Cryogenics

Fundamentals, foundations and applications

Online at: https://doi.org/10.1088/978-0-7503-2757-2

Cryogenics

Fundamentals, foundations and applications

Edited by
Tom Bradshaw
Bennamann Ltd, The Willows, Mill Orchard, East Hanney, Oxon, OX12 0JH, UK

Beth Evans
UKAEA, Culham Science Centre, Abingdon, OX14 3DB, UK

John Vandore
7 Leverton Gardens, Wantage, OX12 9NY, UK

IOP Publishing, Bristol, UK

ISBN 978-0-7503-2757-2 (ebook)
ISBN 978-0-7503-2755-8 (print)
ISBN 978-0-7503-2758-9 (myPrint)
ISBN 978-0-7503-2756-5 (mobi)

DOI 10.1088/978-0-7503-2757-2

Version: 20221201

IOP ebooks

British Library Cataloguing-in-Publication Data: A catalogue record for this book is available from the British Library.

Published by IOP Publishing, wholly owned by The Institute of Physics, London

IOP Publishing, No.2 The Distillery, Glassfields, Avon Street, Bristol, BS2 0GR, UK

US Office: IOP Publishing, Inc., 190 North Independence Mall West, Suite 601, Philadelphia, PA 19106, USA

Contents

Acknowledgement

The editors would like to thank all the contributing authors for their efforts in providing high quality content. In addition, we would like to thank Joe McEntee at the Institute of Physics for guiding the process and keeping us on track. Robert Trevelyan also helped immensely with rights issues. Louise Bradshaw ably assisted with the proof-reading and style.

The Editors

Editor biographies

Tom Bradshaw

Tom Bradshaw is a physicist who completed his PhD in the Low Temperature Group at Lancaster University building a sub-mK refrigerator. This was followed by a Post-Doc at Exeter University on phonon transmission across solid–liquid helium boundaries. He later joined the Science and Technology Facility Council (STFC) Rutherford Appleton Laboratory where he worked on closed-cycle refrigerators for earth observation and astronomy, superconducting magnets and a host of low temperature applications. After 'retirement', he is now Chief Scientific Officer for Bennamann Ltd—a firm working in the renewables sector.

Beth Evans

Beth Evans first worked in cryogenics while a student at CERN After graduating in physics, she gained a PhD in pulse tube refrigeration from the Institute of Cryogenics at the University of Southampton. She then worked at A S Scientific Products Ltd and then at the ISIS Neutron Source at the STFC Rutherford Appleton Laboratory. She is currently the Cryogenics group leader at JET (Joint European Torus) and the Chair of the British Cryogenics Council.

John Vandore

John Vandore is a Trustee of the British Cryogenics Council, and Member of the International Cryogenic Engineering Conference Board. Having studied Engineering and worked in the automotive industry, he first encountered cryogenics running a business (Truflo) making ball valves in Birmingham for application in Liquefied Natural Gas. He went on to other companies in cryogenics including Bestobell Valves, Goddard in USA and Thames Cryogenics. John has worked for a number of years on the Harwell Campus, mainly with the cryogenic community, but also in networks engaged in energy, in food (including the Cold Chain) and new applications of cryogenics such as quantum technology.

Biographies of contributors

Richard A Amos is Associate Professor of Proton Therapy, University College London. Richard trained as a clinical scientist (medical physics) in London with the National Health Service (NHS) before working as a clinical radiotherapy physicist for a number of years in both the UK and Canada. He then spent a couple of years involved in ion-microbeam radiobiological research at the renowned Gray Laboratory Cancer Research Trust in Northwood, UK. In 2002 Richard joined the faculty at Loma Linda University Medical Center in California, the world's first hospital-based proton therapy facility, before moving to The University of Texas MD Anderson Cancer Center in 2005 to help implement and develop a new proton therapy programme, including the first proton pencil beam scanning delivery system in the United States. He returned to London in 2013 to take leadership roles in both the development of a national proton therapy service for the NHS and in developing translational proton therapy physics research. Richard is a Chartered Physicist, a Chartered Scientist, Associate Editor of the British Journal of Radiology, and a Fellow of the Institute of Physics and Engineering in Medicine.

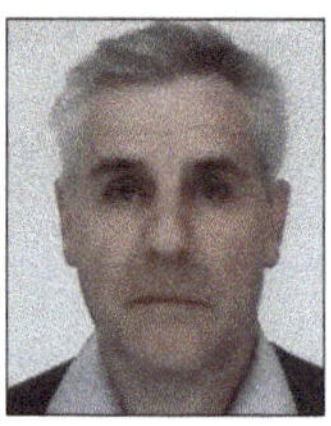

Joseph Boehm is a scientist/engineer who gained his Master's degree at the University of Regensberg in Bavaria and his PhD at the University of Salford in the UK on the topic of 'Open Gradient Magnetic Separation Systems'. He currently works at the UKRI, STFC Rutherford Appleton Laboratory as a Senior Magnet and Cryogenic Systems Engineer. At the Rutherford Appleton Laboratory he has worked on a variety of projects including the magnet and hydrogen systems for the Muon Ionisation Cooling Experiment. Previous work included design, production and installation of MRI magnets at Magnex/Varian, Agilent, and Accel GmbH. He spent many years at Carpco designing and developing magnetic separators, both superconducting and permanent magnet.

Stuart Bradley is Principal Engineer, Power Electronics at WMG, University of Warwick, and has a PhD in Electro-Mechanical Engineering. He works on the development of power electronics, machines and drives for transport, energy and industrial processes. He has worked on wind turbine development for over 20 years, including engineering of generators, drives, blades and foundations.

Gareth Brett joined Highview in 2006 to lead the company in taking liquid air energy storage from concept to market-ready technology. He joined the Board to become full-time CEO in December 2008, before being appointed Vice Chairman in September 2018 and Chief Technology Officer in December 2018.

Gareth has led the company's transition from lab to pilot plant to full-scale demonstrator.

An Electrical Engineering graduate from Bristol University, Gareth began his career with Southern Electric in 1977, becoming Projects Director for its generation subsidiary (SEPG) where he was responsible for development, financing and commercial arrangements on three major UK independent power projects (IPPs). A veteran of the Independent Power Sector, he has subsequently been involved in the development and successful execution of many IPPs in the UK, Europe and the USA from the perspective of owner, financier, operator and customer. Prior to joining Highview full-time as CEO, Gareth was also Head of New Nuclear Build for British Energy. His career has included a five-year period in investment banking and three years based in the USA as President of Entergy Asset Management.

Jack Cahn is the founder and principal researcher at Deep Cryogenics international in Lunenburg, Nova Scotia, Canada. Since 1999, he has used deep cryogenics treatment on machine shop tooling, developed deep cryogenics test procedures for use on JPL's Mars exploration rover, worked with researchers at NIST's cryogenic processing lab, and served as the lead investigator in two US army Cooperative Research and Development Agreements. He is the author of one USPTO-issued patent and five patents-pending.

David Clarke is the Division Head of the STFC UK Central Laser Facility's activities in the Research Complex at Harwell, where he leads teams developing and operating laser-based microscopy and spectroscopy techniques for the scientific community. David is a biochemist and biophysicist who focuses on the development and application of physical techniques for the investigation of biological systems. He is expert in a number of spectroscopy and microscopy methods, including fluorescence (steady-state and time-resolved), circular dichroism, transmission and scanning electron microscopy (including cryo-TEM), and various optical microscopy techniques. A particular area of expertise is the application of kinetic methods combined with synchrotron radiation techniques for the study of macromolecular interactions. This has included the development of fast stopped-flow, temperature-jump, and laser flash photolysis equipment for use with UV and x-ray synchrotron techniques. Currently, David's main area of interest is the development of cryo-techniques for correlative light and electron microscopy.

Louise Cuff is a company owner (Nu Perspectives) and Cryogenic Design Engineer. Louise's experience as a mechanical design engineer lies in cryogenic product development, project management, mechanical design, technical training and the assembly and repair of cryogenic electro-mechanical devices. She has worked alongside a pioneering surgical team to treat cancerous tumours with cryoablation, designed and developed sight saving medical devices and spent two years working in a developing country as an instrument engineer.

In 2008, Louise founded Nu Perspectives Ltd with business partner Gareth. With each year that passes, Louise and Gareth remain dedicated to the work that they do —enjoying a variety of cryogenic projects that continue to challenge and enhance their understanding, their ideas and their expertise.

Robert Done began his career as a mechanical engineering apprentice for UKAEA at Risley in Cheshire. On graduating, he joined the STFC, working first at their Daresbury site in Warrington and eventually moving to the Rutherford Appleton Laboratory in Oxfordshire. For the last 30 years he has been providing the mechanical engineering support for the neutron spallation source at the site, designing equipment which provides the experimental environments for user-provided test samples. Much of this equipment operates at cryogenic temperatures. Rob routinely presents a series of cryogenic training courses for both internal and external clients. He is also a lead auditor for the Council's safety auditing team.

Lionel Duband completed his PhD in Low Temperature Physics at the Centre de Recherche sur les Très Basses Températures (CRTBT) of the CNRS (Fr) in 1987. Since then, his research has been focused on small cryocoolers, mostly for space applications. After three years (1988–90) as a Post-Doc fellow in the Lange's group at the Department of Physics and the Space Sciences Laboratory of the University of Berkeley (USA), he joined the Service des Basses Températures (SBT) of the CEA (FR), where he has worked ever since. He has been project manager for the cooler on board the IRTS satellite and the HERSCHEL satellite, successfully launched in 1995 and 2009, respectively. He is currently the project manager for the development of the so-called 50 mK hybrid cooler, potentially to be flown on board the SPICA and ATHENA satellites. From 2000 to 2013, he led the Cryocoolers and Space Cryogenics laboratory within SBT. In 2013 he was appointed head of SBT which recently became the Département des Systèmes Basses Températures (DSBT).

James E Fesmire is founder and President of Energy Evolution LLC for technology implementation in thermal energy systems and materials. He is also founder of the Cryogenics Test Laboratory at NASA Kennedy Space Center for novel energy technology and materials research. He holds a Master of Science in Mechanical Engineering (Materials Science) from the University of Central Florida and Bachelor of Mechanical Engineering from Auburn University. James has 35+ years of experience in cryogenics and low-temperature problem-solving with specialty in all aspects of liquid hydrogen storage and transfer. His cryogenic systems design work helped to advance the Space Shuttle, the International Space Station, future Moon and Mars exploration, experimental rocket programs, commercial space launch vehicles and facilities, superconducting power, hydrogen storage and transfer, and many commercial and industrial applications. He has extensive publications and patents in thermal insulation systems, novel materials, and cryogenic testing. James is also recipient of NASA medals for Distinguished Service, Exceptional Technology Achievement, and Exceptional Service; R&D 100 award and Space Technology Hall of Fame medal for aerogel insulation technology. He is also an inductee of the NASA Inventors Hall of Fame for developments in cryogenics, materials, and energy technologies.

Eleni Fitsioui studied Chemical Engineering and did her Master's degree in Chemistry with emphasis on Food Chemistry and Technology, at Aristotle University of Thessaloniki (https://www.auth.gr/en/cheng). At Lancaster University she has been completing her PhD in the Chemistry department working on computational chemistry. Eleni enjoys doing different things but has always been interested in the food sector (and worked in a Quality Assurance department in the confectionary industry early in her career). Cryogenics has intrigued Eleni; she has researched many applications of cryogenics with a particular focus on food technologies.

Dr Martin Fray works at the MRC-Harwell Institute which is renowned as a centre of excellence in mouse genetics, supporting a dynamic research programme directed towards in-life studies, from development through to ageing. After his post-graduate studies at the University of Nottingham, Dr Fray worked as a research fellow at the University of Reading investigating the control of ovulation in ruminants. He then moved to the Institute for Animal Health and studied the effects of bovine viral diarrhoea virus (BVDV) on cattle reproduction. In 2001, he moved to the MRC and is currently Deputy Director of the Mary Lyon Centre and also leads the Biological Resources Group. Within this role he has responsibility for the MRC-Harwell Archive, which is one of the largest repositories of mouse germplasm in the world. The MRC-Harwell Archive serves the biomedical community in the UK and abroad. It also acts as the UK node for the European Mouse Mutant Archive (www.infrafrontier.eu).

Jaroslaw Fydrych is a cryogenics engineer at European Spallation Source ERIC. He received his PhD in Technical Sciences with a specialisation in Fluid Mechanics from the Wroclaw University of Science and Technology (WUST) in Poland. He was involved in WUST's collaborations in the field of cryogenics with XFEL, ITER and FAIR projects and he also spent two years at CERN as a Project Associate at Cryogenics Group. Jaroslaw joined ESS in 2013 as the Project Engineer for the ESS Cryogenic Distribution System.

Chris Lawson is a physicist who completed his PhD in the Low Temperature Group at Lancaster University working on a device to measure quantum turbulence in ^{3}He down to 150 μK. Following this, he travelled to The Netherlands, working to deliver a nuclear demagnetisation stage for the dilution refrigerators of Leiden Cryogenics. He is currently responsible for the ultra-low temperature capability of the ISIS Neutron and Muon Source, based at STFC Rutherford Appleton Laboratory.

Professor Graham Machin (FREng) is an NPL Fellow and is its current Head of Temperature Standards. He has >30 years' thermometry experience, published >230 papers and given numerous keynote talks around the world. He is visiting Professor at Strathclyde (Physics) and South Wales (Computing) Universities and Distinguished Visiting Fellow at Valladolid University (Engineering). He represents the UK on the Consultative Committee of Thermometry (CCT) and chairs the CCT working group for non-contact thermometry. He was President of the UK Institute of Measurement and Control (2018–19), chair of the Euramet Technical Committee for Thermometry (2014–18) and served on the EPSRC Physical Sciences Strategic Advisory Team (2014–17). Graham has been awarded the InstMC Callendar medal for 'outstanding contributions to temperature measurement', was elected Honorary Scientist of the Chinese Academy of Sciences (2019) and is a Fellow of the Royal Academy of Engineering (2019).

Current activities include leading the 21-partner Euramet 'Realising the redefined kelvin' project, both NPL's thermometry and metrology for nuclear decommissioning activities, and he is a founder member of the 'NHS Body Temperature Group'. Graham Machin https://orcid.org/0000-0002-8864-6951.

Professor Peter Malkin is a specialist in novel electric power systems for aircraft applications. He has had a long career in industrial R&D with major European companies and was appointed Chief Technologist and Senior Electrical Fellow at Rolls-Royce plc.

Following this he was appointed Professor of Electrical Power at Cranfield University and he is currently Strategic Research Advisor at Newcastle University. His specialist areas are: electrical power systems, superconducting and cryogenic applications, fault management and aircraft applications. He was an examiner on both RAE2008 and REF2014 UK University Assessment exercises and has over 30 patents worldwide.

Marisa Martin-Fernandez is the Group Leader of the STFC Central Laser Facility's Octopus microscopy cluster, where she leads a team developing and operating advanced optical and electron microscopy techniques for the research community. Marisa's research is geared to understanding the structure–function relationships and the behaviour of the human epidermal receptor growth factor family in the cell's physiological environment. The aim is to describe the basic molecular ingredients, the signalling pathways, the supra-molecular structural and spatiotemporal rules regulating signalling outcomes, and the dysregulation of these signals in cancer. Pioneered imaging methods developed are targeted to describing the structure of macromolecular complexes in cells, including ensemble and single molecule imaging, fluorescence resonance energy transfer and polarisation, and fluorescence localisation imaging with photobleaching, to be combined with molecular kinetics and dynamics information derived from multicolour single-particle tracking. These methods, in conjunction with the modelling of macromolecular behaviour at atomic resolution, have revealed the higher order architectures of EGFR macromolecular complexes, and inform how these determine the information output of the system.

Andrew May is a cryogenics engineer in the Accelerator Science and Technology Centre at the STFC Daresbury Laboratory where he works mostly on accelerator cryogenics. He received his PhD, MSc and MEng from the University of Manchester, where he worked on millikelvin systems for CMB polarisation observatories.

Dr Jarek Nowak is a senior lecturer in the Physics Department at Lancaster University and expert in liquid argon detectors for particle physics. He is the leader of the quality control and quality assurance for the cryogenic equipment produced in the UK for the DUNE experiment. Before that, he was a leading figure in the construction, operation and data analysis of particle detectors in the USA and Japan and published over 120 peer-reviewed articles. For the past few years, Dr Nowak has worked on the implementation of his research in industry, including novel approaches to the use of machine learning and cryogenic technologies.

Shrikant Pattalwar is Senior Cryogenics Engineer at UKRI-STFC Daresbury laboratory and leads a team for developing cryogenics technology for particle accelerators for international projects (HL-LHC, ESS, PIP-II etc). He started his career in 1979 at the Tata Institute of Fundamental Research (TIFR) in India as a scientific officer in low temperature experiments. He then moved to the UK in 2001 to work with Oxford Instruments where he led a team for designing custom dilution refrigerators. In 2005 he moved to Daresbury Laboratory where he set up a range of small and large cryogenic facilities primarily for the development of SRF technology.

Toby Peters is a Professor in Cold Economy (University of Birmingham) and Visiting Professor of both the Global Innovation Centre (Kyushu University, Japan) and Centre for Sustainable Road Freight (Heriot-Watt University, UK). He is the Co-Director of the Centre for Sustainable Cooling, Co-inventor of Liquid Air Energy Storage (Highview Power, Dearman Engine) and the architect of the 'Cold Economy' and 'Clean Cooling'. Toby has more than 15 years of experience in energy storage and energy systems, transitioning some nine years ago specifically into clean cooling and the cold economy, including policy and regulatory environments. A cross-disciplinary researcher, uniquely spanning engineering, energy, social sciences, business schools, he works globally with government, multilateral development agencies, banks and NGOs. In 2015 he led the agenda-setting 'Doing Cold Smarter' policy report which was followed by the report 'A Cool World: Defining the Energy Conundrum of Cooling for All' which set out to define for the first time the size of the future global cooling demand if we were to meet the UN Sustainable Development Goals, and its implications for energy systems and emissions. Along with leading research and technology, product development and new system integration globally, his work has included developing a variety of novel system-level approaches to meeting our energy demand. He has also developed new delivery mechanisms for accelerated global deployment, Factory in a Box, and associated business models (Cooling as a Service). Toby is also active in developing the role of clean cooling for sustainable development in low/lower-middle income countries.

Agnieszka Piotrowska (PhD Eng) is a lecturer in the Faculty of Mechanical and Power Engineering at Wroclaw University of Science and Technology (Poland). Cryogenics is her main field of study and research. She is a specialist in risk analysis of cryogenic complex systems and technologies of cryotherapy devices. She gained professional experience working for two companies producing cryotherapy devices. Since April 2020 she has been a member of Commission C2 of the International Institute of Refrigeration.

Russell A S Thomas is Principal Research Scientist at NPL Science Area Leader, Medical Radiation Science, National Physical Laboratory, UK. Fellow of the Institute of Physics and Engineering in Medicine.

Leaving school at 16, Russell had intended to take up an apprenticeship as a silversmith. However, an economic recession put paid to that and instead he trained as an electrical and electronic engineer whilst working on various building sites. An interest in the use of electronics in medicine led him to study for a degree in physics with medical applications at Kings College London but on graduating from there he found himself working for two years in a secure unit for people with violent and challenging behaviour.

In 1994 he was recruited to the Radiation Dosimetry Group at the UK's Primary Standards Institute based in Teddington, UK, originally working on calibration services for photon and electron beam therapy level radiation and being trained as an operator for the NPL Radiation Dynamics Research Linear Accelerator (linac). When it became apparent that this machine needed to be replaced, the decision was taken to do so with a clinical linac and Russell took up a two year secondment at the Royal Surrey County Hospital to learn more about this type of machine. However, from this experience he was also able to gain state registration as a Clinical Scientist and returned to NPL with a much better understanding of the issues faced by colleagues working in the clinical setting.

Since 2000, amongst other areas of research and development associated with the primary standards ionizing radiation, Russell has been involved in the development work of the group in therapy level proton dosimetry. This project has led to the creation of a new primary standard calorimeter specifically for proton beam therapy that can be operated directly in the clinical beam used to treat patients. It is hoped that the introduction of this new primary standard, combined with a new dosimetry code of practice for protons from IPEM, will result in considerably reducing the uncertainty on dose delivered to patients treated with this form of radiotherapy.

Bryan Shaughnessy (STFC RAL Space) manages the Thermal Engineering Group within STFC Rutherford Appleton Laboratory Space department. After a PhD in radiative heat transfer from Cranfield University, he completed a post-doctoral appointment before joining the Thermal Engineering Group in 1999. He has over 20 years of diverse experience as a thermal systems engineer working on high-profile scientific spacecraft missions. He has been involved in all aspects of the spacecraft lifecycle, from development and testing of thermal control systems through to mission operations. A highlight of his career is leading the thermal design of the cryogenic Mid-Infrared Instrument (MIRI) of the James Webb Space Telescope (JWST).

Jon P Trembly works at Air Products Plc and is the Technology Manager for Cryogenic Applications. He obtained his degree in Mechanical and Production Engineering from Kingston University. Jon has been working in the Industrial Gas Business for over 30 years with almost all his entire career focused on low temperature cryogenic applications. Jon joined Air Products UK in 1993 as a research engineer and quickly progressed within the company's technology organisation to the leadership position he holds now.

Having worked in both Europe and North America, Jon now leads an Advanced Technology Team with a focus on developing new and innovative cryogenic applications for Air Products, from his location in the UK.

Viktor Tsepelin is a Reader in Low Temperature Physics at Lancaster University and former EPSRC Advanced Research Fellow (2006–11). He was awarded the IUPAP Young Scientist Prize in 2008 for contributions to the fields of quantum crystals and quantum turbulence. He has more than 20 years of experience working at sub-millikelvin temperatures using nuclear demagnetisation techniques. He has conducted studies of helium-3 crystals down to 500 μK using optical interferometry and routinely uses bolometric and cross-correlation techniques at 150 μK in ^{3}He–B to study quantum turbulence. He has experience of working with a variety of oscillating mechanical structures at low temperatures and investigates the dissipation mechanisms of microelectromechanical systems and nanoelectromechanical systems at ultra-low temperatures.

Tristan Valenzuela is a physicist who completed his PhD in Atomic Physics at the Johannes Guttenberg University Mainz (Germany) measuring the Landé g-factor of the electron bound in a single hydrogen-like oxygen ion (O^{7+}) trapped in a Penning trap, cooled to liquid helium (l-He) temperatures. Later, he joined the Institute of Photonic Sciences (ICFO) in Barcelona where he developed a laser cooling system to trap and control a single atom for quantum information applications. During the last 10 years, he has been working in laser cooling experiments to develop quantum sensors and lately his efforts are focused in translating the laboratory-based Cold Atoms technology to the space environment to enable their use on board spacecraft for the purpose of global surveys.

John Weisend is currently Deputy Head of Accelerator Projects and Group Leader for Specialized Technical Services at the European Spallation Source in Lund, Sweden. He is also an Adjunct Professor of Engineering at Lund University. He received his PhD in Nuclear Engineering and Engineering Physics from the University of Wisconsin—Madison. He has worked at the SSC Laboratory, the Centre D'Etudes Nucleaires Grenoble, the Deutsches Elecktronen-Synchrotron Laboratory (DESY), the Stanford Linear Accelerator Laboratory (SLAC), the National Science Foundation and Michigan State University. Dr Weisend's research interests include He II, cryogenic safety and large-scale accelerator cryogenics. His previous books include: *He is for Helium*, co-authored (with N Filina of Cryogenic) *Two-Phase Flow* and he is editor of the *Handbook of Cryogenic Engineering and of Cryostat Design*. He writes a regular column 'Cryo Bios' for the publication *Cold Facts* and is a member of both the Cryogenic Engineering Conference and International Cryogenic Engineering Conference Boards.

List of contributors

Richard Amos
University College London, Department of Medical Physics and Bioengineering, University College London, Gower Street, London, WC1E 6BT, UK

Joseph Boehm
UKRI, Science Technology and Facilities Council, STFC Rutherford Appleton Laboratory, Harwell Campus, Didcot, OX11 0QX, UK

Stuart Bradley
Ricardo UK Ltd, Midlands Technical Centre, Southam Road, Radford Semele, Leamington Spa, CV31 1FQ, UK

Tom Bradshaw
Bennamann, The Willows, Mill Orchard, East Hanney, Oxon, OX12 0JH, UK

Gareth Brett
Highview Power, Pennine Place, London, WC2H 0HF, UK

Jack Cahn
Deep Cryogenics, Deep Cryogenics International, Lunenburg, Nova Scotia, B0J 2C0, Canada

Dave Clarke
Central Laser Facility, Science and Technology Facilities Council, UKRI, STFC Rutherford Appleton Laboratory, Harwell Campus, Didcot, OX11 0QX, UK

Louise Cuff
Nu Perspectives, Hillview House Leylands Farm Business Park, Colden Common, Winchester, SO21 1TH, UK

Robert Done
UKRI-STFC, Science and Technology Facilities Council, Rutherford Appleton Laboratory, Harwell Campus, Didcot, OX11 0QX, UK

Lionel Duband
Univ. Grenoble Alpes, CEA IRIG DSBT, F-38000 Grenoble, France

Beth Evans
UKAEA, Culham Science Centre, Abingdon, OX14 3DB, UK

James Fesmire
2611 Ivory Way, Titusville, FL 32780, USA

Eleni Fitsiou
Lancaster University, 3 Dutton Drive, Lancaster, LA1 3FT, UK

Martin Fray
The Mary Lyon Centre at MRC-Harwell, Harwell Campus, Oxfordshire, OX11 0RD, UK

Jaroslaw (Jarek) Fydrych
European Spallation Source, PO Box 176, SE-221 00 Lund, Sweden

Chris Lawson
ISIS Neutron and Muon Source, Science and Technology Facilities Council, Rutherford Appleton Laboratory, Harwell Campus, Didcot, OX11 0QX, UK

Graham Machin
National Physical Laboratory, Thermal and Radiometric Metrology Department, National Physical Laboratory, Hampton Rd, Teddington, Middlesex, TW11 0LW

Peter Malkin

Centre for Advanced Electrical Drives, Electrical and Electronic Engineering, School of Engineering, Merz Court, Newcastle University, NE1 7RU, UK

Marisa Martin-Fernandez
Central Laser Facility, Science and Technology Facilities Council, UKRI, STFC Rutherford Appleton Laboratory, Harwell Campus, Didcot, OX11 0QX, UK

Andy May
UKRI-STFC Daresbury Laboratory, Warrington, WA4 4AD, UK

Jarek Nowak
Lancaster University, 18 Mallside Close, Lancaster, LA1 3FZ, UK

Shrikant Patalwaar
UKRI-STFC Daresbury Laboratory, Warrington, WA4 4AD, UK

Toby Peters
Fullers Thatch, Uckham Lane, Battle, TN33 0LY, UK

Agnieszka Piotrowska
Wroclaw University of Science and Technology, Wybrzeze Wyspianskiego 27, 50-370 Wroclaw, Poland

Bryan Shaughnessy
UKRI-STFC, RAL Space, Science and Technology Facilities Council, Rutherford Appleton Laboratory, Harwell Campus, Didcot, OX11 0QX, UK

Russell Thomas
National Physical Laboratory, National Physical Laboratory, Hampton Road, Teddington, Middlesex, TW11 0LW, UK

Jon Trembley
IChemE, Air Products PLC, Hampshire International Business Park, Lime Tree Way, Basingstoke, RG24 8GA, UK

Viktor Tsepelin
Department of Physics, Lancaster University, Bailrigg, Lancaster, LA1 4YW, UK

Tristan Valenzuela
RAL Space, Science and Technology Facilities Council, Rutherford Appleton Laboratory, Harwell Campus, Didcot, OX11 0QX, UK

John Vandore
7 Leverton Gardens, Wantage, OX12 9NY, UK

John Weisend
European Spallation Source, PO Box 176, SE-221 00 Lund, Sweden

IOP Publishing

Cryogenics

Fundamentals, foundations and applications

Tom Bradshaw, Beth Evans and John Vandore

Chapter 1

Fundamentals of cryogenics

Beth Evans, Tom Bradshaw and John Vandore

1.1 Introduction

Cryogenics, or the art/science of maintaining and achieving cold, affects all our lives in one way or another. We hope that in this book you will discover the wide range of benefits that it brings to modern life. At the time of writing, we are witnessing the excitement of new Covid-19 vaccines being deployed—some of which need to be stored at −70 °C or ~200 K. There are also the possibilities being offered by quantum computers, artificial intelligence, food processing, medical imaging, and reducing/slowing climate change effects, many of which depend on cryogenics.

The infographic below gives a non-exhaustive overview of how cryogenics affects our lives. Clearly there are many other applications, but the breadth is clear to see.

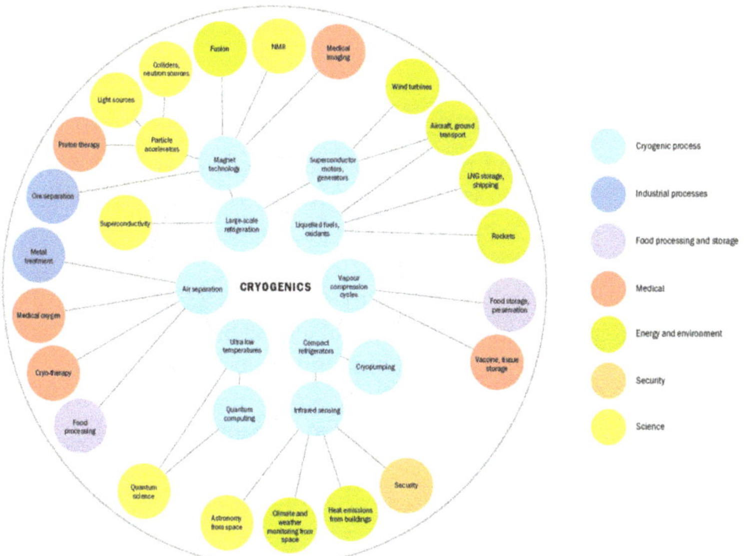

doi:10.1088/978-0-7503-2757-2ch1

1.2 What do we mean by cryogenics?

Much of the existing literature on cryogenics is predicated on the assumption that cryogenics begins below a stated temperature, often 120 K. A distinction is made between cryogenic engineering on the one hand, and low temperature physics on the other—implying a dichotomy that we prefer to avoid in this publication. Indeed, we go further, including reference to laser cooling, a technique important in the emerging field of quantum technology. We do this for two reasons: that there are no benefits to this separation, and we believe the definition of cryogenics should be broad—indeed, many of the changes to thermodynamics and material properties are common across a wide range of temperatures, and not necessarily more exacting the lower in temperature you go! In support of this argument is the eclectic membership of the British Cryogenics Council (https://bcryo.org.uk) which demonstrates how cryogenics is relevant to all industrial sectors.

1.3 Role of cryogenics in the economy

The importance of cryogenics as an enabling technology is often overlooked. In the absence of a 'Standard Industry Classification' (SIC) code for cryogenics, data on production, employment, exports, growth and suchlike are not routinely gathered. There has been just one Economic Impact Study conducted in the UK on cryogenics [https://stfc.ukri.org/files/cryogenics-impact-report/]. This study was conducted in 2015 and concluded that 17% of the UK Economy was 'broadly associated' with cryogenics—in other words, if one took out the cryogenics aspect, almost one fifth of the UK Economy would have to find some other way of operating. This suggests that cryogenics is a more substantial enabling technology than is generally understood.

As a community cryogenics is well organised, not only in the UK but globally. The aforementioned British Cryogenics Council works with the UK Institute of Physics to coordinate and bring together the low temperature community.

1.4 Target audience and how they will make use of this book

This book aims to be a starting point for students, scientists, engineers, technicians and early career researchers looking for a broad-scoped, practically-oriented introduction to cryogenic technology and its applications. The core readership will span academia, research and industry.

A secondary readership will include key stakeholders in the wider 'innovation ecosystem' with an interest in the uptake of cryogenic enabling technologies for research and industrial applications. This could include technology transfer professionals, business development managers, senior management in academia and industry, funding agencies and the financial/investment community.

To this end, we have gathered acknowledged experts who have contributed introductions to their field of expertise, including practical advice and references.

1.5 Physical effects of cryogenics

Lowering temperatures gives rise to a number of physical effects. Material properties are altered. Materials shrink. They change phase: gases condense to liquids, with substantial change in volume; in turn they freeze as solids. Strength is just one other property of solids which alters as they cool down. Lowering temperature results in a lower energy state, with lower molecular vibration. Preservation of organic or biological materials, food or blood and tissue for example, is enhanced at low temperature. The phenomenon of superconductivity is enabled at low temperature too. All these effects and more will be explored in this book.

1.6 Role of cryogenics in medical applications

A visit to a hospital may well require the taking of a tissue sample (biopsy) which could be stored at cryogenic temperatures for future use or reference. Bulk storage tanks of liquid oxygen at hospitals are another connection with function of cryogenics, enabling provision of oxygen to patients where required. Staying in the hospital, but visiting the Magnetic Resonance Imaging (MRI) suite, this equipment would not function without liquid helium to make the superconducting magnets in the MRI scanner work. MRI uses strong magnetic fields created by coils of superconducting wire maintained in this state by liquid helium at 4 K. Superconductors are extraordinary: even good conductors of electricity, like copper, have a resistance which manifests in a loss of energy in the form of heat, but superconductors deliver current without any resistance at all. This means that they can take huge amounts of current and hence deliver high magnetic fields. This ability to produce strong magnetic fields is exploited in a number of applications including high energy particle colliders and magnetic ore separation.

1.7 Role of cryogenics in science

So much science involves low temperatures. Superconducting applications, particularly magnets, are one important field. The Large Hadron Collider at CERN relies on over two thousand superconducting magnets stretching around a ring of 25 km, requiring immense cryogenic systems, to deliver its particle acceleration function. The delivery of new nuclear fusion facilities like ITER will extend the boundaries of cryogenic systems with requirements to cool immense superconducting magnets and achieve the levels of vacuum required in these amazing machines.

Other fields of science exploit the reduced molecular vibration associated with the lower energy states at cryogenic temperatures to conduct various forms of microscopy. In neutron scattering devices, for example, the majority of experiments involve cooling the sample to cryogenic temperatures, to reduce molecular vibration, enabling the capture of a more precise image of the molecular structure.

Cryogenically-produced lower molecular vibration in the cameras or 'detectors' used by astronomers can produce sharper images of the faintest signals captured all the way back to the 'Big Bang.' Thus cryogenics has an application in cooling detectors in astronomy, be they ground-based or housed on a satellite.

1.8 Role of cryogenics in energy and the environment

Energy production is an application for cryogenics with great future potential—but for now it is best exemplified by Liquefied Natural Gas, or LNG. Britain has in recent times been supplied with much natural gas from the North Sea via pipeline but, as the North Sea fields decline, is increasingly being bought from sources like Qatar, which has no pipeline connection to the UK. So, the gas is liquefied, shrinking the volume by a factor of around 700, and enabling shipment via tanker to import terminals in the Thames Estuary or Wales. A similar trade exists on a smaller scale in ethylene, boosted by its availability from shale deposits in the USA.

The merit of contraction from liquefying a gas can be exploited in reverse: there is a segment in energy storage development being embarked upon known as 'Liquid Air Energy Storage.' Using 'spare' energy, air is liquefied and stored. When energy is required, the liquefied gas is vapourised and the expanding gas is used to drive a turbine. Where possible, low grade waste heat (e.g. from a cooling tower) is used for vapourisation, to assist the turnaround energy efficiency. A related development is use of liquefied gases such as nitrogen as a 'fuel' for engines, exploiting the expansive force of vapourisation in a reciprocating engine, in a non-emitting way.

The cryogenic role in enabling superconductivity leads to other energy applications. Permanent magnet technology has been the principal choice in wind turbine generators, up to remarkable ratings of 12 MW. While the industry has exceeded expectations for some with conventional technologies, there is arguably a limit to turbine ratings, perhaps simply for reasons of scale and logistics, where it will be advantageous to incorporate superconducting technology in the generator. This could bring the benefits of lower size, weight, higher efficiency, avoidance of rare earth magnetic materials and of the need for a gearbox.

Global monitoring of the Earth's climate is made possible by Earth observation from satellites in orbit. Measurements of temperature are made using infrared detectors. To reduce noise and to take advantage of windows in the infrared spectrum, detectors and optics are cooled to low temperatures, typically 80 K. The ability to globally monitor sea and land temperatures is vital in weather forecasting and monitoring the effects of climate change.

1.9 Brief resumé of the chapters

Chapter 2 What is cryogenics

This chapter gives a brief introduction to cryogenics. It describes the scope of cryogenics in the economy and its uses in new and developing technologies.

Chapter 3 Cryostat design

This chapter includes the elements required to enable the reader to design cryogenic equipment including thermal insulation and material properties. The fundamentals of thermometry are covered with an introduction to how the temperature scales are specified and a guide to practical thermometry.

Chapter 4 Closed cycle refrigerators
Closed cycle refrigerators, or cryocoolers, are the workhorses of practical cryogenics. This chapter introduces the more commonly used ones, how they work and practical implementation.

Chapter 5 Very low temperature techniques
This chapter is devoted to techniques for achieving very low temperatures (below <1 K). The need for such low temperatures has expanded in recent times due to the requirements for ultra-sensitive astronomy, dark matter detection, quantum computing, cryptography and studies at the atomic scale. The chapter covers dilution refrigerators, adiabatic demagnetisation (electronic and nuclear), sorption coolers and laser cooling.

Chapter 6 Cryogenics in particle accelerators and fusion reactors
Particle accelerators, particularly high energy ones, utilise a range of cryogenic equipment including superconducting magnets and superconducting cavities. This chapter looks at the history and rationale for these devices and a description of the cryogenic distribution systems behind them.

Chapter 7 Cryogenics in energy storage, motive power and renewables
Of the various attributes and manifestations of cryogenics, there are a number which have a topical relationship to the world of energy. One which is already established is the liquefaction of fuel gases, primarily methane as liquefied natural gas (LNG), but also liquid ethylene, to enable transportation.

Cryogenics has many other energy applications, in liquid air energy storage, in electric aircraft, or in wind turbine generators with the incorporation addition of superconducting components. Many of these applications are still in the concept or early development stages and so there is some exciting reading in this chapter.

Chapter 8 Cryogenics in the life sciences and healthcare
This chapter looks at the important contribution cryogenics makes in healthcare and the life sciences. It brings together insights relevant to storing biological samples and vaccines as well as cryosurgery and cryotherapy. It covers specialisations such as cryo electron microscopy and proton beam therapy.

Chapter 9 Industrial applications
Given the endless range of application of the enabling technology which is cryogenics, and its never-ending expansion, this chapter is necessarily selective, covering volatile organic compound [VOC] recovery, superconducting magnet separation, deep cryogenic treatment and cryogenics in the food industry.

IOP Publishing

Cryogenics

Fundamentals, foundations and applications

Tom Bradshaw, Beth Evans and John Vandore

Chapter 2

What is cryogenics?—UK perspective and brief history

Beth Evans, Tom Bradshaw and John Vandore

2.1 Introduction

Cryogenics is usually defined as the art of producing and maintaining cold. There has been a lot of debate over what is defined as a cryogenic temperature, but generally the accepted view is that cryogenics starts where conventional refrigeration stops—although there is often some overlap. Perhaps a clearer way of thinking about the 'start' of cryogenics is as being the temperature below which the permanent gases, such as helium and hydrogen, have their boiling points: −153 °C or 120 K.

The liquefaction points (at standard conditions) of the most commonly used cryogens are shown in figure 2.1. Also shown are the some of the lowest temperatures obtained. In a laboratory setting, one can reach the millikelvin range with a dilution refrigerator, but the current record for the lowest temperature is 100 picokelvin, achieved by cooling the nuclear spins in a piece of rhodium metal.

2.2 Brief history of cryogenics

Cryogenics began with scientists trying to do the then unimaginable—turning air into a liquid. The first documented success of liquefying a constituent of air was in 1877, when Louis Paul Cailletet in France and Raoul Pictet in Switzerland independently managed to produce a mist of liquid oxygen, both publicising results at the same time. In 1883, a team at Cracow university produced small amounts of both liquid oxygen and nitrogen in test tubes (figure 2.2).

Over the coming years liquefied gases were produced at various establishments but these were restricted to small volumes which quickly evaporated due to the heat transfer from the surroundings through the vessel walls. It was not until James Dewar invented a new design of flask in 1892, with a layer of vacuum between a silvered inner vessel and a glass outer vessel, that substantial amounts of cryogens

doi:10.1088/978-0-7503-2757-2ch2

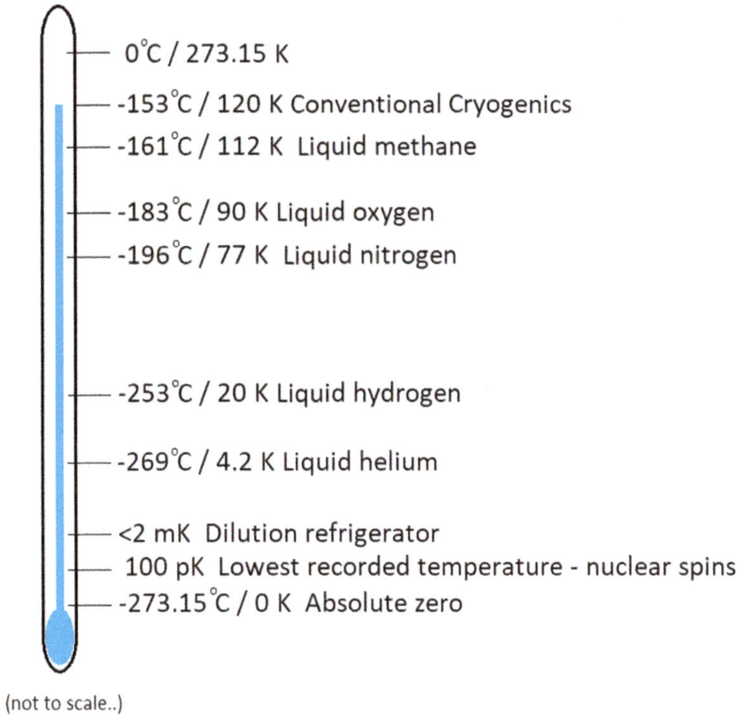

Figure 2.1. Liquefaction points and markers on the temperature scale.

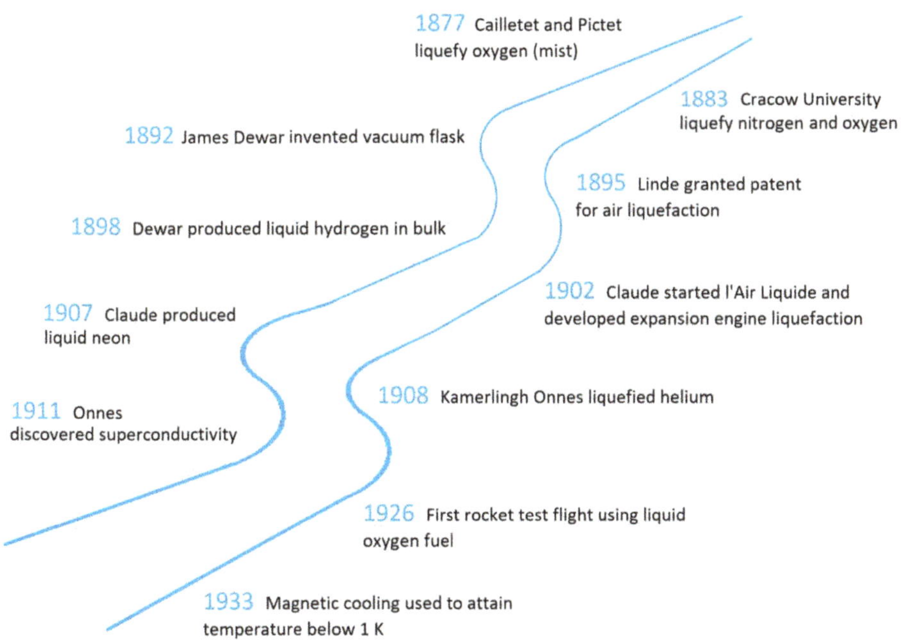

Figure 2.2. Some key points in the history of cryogenics (first half of the 20th century).

Figure 2.3. Dewar lecturing at the Royal Institution. This 'Heike Kamerlingh Onnes - 33 - James Dewar in the Royal Institution in London, around 1900' image has been obtained by the authors from the Wikimedia website, where it is stated to have been released into the public domain. It is included within this article on that basis.

could be generated. This allowed not only Dewar to produce liquid hydrogen in bulk in 1898 but also the beginnings of cryogenic production processes (figure 2.3).

In 1895 Carl von Linde applied for a patent in Germany for his air liquefaction process using Joule–Thomson expansion. A couple of weeks earlier, William Hampson had applied for a patent for a similar cycle in the UK. Hampson went on to produce a design for a hydrogen liquefier which was taken up by BOC.

In 1902, George Claude developed liquefaction using the expansion engine and started l'Air Liquide. He was the first to produce liquid neon, in 1907.

Meanwhile in the Netherlands, Heike Kamerlingh Onnes went on to take two of the greatest steps in the history of cryogenics by becoming the first to liquefy helium in 1908, followed by his discovery of superconductivity in mercury in 1911. He also achieved the then lowest temperature of 1.5 K by reducing the pressure over liquid helium.

Following the First World War the advances in cryogenics became less fundamental and more applied. One of the major developments was the use of cryogens as fuels. This was demonstrated by a successful rocket test flight by Robert H Goddard in the US in 1926.

Another breakthrough came in 1933 with the achievement of sub 1 K temperatures using magnetic cooling, before world events diverted research to more military applications. After the end of the Second World War interest in cryogenics grew significantly, mostly due to the advent of space exploration (figure 2.4).

In 1959 the ship 'Methane Pioneer' delivered the first liquid natural gas (LNG) to the UK from the US. The UK would later become the first country in 1964 to import LNG from the Algerian oil fields in the 'Methane Princess' and later the 'Methane Progress'.

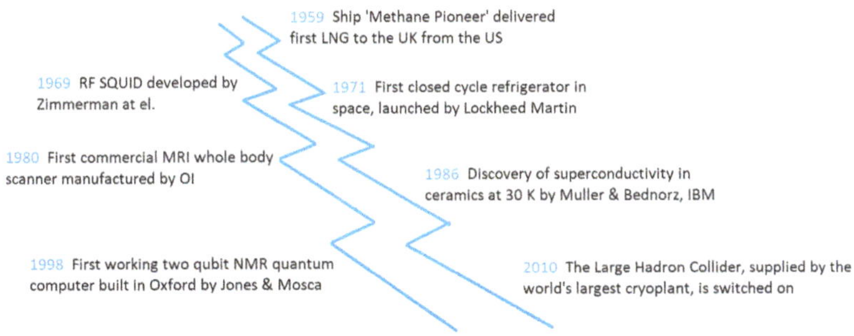

Figure 2.4. Some key points in the history of cryogenics (second half of the 20th century).

In the area of superconductivity, Heinz London postulated flux quantisation in superconductors in 1950. This was followed in 1962 by Brian Josephson developing the theory of tunnelling through a weak link between superconductors, for which he later won the Nobel Prize. This theoretical breakthrough led to the development, and later commercialisation, of the RF SQUID by James Zimmerman together with John Lambe, Arnold Silver, Robert Jaklevic and James Mercereau in 1969. later, in 1987, the 'Woodstock of Physics' took place at an American Physical Society meeting to discuss the discovery of 'high temperature' superconductivity in ceramics at 30 K by K Alex Muller and J Georg Bednorz at IBM's Zurich research laboratory.

In addition to the growing use of liquid oxygen and hydrogen as rocket fuels, in 1971 lockheed Martin Space Company launched the first closed cycle refrigerator into space on board a US Department of Defence Satellite RM-19. The continued use of mechanical coolers in space applications has led to longer lifetimes of systems which are no longer dependent on finite supplies of cryogens.

One of the most beneficial, and now widespread, applications of cryogenics and superconductivity had its beginnings in 1978 when Sir Peter Mansfield developed medical magnetic resonance imaging (MRI) at Nottingham University. He was awarded the Nobel Prize for this in 2003. The production of the world's first commercial MRI body scanners began at Oxford Instruments in 1980 and continues today with Siemens, GE, Phillips and other companies around the world.

At the time of writing, the latest, fast-growing application for low temperatures is quantum technology (see chapter 5). This began in 1998 when the first working two-qubit NMR quantum computer was built and used to solve a mathematical problem in Oxford by Jonathan Jones and Michele Mosca. Today, huge dilution fridges are being used in quantum computers of over 20 qubits made by companies such as IBM and Google. More details can be found in section 5.1.11.

A significant user of cryogenics is the research community, and this has led to the construction of the most impressive cryoplants on the planet. In 2010 the Large Hadron Collider at CERN, Geneva, was first turned on—powered by superconducting magnets it is the largest machine ever built. It is supported by the world's largest cryogenic system, currently supplying over 160 kW cooling power at 4.5 K, distributed around the accelerator's 27 km circumference. Soon ITER, the fusion facility, will have the biggest concentrated cryoplant providing a cooling power of 75 kW at 4.5 K and 1.3 MW at 80 K (see chapter 6).

These great steps in the development of cryogenics and its applications would not have been possible without research in universities and innovation in commercial companies. As we are writing from the UK, we will briefly consider the significant contributions made by researchers based in the UK and the companies which often grew from this research.

2.3 History of cryogenics in the UK

The history of cryogenics in the UK in the early years is mostly centred around the Universities of Oxford and Cambridge.

James Dewar, inventor of the vacuum-insulated flask, spent much of his time entertaining crowds at the Royal Institution but actually held a professorship at Cambridge. The Cavendish Laboratory in Cambridge (figure 2.5) went on to house many significant researchers, not least Peter Kapitsa who constructed his own hydrogen liquefier, infamously housed in a lab with a roof which was designed to 'safely' blow off in the event of an explosion [1]. Kapitsa went on to become director of the Royal Society Mond laboratory from 1930 to 1934, a laboratory for low temperature and high magnetic field physics, named after the benefactor Ludwig Mond. During a visit to see his parents in his native Russia he was prevented from returning to the UK. However, in Russia he went on to develop one of the leading cryogenics laboratories in the world, discovering superfluidity in helium and giving his name to the thermal boundary resistance occurring at the interface between the helium and a solid. Kapitsa was awarded the Nobel Prize in 1978 for his work on helium (figure 2.6).

Meanwhile, at the Clarendon laboratory in Oxford in 1933, while on a short visit from Breslau (now Wroclaw) in Poland, Kurt Mendelssohn became the first person to liquefy helium in the UK. As the Second World War approached, the then head of the laboratory, Frederick Lindemann found funding for a number of refugees including Mendelssohn, Nicholas Kurti and Franz Simon (later known as Simon Francis). Lindemann's team at the Clarendon went on to produce years of low temperature research including important work on superconductors, helium and hydrogen and the adiabatic demagnetisation of paramagnetic salts, producing temperatures below 1 K. It was said that Lindemann, a keen golfer, used to provide the team with old golf balls which they melted down and mixed with lanoline to produce vacuum grease for keeping the laboratory systems leak-tight.

But research and development at low temperature was not restricted to Oxford and Cambridge. William Hampson, who worked at Queens and St John's Hospital in London, has already been mentioned for his patent on an air liquefier. In 1951, at the Atomic Energy Research Establishment in Harwell, the concept of the He3/He4 dilution refrigerator was put forward by Heinz London. The first practical application of the dilution refrigerator was first made in Leiden by Das, Rudolf de Bruyn-Ouboter and Krijn Taconis and it was developed further by Henry Hall at Manchester in the 1960s.

Figure 2.5. The Cavendish Laboratory. This image of Cavendish Laboratory has been obtained by the authors from the Wikimedia website, where it is stated to have been released into the public domain. It is included within this article on that basis.

The Welshman Brian Josephson proposed quantum tunnelling in 1962, work which later led to the development of the Superconducting Quantum Interference Device (SQUID) [2]. He was awarded the Nobel Prize for his work on the 'Josephson effects' in 1973. John Clarke (Cambridge then California) developed this type of technology from its humble beginnings as a SLUG (Superconducting Low indUctance Galvanometer) and a practical device was developed by Jim Zimmerman in the US in 1969. Further advancements and developments were made by Gordon Donaldson at Strathclyde University (work being carried on by Colin Pegrum).

Figure 2.6. Peter Kapitsa. This image has been obtained by the author(s) from the Wikimedia website, where it is stated to have been released into the public domain. It is included within this article on that basis.

The Institute of Cryogenics at Southampton University was formally founded in 1979 although much was happening in research, particularly benefitting industry, before then. Today, important low temperature work continues at Lancaster, Durham, Cardiff, Manchester, London and several other universities.

2.4 Cryogenics industry in the UK

The UK cryogenics industry has grown to be one of the most significant in the world. Two of the most familiar names are BOC and Oxford Instruments (OI).

BOC, originally Brins Oxygen Company, was in fact started by two French brothers in 1886 using barium oxide to produce oxygen—no cryogenics at all. However, as the company grew, production changed to using the Linde-Hampson liquefaction process (figure 2.7).

Over the years, BOC developed and improved methods of cryogen production and storage, insulation, heat exchange and the transport of cryogenic liquids. BOC became part of the Linde group in 2006. BOC/Linde are one of a number of 'gas' companies supplying large amounts of industrial gases as well as liquefying air and producing bulk cryogens.

At the other end of the scale are companies providing 'laboratory' cryogenic equipment. Many of the UK companies providing equipment on this scale can trace their origins back to Oxford Instruments (OI), which grew out of the need of research carried out in the 1950s at Oxford's Clarendon Laboratory. This research required magnets, the strongest of which were powered by a huge generator. With the advent of superconductors, development work on smaller, more powerful magnet coils constructed from superconducting materials began. One pioneering university research officer, Martin Wood, with his wife Audrey, began a company producing superconducting magnets in their house in North Oxford [3]. As the company grew, production moved to the garden shed, then to a series of disused

Figure 2.7. BOC Heylandt plant shop (1930s). Source: Ralph's book [1].

Figure 2.8. Oxford Instrument's first superconducting magnet. Source: Ralph's book [1].

buildings in and around Oxford including a slaughterhouse, a laundry and a boat-builder's shed. Work was begun on the development of magnets for MRI body scanners and the first one was built and tested in Osney Mead, the company's first purpose-built factory. The company grew, extending their range to cryostats and dilution refrigerators, and now with its headquarters in Tubney Woods near Oxford, it has bases in Europe and the US and is listed in the FTSE 250 (figure 2.8).

2.5 Growth of the cryogenic industry in the UK

From the first MRI scanner produced in the OI factory in Oxford a large industry grew, first with Oxford Magnet Technology (OMT), then Siemens which now supplies a third of the world's MRI market. The UK LNG industry has expanded since the first delivery back in 1959 and this is now a substantial industry in its own right—distinct from the cryogenics industry but an important user of cryogenic technology.

A wealth of knowledge and experience is to be found in the UK cryogenics industry, and a large proportion of goods are exported. Dedicated cryogenics companies are also reliant on specialist suppliers, reflected in the corporate membership of the British Cryogenics.

2.6 Cryogenics in the UK economy

Cryogenics has often been the unsung hero of ground-breaking research and technologies, from particle physics to space exploration.

In 2015, the Science and Technology Facilities Council commissioned a study [4] into the importance of cryogenics to the council, and also its impact on the economy as a whole. The study found that cryogenics provides core enabling technologies supporting a diverse range of industries (as illustrated by the infographic in chapter 1) that collectively account for approximately 17% of UK economic output.

Add to this the fact that many large-scale research facilities are dependent on cryogenics to operate instruments from particle accelerators to fusion facilities, as well as smaller laboratory-scale cryostats, meaning a significant proportion of research can only be performed by using low temperatures (figure 2.9).

2.7 The British Cryogenics Council

British Cryogenics Council

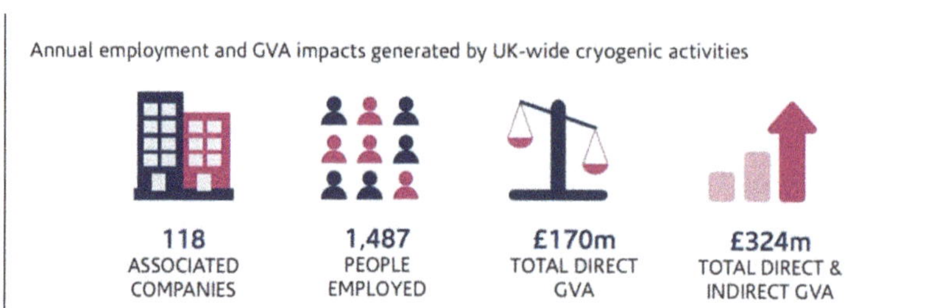

Figure 2.9. Value of cryogenics to the UK economy (2015) (GVA is Gross Value Added).

BRITISH CRYOGENICS COUNCIL

INAUGURAL MEETING

Lecture on 'The Future of Cryogenics in Industry',
by Dr. K. Mendelssohn, F.R.S.

Synopsis

From its initial applications in the chemical industry the large-scale use of low temperature processes and phenomena has already spread into important areas of mechanical and electrical engineering. A survey will be given of these newer applications, including some in the fields of biology and medicine. An indication will be given of the possible extent and speed of some of these developments, and their importance for British industry.

Figure 2.10. Kurt Mendelssohn's inaugural speech (courtesy of BBC Archive) and photograph (image adapted with permission of the National Archives, CC0).

In 1967, the British Cryogenics Council (BCC) was founded to support UK industry in a field that was often overlooked or misunderstood, even then. At that time, cryogenics was merely seen as a form of chemical engineering, supporting low temperature research at universities, but with the increase in space travel the interest in cryogenics was expanding rapidly. However, there was a constant battle for funding for research and development in cryogenics and low temperature and the Council aimed to raise its profile, and to bring academia and industry closer together. It also aimed to improve connections between cryogenics in the UK and the rest of Europe.

The first meeting of the council was held on the 24th May 1967 at the Royal Society in London. The inaugural speech was given by Kurt Mendelssohn on 'The Future of Cryogenics in Industry' [5] which made some interesting predictions. Looking back over 50 years later some have come true while other technologies, such as superconductivity, have still not reached their expected potential (figure 2.10).

Today the BCC has over one hundred corporate members, surprisingly almost equal in number to the Cryogenics Society of America.

The aims of the BCC are still to raise the profile of cryogenics but also to bring academia, research and industry together with networking events. The council continues to provide education about cryogenics, helping students and apprentices, funding conference attendance and subsidising education about cryogenics.

References

[1] Scurlock R G (ed) 1992 *History and Origins of Cryogenics* (Oxford: Clarendon)

[2] Josephson B D 1962 Possible new effects in superconductive tunnelling *Phys. Lett.* **1** 251–3

[3] Wood A 2001 *Magnetic Venture: The Story of Oxford Instruments* (Oxford: Oxford University Press)

[4] Siora G, Webb D, Milway P, Punj P and Baig H 2015 A report for the Science and technology Facilities Council, prepared by Warwick Economics and Development, March 2015

[5] https://bcryo.org.uk/mendelssohn-lecture-may-24th-1967

Chapter 3

Cryostat design

Robert Done, Bryan Shaughnessy, Tom Bradshaw, Graham Machin and Beth Evans

This chapter covers the basics of cryostat design and the analytical tools required to calculate the thermal balance and performance. The materials of construction and their properties are also discussed, with some emphasis on the effective use of multi-layer insulation. Heat exchangers are an important aspect of cryostat design and a simplified introduction is given. Thermometry is often considered a 'bought in item' these days, but it is essential to have a background in how the temperature scales are derived and practical implementation.

3.1 Fundamentals of cryostat design

Rob Done

3.1.1 Introduction

By definition, a cryostat is a specific device which enables the extremely low temperatures employed within cryogenics technology to be both initially achieved and then maintained over a period of time. Cryostats will typically provide some means of mounting samples or component devices within them, so that these items in turn can be cooled to the required cryogenic temperatures. There are two principal methods by which cooling of a cryostat can be achieved: by using the cooling power of liquid cryogens as they evaporate or by the use of direct mechanical coolers.

Whichever of the two cooling techniques is adopted, the principles of reducing the temperature within a cryostat follow the fundamental laws of thermodynamics whereby a warm object will always naturally transfer some of its heat to a colder object until a temperature equilibrium is achieved [1]. A good example of this process is when ice is added to a beverage. Over time, the ice melts due to heat transfer from the surrounding warmer liquid. The final temperature balance of the

doi:10.1088/978-0-7503-2757-2ch3

drink will not be as low as the solidification temperature of the ice, but it will be noticeably lower than its original temperature.

Cryostats work in the same way only, instead of using ice, they have much colder coolant systems which absorb the heat preferentially from the inside of the assembly, thereby cooling the working area.

3.1.2 Cooling methods

Using the cooling power of liquid cryogens

Liquid cryogen cryostats function by using the same principles employed many years ago when blocks of ice were used to keep food cold whenever it was being transported or stored. In this situation, the heat from the surrounding environment gradually melts the ice until it has turned to water. During that melting period, the food itself will be kept at a constant low temperature as the phase change process of turning solid water into liquid will extract heat (figure 3.1). Inside a cryostat, instead of ice a liquid cryogen is used, which on being heated by the ambient temperature around it, will evaporate slowly, thereby maintaining the bulk temperature of the liquid at its boiling point.

Using direct mechanical coolers [2]

Mechanical coolers are constructed from two main components; a room temperature compressor and an expansion chamber which can form an integral part of the cryostat (figure 3.2). The function of the compressor is to take a low pressure gas and compress it up to a high pressure expelling the heat of compression to ambient as it does so. The valves control the pressure in the cold finger and the displacer shuttles the gas from the warm part to the cold part. The displacers and the valve opening are timed in the following manner:

(a) The displacer is moved to the cold end expelling gas from that region and then the high pressure valve is opened;
(b) The displacer moves to the warm end admitting gas to the cold end at high pressure, the high pressure valve is closed;
(c) The valve to the low pressure side is then opened, the gas in the cold end expands and cools;
(d) The low pressure side valve is closed and the displacer moves to the cold end.

The cycle is then repeated—on each cycle the gas is expanded and cooled at the cold end and the compression is done at ambient.

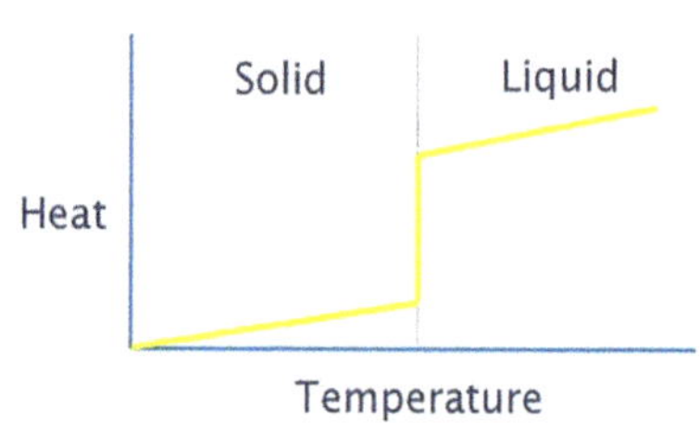

Figure 3.1. Showing the heat needed to change a solid into a liquid at a constant temperature.

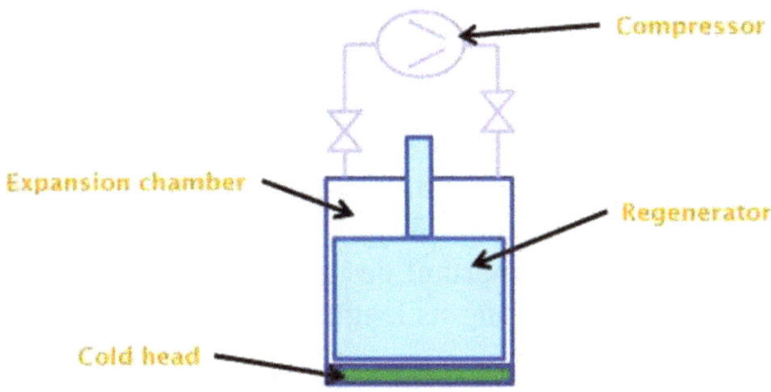

Figure 3.2. The components of a mechanical cooler.

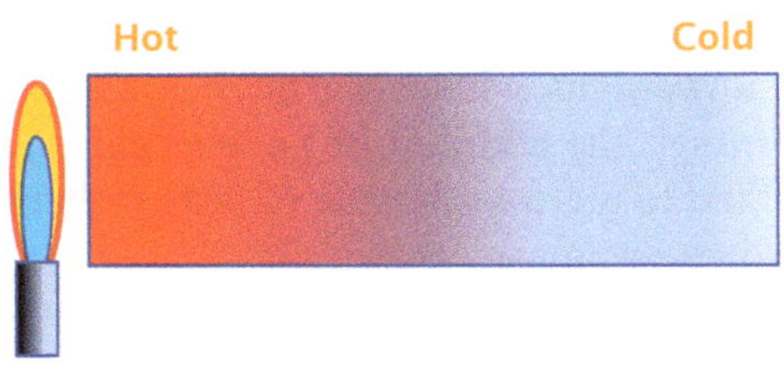

Figure 3.3. Steady-state conduction.

Integral to the displacer is a regenerator which acts as a store of cold. The regenerator is a high heat capacity heat exchanger that has to satisfy conflicting requirements of high heat transfer to the storage material and low pressure drop. When the gas is shuttled from the cold part to the warm part the cold is stored in the regenerator ready to pre-cool the gas before it enters the cold end on the next cycle. In practice the valves are usually combined in a rotary valve to aid the timing.

3.1.3 Heat gains

Once a cryostat is sitting at its base temperature, the ambient heat will work to bring it back to room temperature conditions. There are three routes by which this heat can enter the cryostat and the design of any cryostat assembly needs to ensure adequate provisions are in place to mitigate this heat gain.

Heat gains due to conduction [3]
Heat can be transferred through all media (solids, liquids and gases) by conduction (figure 3.3). Some materials have better heat conductive properties than others. Those with low heat conduction coefficients are considered insulators and these would include materials such as plastics and ceramics. The thermal conductivity of materials generally reduces as their temperature is lowered. Thermal conduction cannot take place through a vacuum.

Generally, cryostats are manufactured as a double vessel design, with one located inside the other. The outer vessel is then evacuated, with the vacuum acting as a thermal insulator to eliminate the heat conduction coming in from its surroundings.

The inner vessel is supported within the outer vessel by small components made from a suitably low-conductive material which must also have adequate structural strength.

The other main source of heat entering a cryostat through conduction is via the mechanism used to load samples into the cryostat. Therefore, this device is normally constructed from a material such as stainless steel, which has relatively low thermal conductivity. In addition, the device will typically be a thin-walled construction and made as long as possible; two important design features which will effectively limit the amount of heat conducted along its length.

Heat gains due to convection [4]

Another means of transferring heat, applicable to liquids and gases, is convection. It is brought about due to the movement of the molecules within fluids as temperature changes cause warmer fluid to circulate into cooler regions. Convection can occur naturally since warmer fluids become lighter in mass and, therefore, they will preferentially migrate upwards in whatever container they are held (figure 3.4). It can also be created through the addition of an external force, such as a fan or a pump, circulating the fluid to increase the rate of heat exchange. Convection cannot take place in solids or in a vacuum.

Convection tends not to be a major problem in most cryostat designs as many employ vacuum insulation between the inner and outer vessels. This eliminates any heat transfer from convection between the vessels. Even in the sample chamber itself, heat convection is negligible as these tend to be relatively small in volume and are usually separated from the rest of the sample feed-pipe by a series of baffles. These baffles effectively produce a series of smaller intermediate chambers within which the effects of convection are reduced to insignificant amounts.

Heat gains due to radiation [5]

Thermal radiation is an energy emitted by any object with a temperature above absolute zero (figure 3.5). It can be transferred through solids, liquids, gases and even a vacuum. The obvious example of this being the radiated heat which the Earth receives from the Sun through the vacuum of space. Thermal radiation can be a significant factor where a large temperature difference exists between one object and another, such as inside a cryostat. A key factor in the effectiveness of an object to emit thermal radiation is a characteristic known as its emissivity. This is in effect a

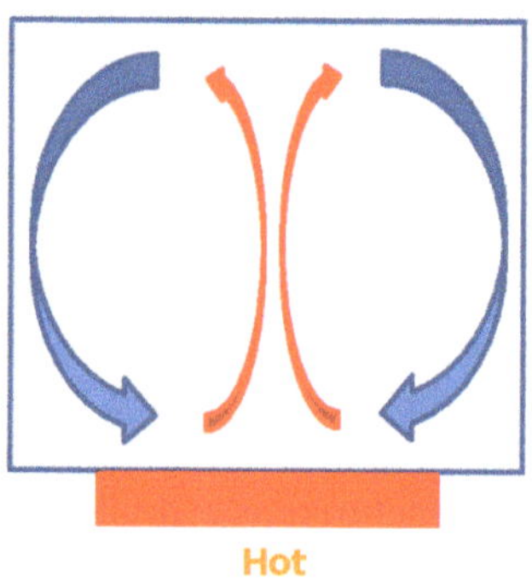

Figure 3.4. Natural convection through a fluid.

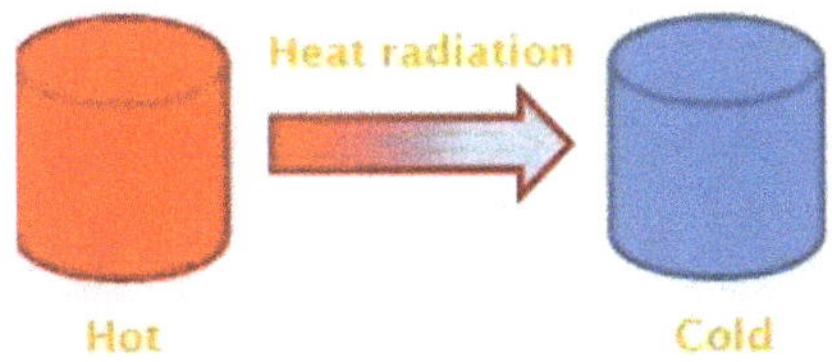

Figure 3.5. Heat radiation between two objects.

factor specific to the material of the object, but emissivity can be changed to some extent by modifying the surface finish of the object.

To mitigate the heat gains through radiation inside a cryostat, several mechanisms are employed. The main method is by adding a so-called radiation shield into the assembly, setting this at as low a temperature as possible. This has the effect of reducing the immediate temperature difference between the outside and the inside of the cryostat and therefore decreasing the amount of thermal radiation entering the system. The other mechanism involves choosing a material with a low emissivity value for this radiation shield, so that as much heat as possible is prevented from entering the cryostat. Many cryostats also make use of multi-layer insulation, which is principally a set of thin insulating sheets layered over each other and coated with a reflective material which, when combined, act like a series of radiation shields.

3.1.4 Types of cryostats

Closed-cycle cryostats [6]

A closed-cycle cryostat is fundamentally a cryostat with a mechanical cooler. A mechanical cooler (figure 3.6) is fundamentally a chamber into which helium gas is introduced. Through a process of expansion and regeneration inside this chamber, the helium gas is cooled. This cold gas is then used to extract heat from a metallic cold-plate attached to the chamber, regaining temperature as it does so. Finally, an external mechanical compressor, which is itself normally water-cooled, draws this warmer gas out of the chamber, allowing the whole process to be repeated until the cold-plate is operating at cryogenic temperatures. Objects can be cooled in this type of cryostat by attaching them to the cold-plate. Closed-cycle mechanical coolers are commercially available which can typically attain 4 K in temperature. The temperature of the object being cooled can be further controlled by attaching a heater and thermocouple to the object which are subsequently operated through a temperature control loop.

The main drawback of using this type of cryostat is that, due to their internal mechanism, they cause vibrations. These movements can affect not only the object mounted to the cold-plate but any other equipment the cryostat is fitted to. On the other hand, they do not need to be refilled with helium as other types of cryostats do, as the single charge of gas is recycled internally. Likewise, as they do not contain a cryogen bath, this makes them intrinsically safer to use. In essence, closed-cycle cryostats can run uninterrupted for extended periods of time and in remote locations without needing maintenance. They can also be fitted in any orientation to suit the cryogenic application. Generally, they are more expensive to buy than other types of

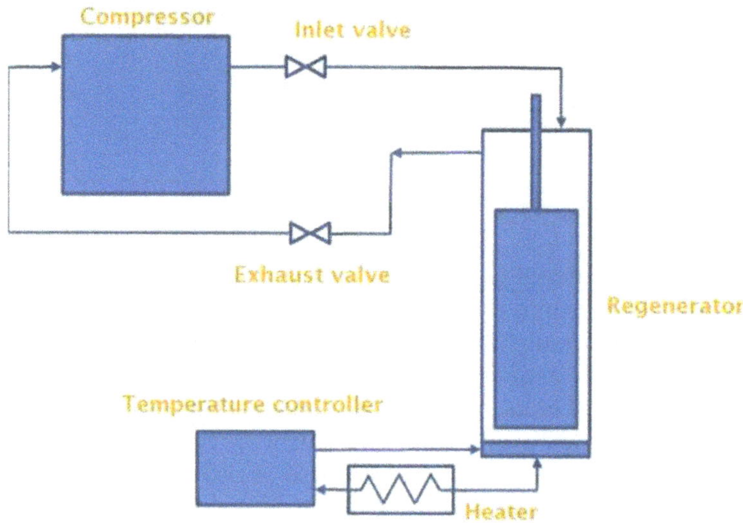

Figure 3.6. Schematic of a closed-cycle cryostat.

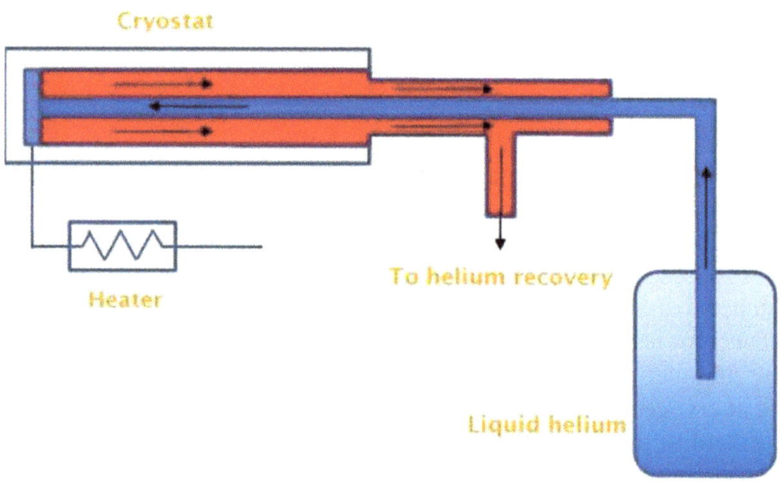

Figure 3.7. Schematic of a continuous-flow cryostat.

cryostat and they use a relatively large amount of electricity to operate. These costs can however be balanced by the fact they do not consume continuous amounts of cryogens.

Continuous-flow cryostats [7]
Continuous-flow cryostats (figure 3.7) consist of a heat exchanger cooling system which utilises the enormous cooling power of liquid cryogens, typically liquid helium and liquid nitrogen. As the cryogens flow into the cryostat they will gain heat from an internal heat exchanger which houses the object being cooled. This heating

transforms the cryogen into cold gas which subsequently flows through a further heat exchanger, normally attached to a set of radiation shields. The gas finally exits the cryostat and is automatically replenished by a steady flow of cryogens from a storage vessel. Temperature control of the object being cooled is provided by controlling the flow rate of the cryogen entering the system, together with a small heater and temperature sensor being fitted to the object.

As continuous-flow cryostats contain nothing more than the liquid cryogen cooling circuit, they tend to be more compact, lower in mass and smaller than other types of cryostats. Their operating limit is governed by the volume of cryogens supplied from the associated storage vessel. They also suffer less accompanying vibration than the closed-cycle cryostats as they have no internal moving parts, so can be utilised where stable conditions are required, such as ground-based astronomy detectors. As the supply of liquid cryogens is normally automated, they do not require operator intervention and would fail-safe in the event of a power failure. Continuous-flow cryostats can, however, consume large quantities of liquid cryogens during their operating period, the expense of which needs to be considered as part of their overall operating costs. To mitigate this to some extent, as the scarcity of liquid helium in particular has become more pronounced, many facilities operate a capture and recover system of the helium gas as it is released from continuous-flow cryostats. These facilities can themselves be costly to initially set up and run.

Liquid bath cryostats [8]

In liquid bath cryostats (figure 3.8), an integral cold-plate is arranged such that it is in thermal contact with a liquid helium storage bath. This cold-plate is part of the inner vacuum chamber of the cryostat and heat from it is extracted by the liquid helium which, as a result, will boil away as its temperature is increased. The rate of this evaporation is controlled with the use of an internal helium cold valve.

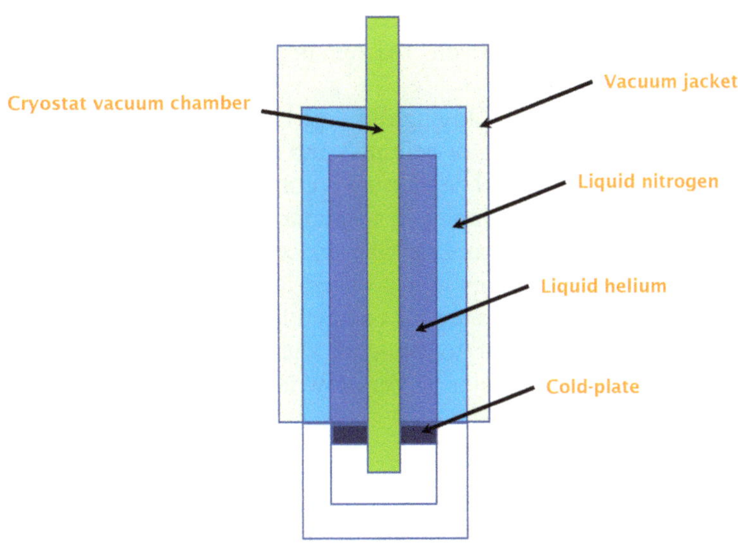

Figure 3.8. Schematic of a liquid bath cryostat.

Generally, the liquid helium bath is surrounded by a secondary bath filled with liquid nitrogen which is used to cool radiation shields around the cold-plate assembly. Both cryogens need to be replenished as they boil away, where the rate of depletion is dependent on the size of the cryostat. The object to be cooled is inserted into the cryostat vacuum chamber, normally attached to the end of a stick assembly. The object is typically surrounded by a low-pressure helium exchange gas inside this chamber which provides the means of extracting heat from the object through the cold-plate. The object temperature is controlled by adjusting the flow of liquid helium through the cold valve as well as attaching heaters and temperature sensors directly to it. Liquid bath cryostats can operate in the temperature range of 1.5 K to 300 K.

Liquid bath cryostats are relatively easy to operate and provide an enormous amount of cooling power for cryogenic applications, allowing large objects to be cooled to extremely low temperatures. They can, however, be quite bulky in size as they need to allow for internal storage of the cryogens and, unless special constructions are employed to retain the cryogens, they can only be used in a vertical orientation. They need to be replenished periodically for long-term applications, but obviously do not have to be continually connected to cryogen transfer-lines as in the continuous-flow cryostats. As ever, availability of cryogens dictates the economics of using liquid bath cryostats, although recovery systems can go some way to mitigate this expense.

Multi-stage cryostats

If cryogenic temperatures are required below that of liquid helium (4 K), it is necessary to provide additional cooling stages to a cryostat assembly. Temperatures in the region of 1 K or below can be attained by linking the cold-plate of a cryostat to a device known as a 1-Kelvin pot. This is a small vessel which is filled specifically with helium-3 isotope, connected to a vacuum pump. By pumping on this vessel, the vapour pressure above the liquid helium surface is lowered sufficiently to allow energetic molecules to be evaporated from the liquid. This is a process known as evaporative cooling and, as a result, the remaining liquid will cool sufficiently to temperatures down to 1 K.

An example of a cryostat employing this type of cooling is a dilution refrigerator (figure 3.9). Here, a combination of helium-3 and helium-4 is mixed to form a phase boundary in a mixing chamber. The mixture is condensed using a 1-Kelvin pot, with subsequent vacuum pumping employed to manage the evaporative cooling of the system. The lower temperature cooling takes place as the helium-3 moves across the phase boundary. In these devices it is possible to achieve temperatures as low as 1 mK. Dry dilution refrigerators are also available where the cryogens are replaced by a cryocooler which is used to pre-cool the helium-3/helium-4 mix. This is described in more detail in chapter 5 ‘Very low temperature techniques’.

3.1.5 Cryostat construction

When considering the construction of any type of cryostat, there are some common principles for how they are both designed and manufactured. By requirement, they are made from materials suitable for the application, invariably metals, which must

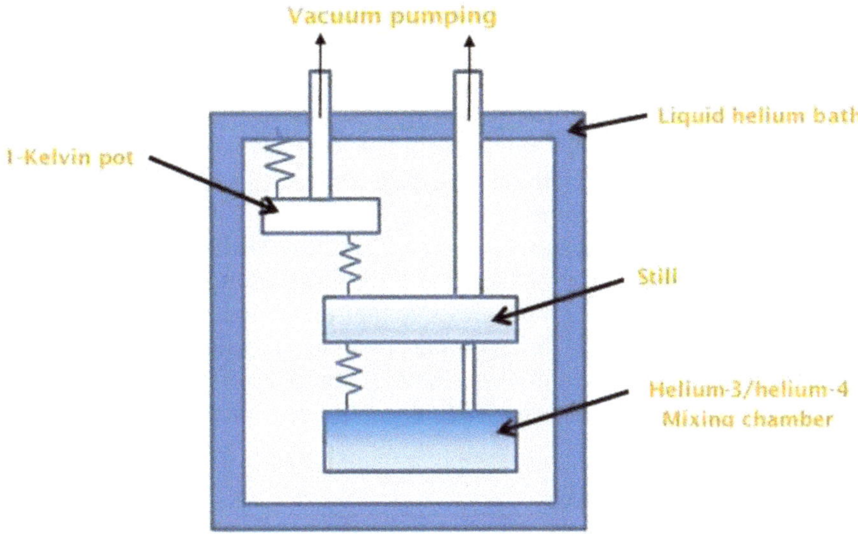

Figure 3.9. Schematic of a dilution refrigerator.

be fabricated in such a way as to be both structurally stable and safe. Cryostats typically consist of chambers which are either pressurised or evacuated and these need to be designed to certain standards so that they do not fail during use. Vacuum vessels in particular are prone to the phenomenon of buckling and collapse, especially where thin-wall sections of materials have been selected. As such, the design of these components must comply with either the Pressure Equipment Regulations in the UK, the Pressure Equipment Directive in the EU, or the ASME Boiler and Pressure Vessel Code in the USA. Any cryostat containing liquid cryogens must also be fitted with a bursting disc or a pressure relief valve to protect against over-pressurisation. This can occur in the event of a leakage of these cryogens into the outer vacuum chamber of the cryostat, whereby the liquid will rapidly evaporate, building up pressure in the chamber.

Most cryostats are of a cylindrical construction, made from stainless steel sheet with its excellent corrosion resistance properties, which has been formed into the tubular shape and welded together. However, this inherent corrosion resistance can be greatly diminished in the welded regions by the precipitation of chromium carbides at the grain boundaries. This often causes intergranular attack, which can eventually result in weld failure. One method of preventing this problem is through the specific use of low carbon grades of stainless steel such as grade 316L (1.4404) for all construction applications as they are less likely to suffer from the formation of the chromium carbides during welding [9].

Tungsten Inert Gas (TIG) and Metal Inert Gas (MIG) welding are the most common methods employed. As the cylinders need to comply with the relevant regulatory standards for pressurised vessels, non-destructive testing of these welds is often necessary. Sound welding procedures and suitable qualifications being held by the persons undertaking them are required as part of this compliance obligation.

For materials used within the body of the cryostat, especially where the working temperatures are likely to be in the cryogenic region, the selection of the correct grades of materials becomes critical. The so-called ferritic and martensitic grades of stainless steel, for example, become brittle as the surrounding temperature is reduced. Typical ferritic grades would include 430 (1.4016) and martensitic grades 410 (1.4408). The austenitic grades however, such as 304 (1.4301) and 316 (1.4401), remain tough at cryogenic temperatures [10].

Some materials, such as plastics, become incredibly brittle at cryogenic temperatures. As they contract in size due to the resultant change in temperature, the internal stresses increase to such a high value that the materials shatter. In addition, the rapid cooling or heating of materials used in cryostats can also introduce undesirable internal stresses. For this reason, components should be both cooled down at a slow rate and heated up gradually to prevent these detrimental effects.

3.1.6 Sealing and leak detection

All cryostats will be made up of vacuum or pressure components which will require some form of seal arrangement to maintain integrity of the overall circuit. Sealing at low temperatures can be a problem as many commonly used sealing materials, such as polymers, become brittle at cryogenic temperatures. Metal seals of copper and aluminium are frequently used, but these tend to require thick mating flanges to maintain leak tightness. Alternatively, the soft metal indium forms a reliable seal at cryogenic temperatures in most applications without the need of large flange arrangements.

No vacuum or pressure system is absolutely leak-tight, so it is common for the integrity of cryostat assemblies to be defined in terms of a satisfactory leak rate. In any cryostat vacuum system, there are two sources central to this leak rate; the finite leak and the so-called outgassing from the materials making up the vacuum circuit. These can be easily distinguished as the leak will remain a constant over time, whereas the outgassing rate will eventually tend towards a zero value as the amount of gas released from the materials becomes exhausted (figure 3.10).

The method of testing for leaks in cryostats, and thereby defining a leak rate, are varied but typically work on one of two principles. The first is where the system is

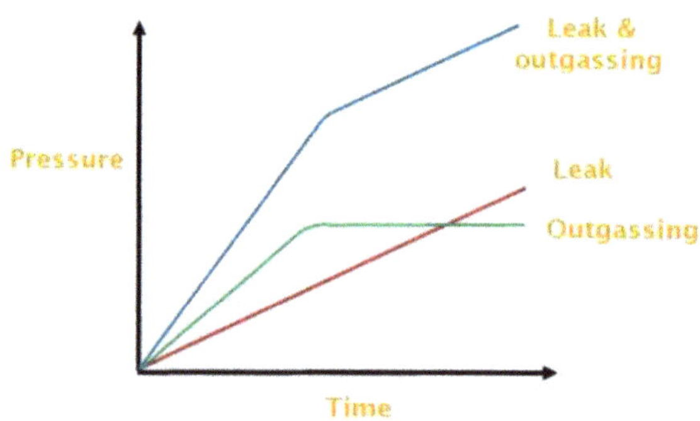

Figure 3.10. Leaks and outgassing in a cryostat system.

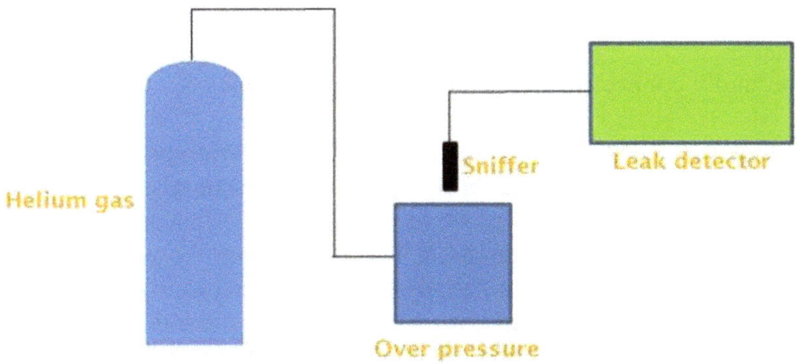

Figure 3.11. Sniffer leak detection method.

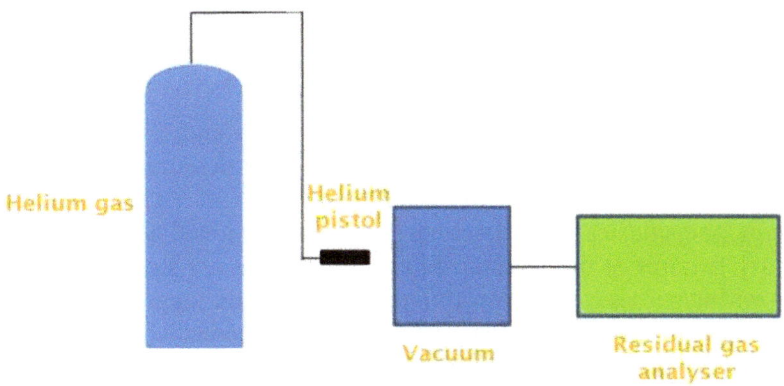

Figure 3.12. Vacuum leak detection method.

subjected to an over-pressure using helium gas. A sniffer device, linked to a helium mass spectrometer, is then employed to detect any leakage of gas from the system (figure 3.11). The second method is where the circuit is evacuated and linked directly to a residual gas analyser. Helium gas is then introduced externally to the cryostat parts and any ingress of the gas through leakages is detected by the analyser (figure 3.12). Both methods can determine how much gas is detected over time which is converted into the defining leak rate value. Helium gas is used due to its small molecule size, which can permeate through the smallest of leakage points. Leak rates are therefore generally quoted in terms of a 'standard helium leak rate' in mbar l/s [11].

3.2 Thermal balance and insulation techniques

Bryan Shaughnessy

3.2.1 Introduction

Cryostats must work within the capabilities of their cooling systems (see previous section). Put simply, if the heat loads are too high for the cooling system then the required temperatures will not be achieved. Appropriate steps are needed to:

1. thermally insulate and isolate where required to minimise unwanted heat loads;
2. thermally connect those regions that must be cooled.

As well as an understanding of appropriate material properties and design practice (such as avoiding thermal bridges), a robust design relies on careful process, an understanding of heat flow paths and management of uncertainties.

Thermal balance is a term often used. This defines that the system is operating at a thermodynamic equilibrium (i.e., steady-state) with heat sources/sinks and temperatures being stable with respect to time. Many cryostats are indeed designed to work at steady-state and this approach simplifies the analysis and testing of systems. Nevertheless, definition of an efficient design for steady-state operation must also consider the impact on the transient performance, such as cool-down and warm-up times.

3.2.2 Heat flows

The designer must understand all the heat flows into and out of the system. At steady-state these will sum to zero. These heat flows determine the temperatures achieved and the effectiveness of the system.

The term 'parasitic heat leak' describes sources of unwanted heat that are often due to necessary features of the cryostat. Examples of these are structural supports and electrical wiring/harnesses between different temperature stages of the equipment. These can be minimised by selecting low thermal conductivity materials and by increasing the ratio of length to cross-sectional area of the part. Electrical harnesses often have a copper conductor core, which has a very high thermal conductivity. Stainless steel, phosphor-bronze and manganin are amongst the options for a lower thermal conductivity core [12]. When selecting alternative harness materials, the designer should be aware that, due to differences in material resistivity, there might be (i) voltage drops that affect instrument performance and (ii) ohmic dissipation that adds to parasitic heat loads. Further, groups of harness wires may be twisted together, which can affect the overall length required for heat conduction calculations. The effect of any electrical shielding (e.g., braids) must also be considered.

Cryostats are usually within a very low-pressure environment, meaning that only thermal radiation needs to be considered in addition to thermal conduction (very sensitive equipment, or those that cannot achieve a low pressure, may need to consider heat transfer through the residual gas). Thermal radiation will be an unavoidable heat path between many surfaces within a cryostat. This is generally minimised by using low-emissivity (high reflectivity) finishes. However, high-emissivity (i.e., black) surfaces are sometimes required to reduce optical reflections (stray-light) in measurement systems or to promote radiative cooling of enclosed equipment.

Multi-layer insulation (MLI) is often used within cryostats and is discussed in more detail in the next section. MLI has a number of synonyms, including super-insulation, insulation blankets and reflective insulation. The heat leak through MLI when used in a vacuum is a combination of both thermal radiation and thermal conduction (due to contact between the individual layers). Optimal MLI performance requires very careful design, production and integration. Nevertheless, there is considerable uncertainty in MLI performance, which must be accounted for in the design process.

It is useful to show a breakdown of heat flows between different regions of the equipment. There is no prescribed way of doing this—it should however be meaningful in the context of the system being designed. Examples are shown in table 3.1 (for the James Webb Space Telescope (JWST) Mid InfraRed Instrument (MIRI)) and figure 3.13 (for the Planck spacecraft). Importantly these representations show the relative contributions of different features of the equipment. They identify those regions critical to the overall heat load. Thus, they show where design effort should be focussed to ensure the overall heat load is compatible with the capability of the cooling system.

Table 3.1. Example of a steady-state heat load breakdown table for the JWST MIRI; reprinted by permission of the American Institute of Aeronautics and Astronautics, Inc [15].

	Calculated heat load (mW)	Uncertainty (%)	Uncertainty (mW)
Conducted			
ISIM to STRUTS	7.581	+0/0	0/0
ISIM to STRUT SLI	3.723	+0/0	0/0
ISIM to PURGE PIPE	0.682	+0/0	0/0
ISIM to ICE HARNESS	3.919	+0/0	0/0
ISIM to FPH HARNESS	1.686	+0/0	0/0
ISIM to IRSU HARNESS	0.239	+0/0	0/0
Radiated			
ISIM to STRUT SLI	0.632	+60/60	+0.379/0.379
ISIM to BODY MLI	1.556	+60/60	+0.934/0.934
ISIM to APERTURE EDGE	0.652	+5/−5	+0.033/−0.033
ISIM through APERTURE	1.863	+5/−5	+0.093/−0.093
SHIELD to STRUT SLI	−0.368	+60/60	−0.221/−0.221
SHIELD to BODY MLI	1.175	+60/60	+0.705/0.705
SHIELD to APERTURE EDGE	0.001	+5/−5	0/0
SHIELD through APERTURE	0.0005	+5/−5	0/0
Dissipated			
Internal dissipations	10.4630	+15/−8	+1.607/−0.796
		Sub-total (RSS):	+2/1.5
		Heat load measurement:	+4/4
Total nominal heat load:	33.8	Overall uncertainty	+6.0/−5.5

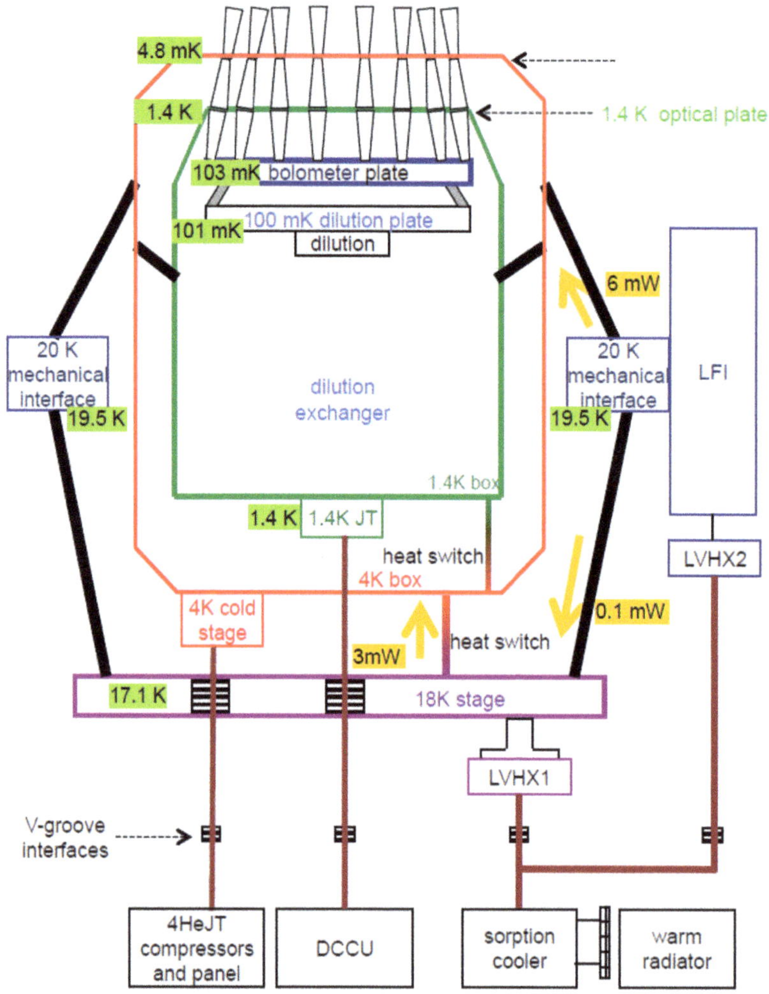

Figure 3.13. Example of a heat flow diagram for the Planck High Frequency Instrument (HFI) 4 K system [16], reproduced with permission, copyright ESO.

Table 3.1 includes estimates of uncertainty in the tabulated heat loads. At early stages of design, this approach helps identify whether specific testing (or design change) is needed to reduce the risk to the overall heat load. For complex systems, it may be useful to use such data to create heat load budgets or requirements for individual sub-systems. For example, defining heat leak requirements to a harness supplier.

The table has many acronyms but these just refer to different parts of the spacecraft. In this example, conduction, radiation and dissipation are considered. Other examples may include things like fluid flow, refrigeration etc.

3.2.3 Thermal modelling

Thermal modelling software is an aid to the design engineer which can allow for the precise calculations that help to justify and verify the design that the engineer has developed.

At the outset, the thermal engineer should review the requirements and any conceptual designs that may exist. Based on this a thermal design concept can be established, by considering good practice in defining what items should be thermally isolated and what should be thermally coupled.

Once a concept is designed, thermal software models can then be created, which allow the design to be further developed. On many projects this will go through a small number of phases as the overall design is refined. The thermal engineer will use the analysis results to justify the development of the design and any changes that are required as the overall design evolves.

Careful testing of the overall system and using these data to improve modelling (thermal model correlation) is recommended. On space programmes, a representative 'prototype' is often tested in this way to verify the design before the 'flight' item is produced. Updating models based on test data of the final deliverable is very useful because that model can then be used as a diagnostic tool.

3.2.4 Managing uncertainty

Managing uncertainty is crucial to the successful design of cryostats. Ultimately, the cooling system will only be able to remove a fixed amount of heat (at a certain temperature). If uncertainties in heat load contributions add up, then there is a risk that the system will not reach temperature. Further, if there are unknowns within the construction of the cryostat, crucial components may not achieve the desired temperature, even if the overall heat load is within specification.

Such risks are mitigated by:

1. Having an understanding of the relative uncertainties of different features within the cryostat—and calculating what the impact is on key performance criteria (such as cooling system capability, heater power for temperature control);
2. Having an awareness of assembly issues that may affect thermal performance;
3. Calculating the impact of uncertainty on both temperature and heat loads;
4. Using outcomes of an uncertainty analysis to define where specific testing, property measurement or assembly procedures are required to reduce risk;
5. Allocating a margin on cooling system capability and calculated heat loads, margin can be reduced as the design matures.

Usually, the overall uncertainty is calculated through a sensitivity analysis of contributing factors (e.g., as shown in table 3.1). Often the difficulty is identifying those contributing factors and defining sensible parameter ranges. For cryogenic systems, the parameter uncertainties must usually be determined on a case-by-case basis. Typical sources of uncertainty include:

- Model parameters—discretisation, completeness of model representation, solver parameters, convergence;
- Environmental parameters—understanding of the boundary conditions;

- Material properties—these can be based on the spread of known property data for similar materials;
- Dimensions—particularly of critical items like thermal straps;
- Heat dissipations;
- Heat transfer across bolted/pressed joints—can be a source of considerable uncertainty. A parameter uncertainty of ±50% is not uncommon;
- MLI performance—there can be considerable uncertainty in MLI performance. A parameter uncertainty of ±50% is not uncommon.

The engineer should be aware of potential assembly issues that can affect the thermal performance. Small spaces can increase the risk of thermal bridging between temperature stages. This can occur, for example, if there is no clear definition of the harness routing and it inadvertently bridges two radiative shields. If assembly procedures are not well defined, then further uncertainty may be introduced. Prime examples of these relate to pre-loading and torquing of mechanical fasteners and integration steps used for MLI.

3.3 Insulation and isolation

Bryan Shaughnessy

3.3.1 Introduction

In general, solid insulation such as foam is used on the external (airside) surfaces of equipment and radiative isolation or insulation is used within the vacuum vessel. Radiative isolation implies the use of low-emissivity surfaces or shields. Radiative insulation implies the use of MLI, which is discussed in more detail below. If implemented properly, MLI offers a high level of insulation down to moderate cryogenic temperatures (~40 K).

3.3.2 Multi-layer insulation (MLI)

Overview

MLI is the most common form of insulation used on spacecraft and is frequently used as insulation within cryostats. Although it has the potential to provide high levels of insulation, careful attention must be given to the design and assembly to ensure that no unnecessary heat leaks are introduced. Further details of MLI design for space applications are given in the reference section [24].

Petersen is credited with defining MLI as we know it now [22]—and much has been published since. In its basic form, it comprises a stack of separated layers of low-emissivity (reflective) material (figure 3.14). The low-emissivity surfaces minimise radiative heat transfer and reducing the contact between each layer minimises conductive heat transfer. Use in a vacuum minimises gas conductive and convective heat transfer.

As the blankets are used in a vacuum, a mechanism is required for venting when the environment pressure is reduced. Often blankets use perforated materials but it is

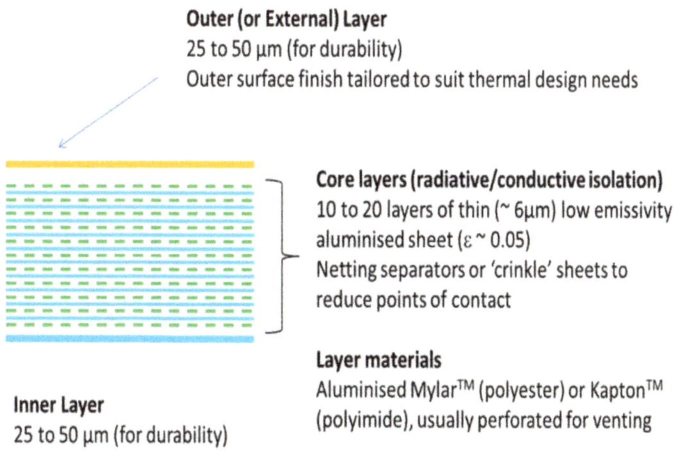

Figure 3.14. Example of multi-layer insulation construction.

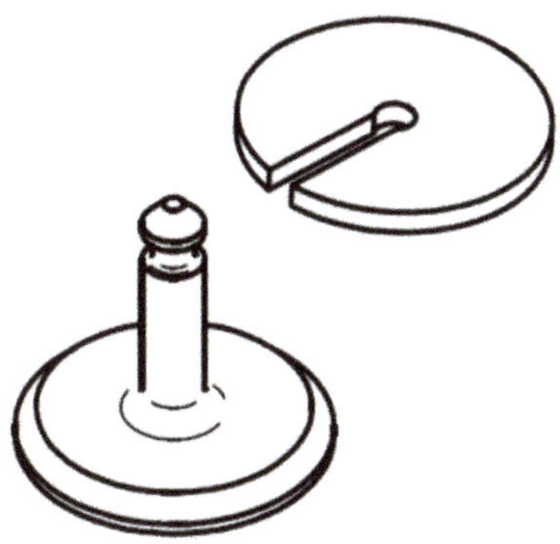

Figure 3.15. A RAL Space MLI attachment stud and clip.

also possible to include dedicated gaps or holes for venting. It is good practice to allow the blankets to vent away from any contamination-sensitive equipment.

The edges of the blankets are sealed by folding over the outer layer and securing with an appropriate tape. An alternative approach is to stitch this overlap.

There are a number of methods of attaching MLI blankets. A standard approach is to use studs and clips (figure 3.15). These can be used over the wide temperature range experienced by cryostats. The studs are attached to either the equipment or another blanket. Holes are punched through the blankets so that they can be placed onto the studs and then clipped in position. This is advantageous, because blankets may then be removed and refitted with relative ease. For cryostats, studs will usually be made from an aluminium alloy or polytetrafluoroethylene (PTFE). Differential thermal expansion is a consideration when selecting the stud material and the attachment method. Studs on equipment are usually bonded with a suitable epoxy, although mechanical fastening is feasible. Where heat flow requirements preclude

Figure 3.16. Sketch of two-part stud.

the use of aluminium, a two-part stud construction may be used, comprising an aluminium base with a PTFE post (figures 3.15 and 3.16).

Other methods of attaching MLI are (1) 'Hook and pile' (e.g., VELCRO® brand), (2) lacing and (3) tape:

1. 'Hook and pile' fasteners are available in organic and metallic materials [24] and are usually sewn or bonded in position. The temperature limits of the fasteners and any adhesives would need to be considered before use because some sources show they are not suitable for cryogenic applications [24]. A further consideration with this type of attachment is the process of removal. To avoid tearing the blanket, care should be taken to separate at the fastener, rather than pulling at the blanket to release the fastener.
2. Lacing cord is an additional option for connecting adjacent blankets or securing blankets to features on the hardware. Grommet reinforcements are required to avoid the cord cutting into the blanket. The lace ends should be tied in a square knot—and staking with epoxy to prevent the knot unravelling is recommended.
3. Taping can also be used for securing blankets, although it is not recommended as the primary method of attachment. An example would be closing gaps at instrument ports or between adjacent blankets (when the thermal design requires that views though gaps should be minimised). Removal of the blankets is of course an important consideration. Tape will usually need to be cut very carefully to be able to separate blankets—and it is unlikely that the tape remaining on the instrument or blanket can be removed. A final observation is that some tapes do not stick to themselves as well as they do to other materials. This should be tested as part of the design process.

The choice of layer materials for MLI depends on factors such as temperature range, thermo-optical properties and electrical surface resistance. For cryogenic applications, it is likely that aluminised polyester (e.g., Mylar™) would be used, separated by a polyester netting (e.g., Dacron™). The aluminised coatings are typically 300–1000 Å (0.03–0.1 μm) thick. At colder temperatures, such coatings will appear more transparent to thermal radiation [17]. Therefore, it is prudent to

Figure 3.17. Specialist MLI on the JWST MIRI (Credit: STFC, RAL Space).

consider shields or options that can use a thicker coating. For example, specialist insulation was developed for the JWST MIRI, including layers with 5000 Å (0.5 μm) aluminium to operate at about 7 K in a 20 K environment (figure 3.17) [12].

The support struts in figure 3.17 are wrapped in a low-emissivity single-layer insulation (SLI). Aluminised coating can contribute to the parasitic heat load into the cooling system. However, the thermal conductivity of film coatings can be less than for the bulk material. This is because the mean free path of electrons, l_e, may be much larger than the thickness of the aluminium coating on the MLI material. The effective conductivity of the aluminium coating can be determined from the approximate relationship [23]:

$$k_{\text{coating}} = k_{\text{bulk}}\left(\frac{3}{4}\right)\left(\frac{t}{l_e}\right)\ln\left(\frac{l_e}{t}\right)$$

where t is the thickness of the coating.

The mean free path of electrons is proportional to $1/T^3$. At room temperature (300 K) the mean free path of electrons for pure aluminium is about 0.04 microns, therefore:

$$l_e(T) = 0.04\left(\frac{300^3}{T^3}\right)$$

Design, construction and fitting

Generally, MLI should be designed using templates, not unlike those used for clothing. These are made from a semi-transparent drafting film to represent the individual blankets and will contain much of the information needed for manufacture and fitting. A computer aided design (CAD) package can be used to develop the template geometry, although they can also be developed by hand on the actual hardware. It is useful for the CAD engineer to be familiar with the constraints of

MLI handling and production, so that they can develop a geometry that is easily produced from flat sheets of material.

Once a design has been developed it is very good practice to do a fit-check of the templates to the hardware. CAD designs may not always capture the exact location of some features, such as routing of a harness. Further, it can be difficult to represent the pliability of blankets in the CAD model. Aside from confirming that the planned design fits the geometry of the hardware, the fit-check confirms there is access to underlying features where needed (e.g., harness connectors) and that attachment points are accessible. The fully assembled hardware might not be present, so some allowance must be made for items that are not present at the time of the fit-check (harness bundles for example). The fit-check should also be used to confirm the best sequence for integrating the individual blankets. Any changes are incorporated into the CAD model, from which production templates can be produced.

The first step in manufacturing blankets is to produce a 'lay-up' of material. This is the stack of thinner layers and netting that form the bulk of the blanket. This can be done by hand, but there are some suppliers that provide a pre-laid MLI blanket that can then be cut to shape. At this stage, the thicker inner layer is also added. Individual blankets are cut from the lay-up either by hand or by machine. The outer layer is added last as this is usually oversized so that its perimeter can be folded over to seal the edge of the blanket.

Recommendations for good practice in MLI design, manufacture and integration are:

- Keep blankets as large as practical (for thermal efficiency), however split into smaller blankets if it helps fit or integration;
- Keep blanket shapes, geometry, and cut-outs simple;
  - Excessive folds and seams reduce performance and increase manufacturing time/cost;
  - Rectangular cut-outs are easier to prepare than circular
- Design to keep blankets and layers 'loose';
  - Consider use of formers during MLI production for blankets that fit around edges or cylinders;
- Aim for adjacent blankets to overlap each other by a minimum 50 mm;
- Optimal performance relies on correct integration (e.g., ensuring adjacent blankets overlap correctly). A precise integration procedure should be defined and followed.

The insulation performance of MLI can be further improved by careful attention to design and integration. For example, blankets will usually have a 'cold' and a 'warm' side, with the 'cold' side being adjacent to the cryostat. A conventional design will have adjacent blankets overlapping each other so that the cold side of one blanket will be in contact with the warm side of the other. An improvement is to look at alternative overlap designs. Some ideas are shown in figure 3.18. A development of these ideas is to wrap hardware in several layers of individual blankets, so that joins between blankets are covered completely by blankets above (figure 3.19).

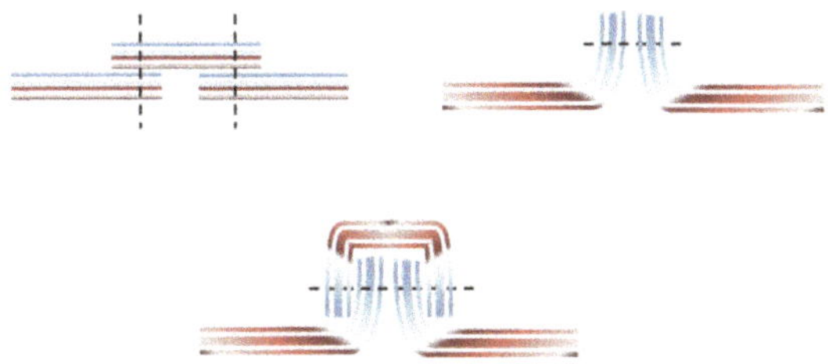

Figure 3.18. Options for improving joints between adjacent MLI blankets.

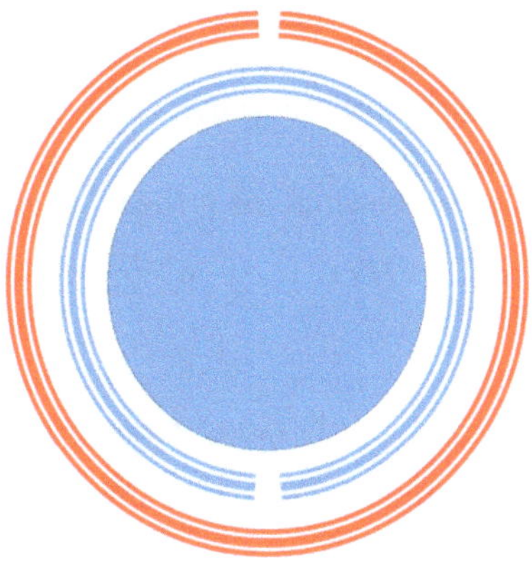

Figure 3.19. Schematic of double blanket wrap with non-aligned overlaps.

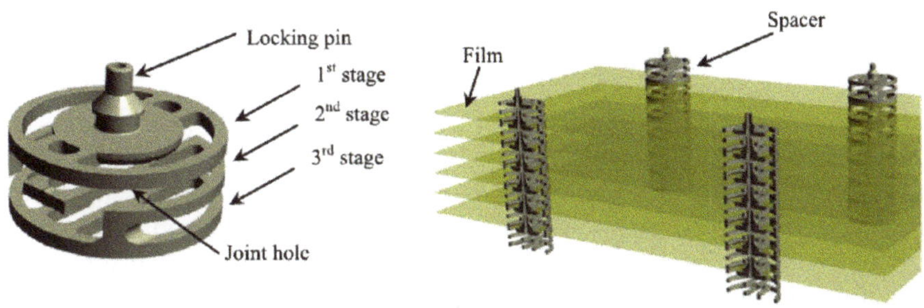

Figure 3.20. JAXA concept for layer supports. Reprinted from [13], copyright (2014), with permission from Elsevier.

Researchers at the Japanese Space Agency (JAXA) have presented a solution that uses thermally isolating stand-offs to individually support each reflective layer (figure 3.20) [13]. This approach offers a superior performance but may be more time consuming to assemble onto the hardware.

Thermal performance

MLI performance describes the heat flow between the hot and cold sides of blanket. This varies across the surface of an individual blanket and from blanket to blanket.

However, it is usual practice to define the typical performance of individual blankets or a set of blankets.

Heat flow across blankets in a vacuum is a combination of conduction and radiation. MLI thermal performance is represented as conductive couplings, radiative couplings or a combination of the two. Representation as a radiative coupling is common, using the effective emissivity to define the performance of the blanket:

$$Q = \varepsilon_{\text{eff}} A\sigma(T_{\text{outer}}^4 - T_{\text{inner}}^4)$$

where ε_{eff} is the effective emissivity and σ is the Stefan–Boltzmann constant.

This is not saying that the heat transfer is purely radiative. It assumes that the combined effect or conductive and radiative heat transfer can be adequately represented in terms of a radiative coupling. The opposite would be true if a conductive representation were used.

Performance is sometimes represented as a 'heat transfer coefficient':

$$HTC = \varepsilon_{\text{eff}} 4\sigma T_m^3$$

where

$$4T_m^3 = (T_{\text{outer}}^2 + T_{\text{inner}}^2)(T_{\text{outer}} + T_{\text{inner}})$$

Figure 3.21 shows some examples of effective emissivity of different blankets as a function of number of layers and temperature. The following observations are made:

- There is a large spread in performance for different blankets;
- Increasing the number of layers improves performance up to a point. After about 20 layers there is very little further improvement;
- The performance decreases with temperature. Therefore, more attention must be paid to the design and application of MLI in cryogenic systems.

Contamination control

Contamination is an important consideration for the thermal performance of radiative shields, MLI and anything else that relies on maintaining a certain thermo-optical property. Steps must also be taken to avoid introducing sources of contamination that may be detrimental to other features of the cryostat.

Contamination can be considered either particulate or molecular in nature. Particulates can arise from the manufacturing or assembly environment (e.g., dust) and may include fibres generated when cutting MLI materials. In extreme cases, particulates can partially obscure and thus alter the radiative properties of a surface. If the percentage area coverage (PAC) of particulates is known, the impact on the emissivity of a surface can be estimated using a weighted average, assuming a high emissivity for the particulates.

Molecular contamination can arise from outgassing of materials when under vacuum as well as molecules present during production. Under vacuum, polymers in particular outgas water and other species, which can potentially freeze onto cold surfaces and increase their emissivity (see references for more detail [29]). During

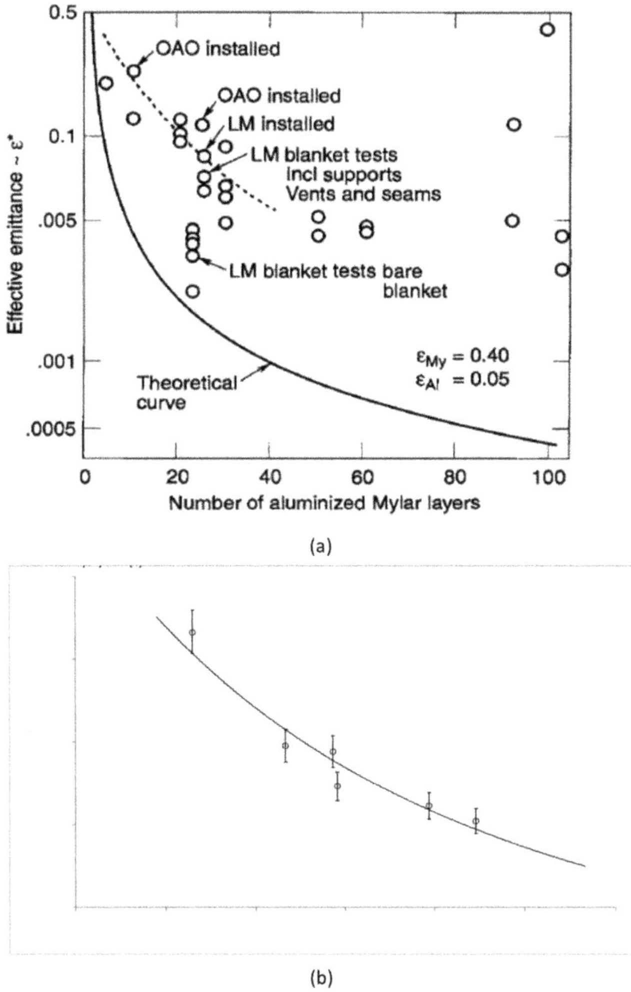

Figure 3.21. Examples of the wide variation in MLI performance (a) from [37] and (b) RAL Space measured data. Courtesy of and reprinted by permission of the Aerospace Corporation.

production, molecular contamination can be introduced by factors such as inappropriate materials, dirty tools or work area, cutting fluids used in manufacture, poor cleaning and poor handling (e.g., finger grease).

Although tailored to the design of space missions, the reference section provides useful background to cleanliness and contamination control that could be adapted to cryostat design (see [25, 27]). Good practice is to define a 'Cleanliness and Contamination Control Plan' that defines the controls required to limit contamination to permissible levels. There are a number of measures that can be taken:

- Controlling the cleanliness of the work environment (e.g., working in an appropriate class of clean room);
- Dress codes and definition of correct gowning and de-gowning process;

- Definition and selection of approved materials for production, consumables and tooling (e.g., those with low outgassing properties);
- Definition of processes to be used (e.g., correct methods for cleaning);
- Monitoring of local contamination levels using witness plates;
- Inspection of materials for contamination and cleaning. Appropriate storage after cleaning;
- Avoid materials with poor outgassing properties. Some outgassing data is available online [26]. Material formulations can change over time, therefore it is good practice get up-to-date measurements for critical applications;
- Condition materials—heat while under vacuum to reduce outgassing potential (i.e., a vacuum bake-out or 'bake'). Monitoring during bake-out. Appropriate storage after bake-out (in a 'dry' environment);
- Design in adequate venting paths to allow the outgassed products to escape;
- Shield sensitive items from outgassing;
- Design so that outgassing materials are maintained at a low temperature and sensitive surfaces at a high temperature—or design so that they can be heated to decontaminate;
- Include 'getter' surfaces to collect the volatiles (e.g., zeolite).

3.4 Material properties

Bryan Shaughnessy

3.4.1 Introduction

This section considers properties that define heat flows between and across solid materials, as well as the rate at which materials respond to changes in heat input.

Good practice is to apply judgement and caution to the use of material property data. Appropriate levels of uncertainties should be defined for the data being used. Literature may report different properties for seemingly identical materials. For physical properties, this may be due to differences or changes in material composition. Even slight differences to the composition of pure metals can have a significant effect on thermal conductivity at cryogenic temperatures. Further, property measurement methods will have an inherent uncertainty.

3.4.2 Thermal conductive properties

Thermal conductivity

Thermal conductivity is a measure of the ability of a material to conduct heat (allowing heat flow to be calculated). It is usually denoted by the symbol k or λ.

Thermal conduction in materials is due to (i) migration of free electrons and (ii) lattice vibrations (phonons). For pure metals, the contribution due to free electrons dominates. According to the Wiedemann–Franz law, this contribution is approximately inversely proportional to resistivity. As resistivity decreases, the thermal conductivity increases. For non-metallic (insulating) materials, the conductivity is usually due to lattice vibrations (the free electron contribution is much smaller).

The lattice vibrations are reduced by lower temperatures, resulting in a decreasing thermal conductivity. Thermal conduction by lattice vibrations (phonons) is also a significant contributor to overall conduction in non-metals and alloys.

Fourier's law describes conductive heat transfer. For one-dimensional heat transfer in the x-axis direction:

$$q_x = -k\frac{dT}{dx}$$

Here, q_x is the heat flux, having units of W m^{-2}.

The SI units of conductivity are W $(mK)^{-1}$. However, different units may be used in reference sources. Conversions between typical units are given in table 3.2.

The conductive heat transfer across a material is defined:

$$Q_{12} = k\left(\frac{A}{L}\right)(T_1 - T_2)$$

Q (W)	Heat rate (or load, or power)
k (W $(mK)^{-1}$)	Thermal conductivity
A (m^2)	Surface area
L (m)	Thickness
T (°C or K)	Temperature

This formulation assumes one-dimensional conduction between two faces across a constant cross-sectional area. An example of this is the heat flow from one end of a rod to the other.

Typical properties

Examples of room temperature properties for commonly used materials are given in table 3.3. However, interesting things happen to material properties as they are cooled through the cryogenic temperature range. There are various data available online and in the literature. A small selection are cited here [18–20, 30–33].

Pure metals have a significant increase in thermal conductivity as temperature reduces. Pure aluminium and copper are regularly used for thermal straps (figure 3.22). A peak conductivity for these is reached at temperatures below 30 K. Thermal conductivity at cryogenic temperatures varies by many orders of magnitude depending on level of purity and treatment. Metal purities often referred to as numbers of 'nines'

Table 3.2. Conversion between typical units of thermal conductivity.

	W $(mK)^{-1}$	mW $(mK)^{-1}$	W $(cmK)^{-1}$	BTU/h·ft·°F
1 W $(mK)^{-1}$ =	–	1000	0.01	0.578
1 mW $(mK)^{-1}$ =	0.001	–	10^{-5}	0.578×10^{-3}
1 W $(cmK)^{-1}$ =	100	100 000	–	57.8
1 BTU/h·ft·°F =	1.73	1730	0.0173	–

Table 3.3. Typical room temperature values of thermal conductivity.

Material	Thermal Conductivity, W $(mK)^{-1}$
Copper (99.95% Pure)	400
Aluminium (99.99% Pure)	235
Aluminium Alloy 1100	212
Aluminium Alloy 6061	155
Aluminium Alloy 5083	117
Manganin	22
Stainless Steel 304	15
Titanium alloy Ti_6Al_4V	6.7
Delrin™	0.3–0.37
PEEK™	0.35
Vespel™	0.25–1.73
PTFE™	0.25

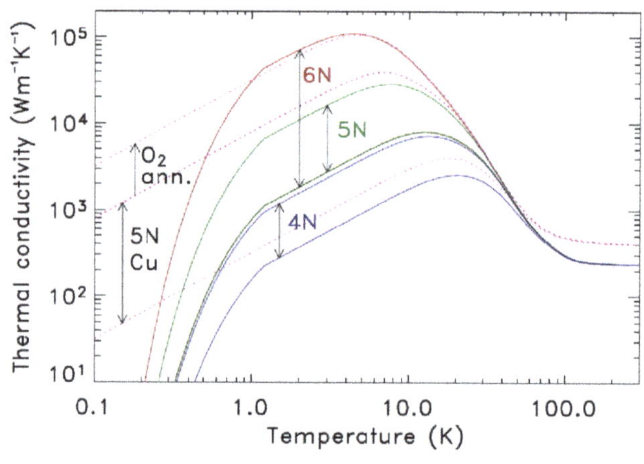

Figure 3.22. Cryogenic thermal conductivity of annealed aluminium and copper. Solid lines—annealed aluminium with 4N, 5N and 6N purity. Dashed lines—vacuum and oxygen annealed 5N purity copper. Reprinted from [31], Copyright (2005), with permission from Elsevier.

or 'Ns' (e.g., 4N, 5N and 6N). For example, 99.999% pure aluminium is referred to as 'five-nines' or 5N aluminium. Purity is sometimes represented (e.g., for copper) in terms of the residual resistivity ratio (RRR), the ratio of the electrical resistivity of a material at room temperature to that at close to 0 K.

The thermal conductivity of alloys and non-metals will decrease with temperature (figure 3.23). These thermal conductivity characteristics can be exploited to the benefit of cryogenic systems. For example, the decrease in conductivity of isolating components. The increase in conductivity of metals means that thinner (lower mass) straps can be used—however, it is worth noting the conductivity is lower when warm, so there may be a trade-off with cool-down time.

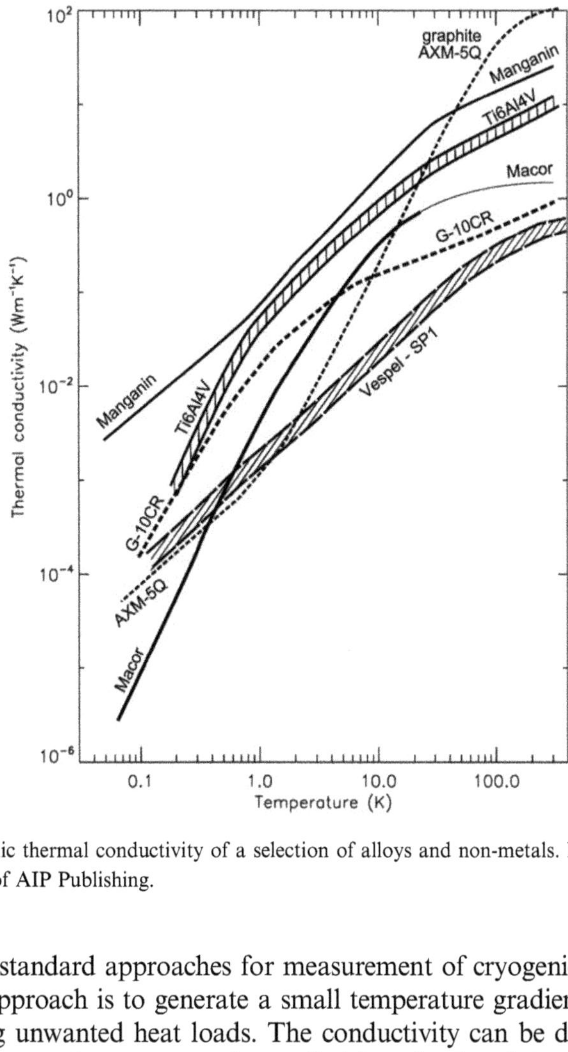

Figure 3.23. Cryogenic thermal conductivity of a selection of alloys and non-metals. Reproduced from [30], with the permission of AIP Publishing.

There are no standard approaches for measurement of cryogenic thermal properties. The basic approach is to generate a small temperature gradient across samples, while minimising unwanted heat loads. The conductivity can be determined from a knowledge of the sample geometry, the applied heat load and the measured temperature difference. Due to the wide variation in material conductivities, bespoke test measurements are often required. An example is given in the reference section [14]. At sufficiently low temperatures, the thermal conductivity can be determined from the electrical resistivity using the Wiedemann–Franz law. This has been shown to be a good approximation for most pure metals and is useful because the electrical measurements are easier to make. More information on this is given in reference [19].

Conductivity integral

In cryogenic calculations, there may be a large temperature gradient across the material. In this case, the conductivity will also change as a function of temperature though the material. The conductive heat transfer is determined using:

$$k_T = \int_0^T k(T)dT$$

By taking data for the two temperature points, the conductive heat transfer is calculated:

$$Q_{12} = \left(\frac{A}{L}\right)\left(\int_0^{T_1} k(T)dT - \int_0^{T_2} k(T)dT\right)$$

$$= \left(\frac{A}{L}\right)(k_{T_1} - k_{T_2})$$

Figure 3.24 gives an example of conductivity integral data.

Thermal radiative properties

Radiative properties define the heat flow between surfaces—the ability of a surface to absorb and emit thermal radiation.

Thermal radiation is defined as the radiation emitted due to the thermal agitation of the constituent particles of a material. A convenient reference point for thermal radiation is the 'blackbody', which refers to an idealised surface that completely absorbs all incident radiation. Real surfaces (as used in engineering applications) are defined relative to the blackbody. They are neither perfect emitters nor absorbers of radiation. The emissivity, ε, describes the ability of a surface to radiate and is defined as the radiation emitted by a surface to that emitted by a blackbody at the same temperature. (The term emittance is also used. The ending '-ivity' is sometimes used

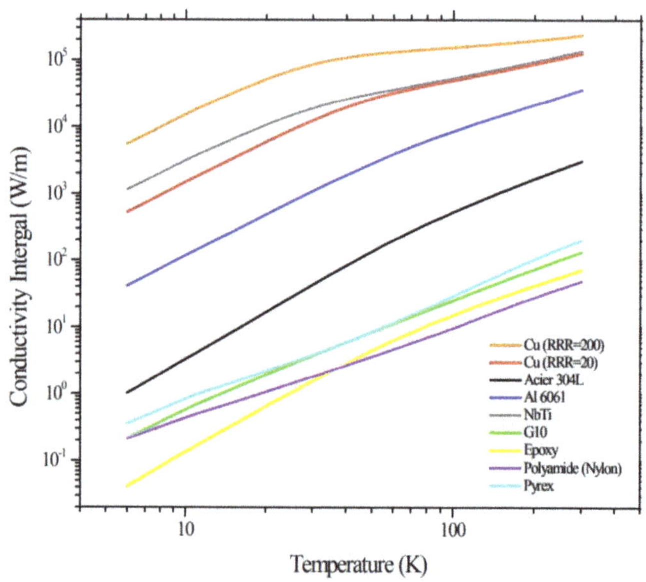

Figure 3.24. Example of thermal conductivity integral data [20]. Produced using CryoComp software, www.eckelsengineering.com. With permission of Lakeshore Cryotronics.

to identify optically smooth materials, with '-ance' being used for 'real' surfaces [28].) The emissivity has a value between zero and unity and is dimensionless. The emissivity is considered a true property of a material as it is not dependent on incident radiation [34].

Typically, non-conductors have a relatively high emissivity that is constant over a large range of emitted directions (but decreases rapidly to zero for shallow emission angles). Conductors tend to have a lower emissivity. The variation of emissivity is also dependent on the condition of the material surface and can vary erratically if the surface roughness scale is similar to, or greater than, the wavelength of the emitted radiation. Increasing surface temperature will tend to raise the emissivity of conductors. However, the emissivity of non-conductors does not have a consistent correlation with temperature [34].

The ability of a surface to absorb radiation is defined by the absorptivity. According to Kirchoff's law of thermal radiation, the absorptivity and emissivity at any specific wavelength are equal [28].

$$\varepsilon_\lambda = \alpha_\lambda$$

Although emissivity and absorptivity can vary with wavelength, a standard thermal engineering assumption is to consider their values integrated over the entire spectrum. This is the 'grey body' assumption.

$$\varepsilon = \alpha$$

For radiative heat transfer between a surface, A1, enclosed by 'large' surroundings:

$$Q_{12} = A_1 \varepsilon_1 \sigma (T_1^4 - T_2^4)$$

Q = Heat rate (or load, or power) (W)
σ = Stefan–Boltzmann constant (5.67×10^{-8} W m^{-2} K^{-4})
T = Surface temperature (K)

Measured emissivity values may be described as 'normal' and 'hemispherical'. This relates to the test measurement approach. However, this distinction is not always made in the literature, which means that the engineer must make some judgement on how to use the data. Normal emissivity is the value measured in a single, near-normal, direction (usually derived via reflectivity measurements). Hemispherical emissivity describes the value integrated in all directions from the surface. Normal and hemispherical emissivity values are not necessarily the same and this should be considered in the analysis. Hemispherical emissivity is often preferable for radiative heat transfer calculations and an estimate based on the normal emissivity may be needed. Alternatively, the variation could be bounded by sensitivity analysis.

For low-emissivity metallic surfaces, the normal emissivity may be less than the hemispherical emissivity by about 10% to 15%. At high values, the normal emissivity may over-predict the hemispherical emissivity by around 5%.

Some typical values of emissivity at room temperature are given in table 3.4. Some measurements made for the James Webb Space Telescope are given in [21].

Table 3.4. Typical values of normal emissivity.

Material	Normal emissivity
Black paints/coatings	0.75–0.95
Anodized titanium	0.44–0.48
Alodyne aluminium	0.1–0.15
Vacuum deposited aluminium	0.02–0.07
Machined stainless steel	0.14
Polished stainless steel	0.11
Beryllium copper oxidised	0.09
Polished aluminium or copper	0.03

Typically, low-emissivity (i.e., reflective) materials are used when radiative isolation is required. Examples of this are radiative shields or the sheets within MLI. High-emissivity materials are used when a strong thermal coupling is required, for example the black coatings of cooling shrouds in a test chamber.

Special care is required in specifying low-emissivity materials because of the risk that their finish may be compromised by contamination. This is a specific consideration in cryogenic applications because (i) the materials are usually selected as a key component of the thermal design, and (ii) the cold temperatures mean that surfaces are prone to collecting contamination 'frozen out' from the environment.

Variation of emissivity with temperature is also a consideration. However, as radiative heat transfer calculations are computationally intensive it may not be appropriate to use a temperature dependent emissivity in cool-down or warm-up simulations. Instead, a judgement is needed on whether a 'worst case' value should be selected. For steady-state calculations, the appropriate temperature dependent value should be used.

Frolec *et al* describe a test method for cryogenic emissivity and present a selection of data [36]. An example is shown in figure 3.25. Figure 3.26 shows data for Chemglaze z306 (also called Aeroglaze), a black paint commonly used in aerospace. The room temperature emissivity is around 0.9 [35]. This data shows a significant drop in emissivity below 100 K.

Alternatively, some radiative properties can be predicted from theory. Those of optically smooth surfaces can be predicted from electromagnetic wave theory: the Drude model, Fresnel's equations or Hagen–Rubens relation accurately predict the wavelength, directional and temperature dependent properties of smooth metals if the optical properties are known [34].

Specific heat capacity

The specific heat capacity defines the rate at which a substance changes temperature in response to a given heat input (or output). It is the heat capacity of a material divided by its mass. Its SI unit is J (kg K)$^{-1}$.

The following equation relates the rate of temperature change to the heating/cooling load applied:

$$Q = mc_p\frac{dT}{dt}$$

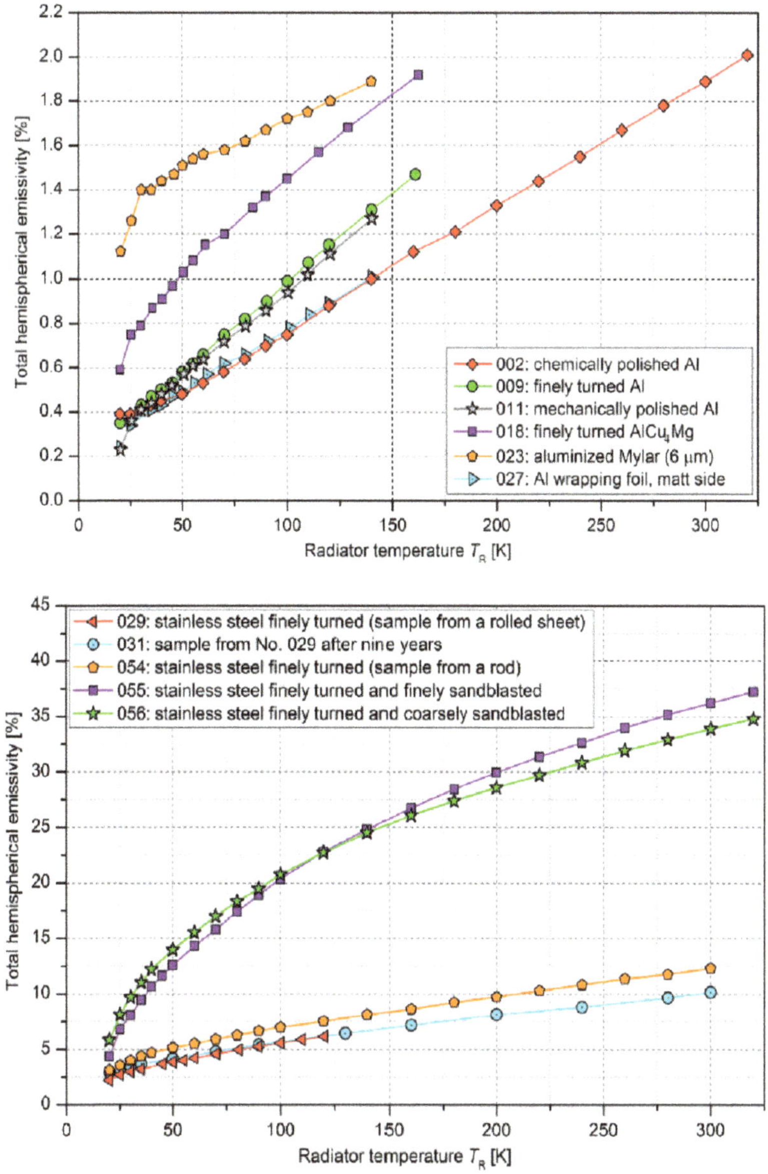

Figure 3.25. Temperature dependent emissivity for aluminium and stainless steel finishes [36], copyright (2019), with permission from Elsevier.

Q = heat rate (or load, or power, or cooling) (W);
m = mass (kg);
c_p = specific heat capacity (J (kg K)$^{-1}$).

Lattice vibration excitations are the main contribution to the heat capacity of solids at all except the lowest temperatures. This is well described by the Debye model.

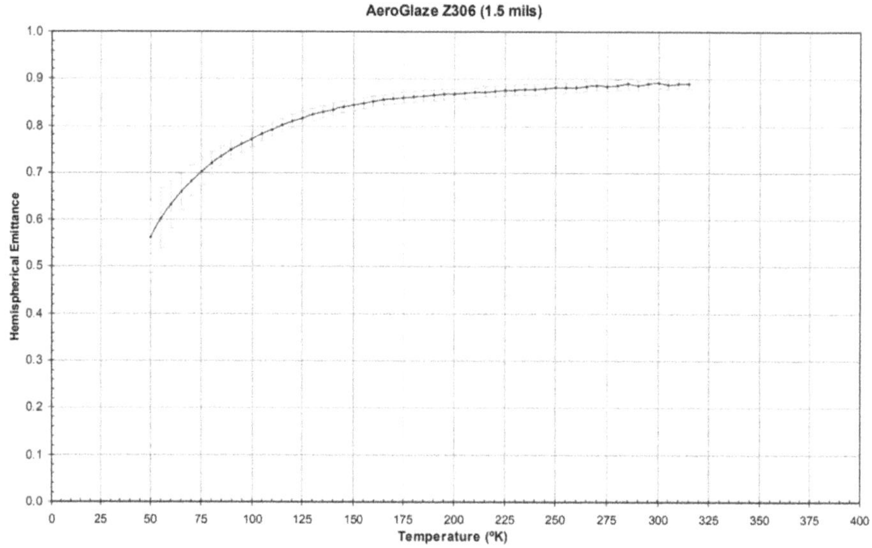

Figure 3.26. Example temperature dependent emissivity for Chemglaze z306 black paint [35]. Courtesy of NASA.

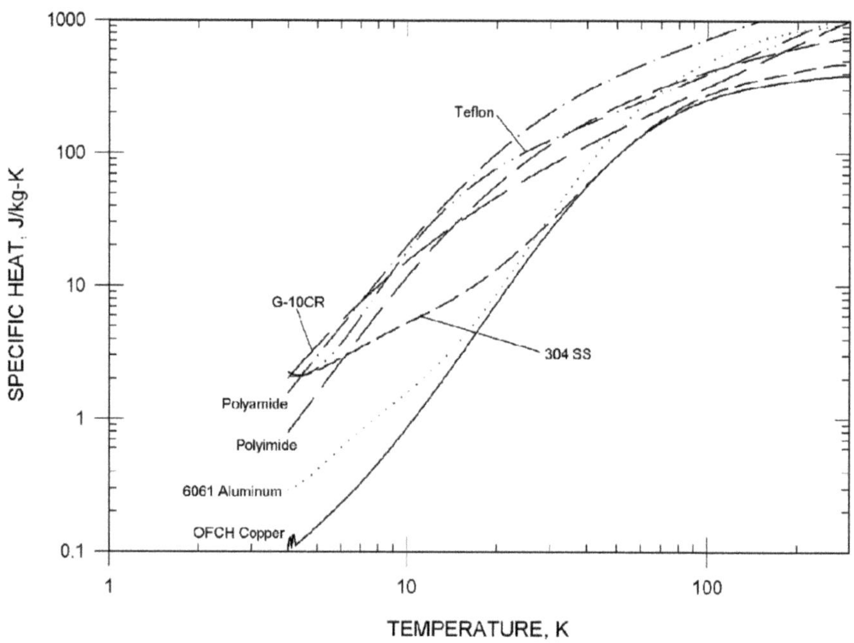

Figure 3.27. Specific heat capacity of a selection of materials [18], copyright (2002) Kluwer Academic Publishers, with permission of Springer.

As with temperature dependent thermal conductivity, it may be necessary to integrate the specific heat capacity over the temperature range of interest:

$$\int c_p = \int_{T_1}^{T_2} c_p(T)dT$$

A selection of data is shown in figure 3.27.

3.5 Heat exchangers

Tom Bradshaw

3.5.1 Introduction

When first setting out to design a heat exchanger for a specific application it is often daunting and difficult to know where to start. This section is a very brief introduction to the subject. Other books go into the subject in more detail [38, 39] and cover more complicated geometries. Both of the books cited are highly recommended.

In the analysis of heat exchangers and hydrodynamic flow, dimensionless groups are often used. These enable experimental data and theoretical descriptions to be applied across all fluids. The actual fluid used, be it a gas or liquid becomes unimportant. You will come across Reynolds numbers, Prandtl numbers, Stanton numbers, Colburn factors, Froude numbers etc. A few are used in this section and are described in table 3.5.

Table 3.5. Dimensionless groups used in this section.

Reynolds Number	$Re = \frac{\rho v d}{\eta}$	Ratio of inertial and viscous forces in fluid flow—an indicator of laminar or turbulent conditions
Nusselt Number	$Nu = \frac{Hd}{k}$	Ratio of convection and conduction in a heat transfer problem.
Prandtl Number	$Pr = \frac{C_p \eta}{K}$	Ratio of momentum diffusivity and thermal diffusivity. Helps in reducing data for various fluids.
NTU	NTU = AU/Cmin	Strictly a dimensionless parameter, it combines heat transfer, transfer area fluid flow rate and heat capacity.

Nomenclature

ρ	Density [kg/m^3]
v	Velocity [m/s]
d	Hydraulic diameter [m]
η	Viscosity [Pa s]
H	Heat transfer coefficient [W/m^{2}K]
κ	Thermal conductivity [W/m K]
A	Area for heat exchange [m^2]
U	Overall heat transfer coefficient [W/m^{2}K]
Cmin	Lower heat capacity x mass flow product [W/K]
Cp	Isobaric heat capacity [J/kg K]

The analysis of a heat exchanger takes on the following rough sequence:

(a) Determination of the hydraulic diameter for the ducts in the heat exchanger;
(b) Evaluation of the Number of Transfer Units (NTU);
(c) Calculation of the thermal conductance of the streams to the wall and the wall itself;
(d) Calculation of the heat exchanger effectiveness;
(e) Calculation of the temperatures.

Hydraulic diameter

The hydraulic diameter, dh, is used as a representative length for a specific geometry. A definition for a duct is:

$$d_h = \frac{4A}{P}$$

where A is the flow area and P is the wetted perimeter. It can be seen that the hydraulic diameter for a round pipe is simply the diameter.

Evaluation of the number of transfer units

Using the NTU effectiveness method from Kays and London [38]:

$$NTU = AU/C\mathrm{min}$$

where A is the area for heat exchange between the two streams and U is the overall heat transfer coefficient.The product AU is therefore the thermal conductance. Of the two streams in the heat exchanger, Cmin is the heat capacity [J/kg K] × Mass flow [kg/s] of the flow having the lowest product.The NTU is a dimensionless number.Note that other methods are available for solving heat exchanger problems, but the NTU-effectiveness method is the most universal.

Calculation of the thermal conductance

The basic geometry and nomenclature for countercurrent and parallel flow heat exchangers is shown in figure 3.28.

Typically, there is a chain of thermal resistances that link the hot stream and the cold stream. These are the resistances between the streams and the heat exchanger walls (which depend on flow) and the wall thermal resistance. This is shown pictorially in figure 3.29.

The thermal resistance between the fluid streams is calculated by first calculating the Reynolds number to establish the flow conditions (laminar or turbulent flow—see figure 3.30):

$$Re = \frac{\rho v d}{\eta}$$

where ρ = density [kg m^{-3}], v is the fluid velocity [m s^{-1}] and η the viscosity [Pa s]. For a Reynold's number ≈<2500 the flow is said to be laminar and for greater than ≈>2500 turbulent. The problem is that this transition is not well defined and can depend on the specific flow geometry. In practice, this region is usually avoided as

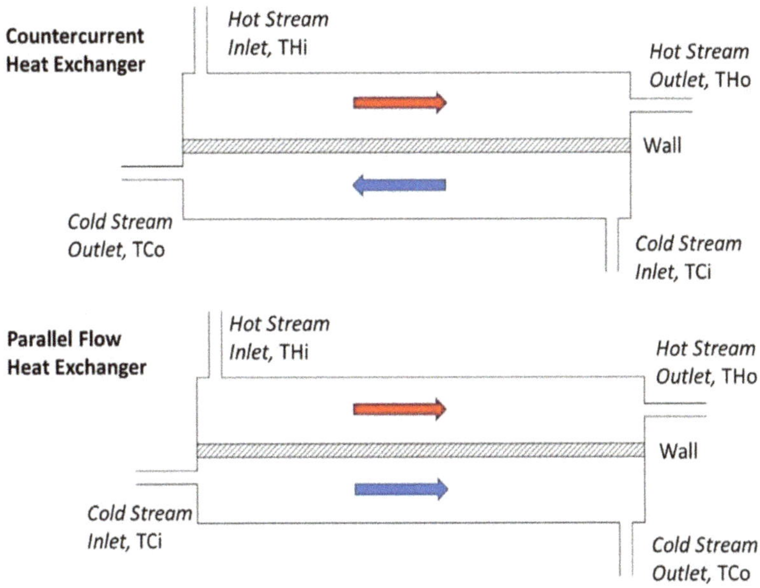

Figure 3.28. Countercurrent and parallel flow heat exchangers.

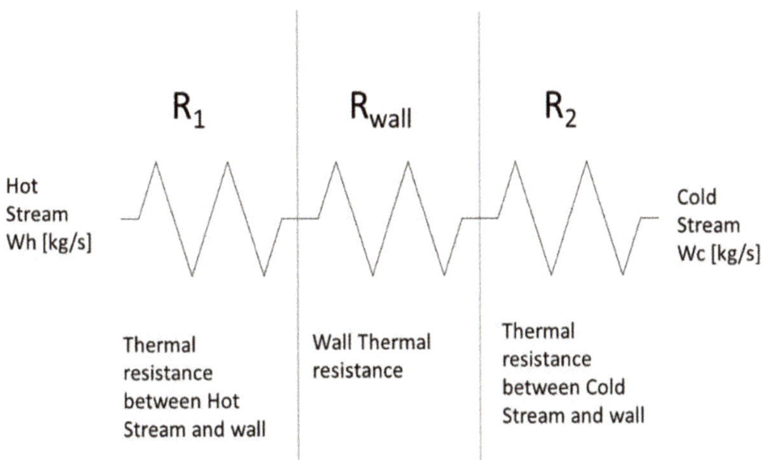

Figure 3.29. Thermal resistances in a heat exchanger.

the behaviour is unpredictable. Having established the flow regime, then the heat transfer coefficients can be calculated.

The Nusselt number, *Nu*, is the ratio of the convective to conductive heat transfer:

$$Nu = \frac{Hd}{k}$$

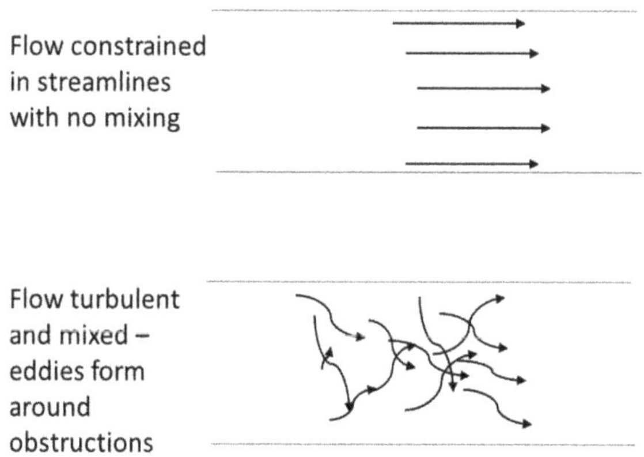

Figure 3.30. Laminar and turbulent flow.

where H is the heat transfer coefficient that we are trying to evaluate [W m^{-2} K^{-1}], d the hydraulic diameter and k the thermal conductivity [W m^{-1} K^{-1}].

The Nusselt number for laminar flow in a plain tube with constant heat flux it is 4.364 [38]. For turbulent flow the following form is appropriate for fluids with a Prandtl number of 0.7.

$$Nu = 0.0244 Re^{0.775}$$

Common fluids (e.g., air and other gases) have a Prandtl number of approximately 0.7 but this *must* be checked for any given situation. For other Prandtl numbers consult appropriate references (e.g., reference [38] p 108 figure 6-7). The Prandtl number is defined by:

$$Pr = \frac{C_p \eta}{K}$$

where C_p is the heat capacity [J (kg K)$^{-1}$], η the viscosity [Pa s] and K the thermal conductivity [W (m K)$^{-1}$].

Knowing the Nusselt number the Heat transfer coefficient, H [W m^{-2} K^{-1}], can then be calculated. The thermal resistance (R_1 or R_2 above) is $1/(AH)$ where A is the heat transfer area. The wall thermal resistance is given by:

$$R_{\text{wall}} = \frac{\text{Wall thickness}}{K_{\text{wall}} A}$$

The total thermal resistance is therefore:

$$R_{\text{total}} = R_1 + R_{\text{wall}} + R_2$$

The overall conductance, AU is simply $1/R_{\text{total}}$ which is used in the calculation of the NTU as described earlier.

Calculation of the heat exchanger effectiveness

The two streams in a heat exchanger can have different flows and heat capacities. The products of the mass flow multiplied by the heat capacity will therefore have two values. The maximum power that can be transferred in the heat exchanger is determined by stream having the lowest W^*c_p product.

Note that C is:

$$C = W\ c_p$$

where W is the mass flow in kg s^{-1} and c_p is the heat capacity in J $(kg\ K)^{-1}$.

The flow has two streams—the hot and the cold—for a counterflow heat exchanger the effectiveness is given by:

$$\varepsilon = \frac{1 - \beta}{1 - \left(\frac{C_{min}}{C_{max}}\right)\beta}$$

where

$$\beta = e^{-Ntu*\left(1-\frac{Cmin}{Cmax}\right)}$$

For a parallel flow heat exchanger, the corresponding effectiveness relationship is:

$$\varepsilon = \frac{1 - \beta}{1 + \left(\frac{C_{min}}{C_{max}}\right)\beta}$$

where

$$\beta = e^{-Ntu*\left(1+\frac{Cmin}{Cmax}\right)}$$

The maximum heat that can be transferred between the streams depends on the minimum capacity stream.

$$q\text{max} = C\text{min}^*(\text{Thi} - \text{Tci})$$

The effectiveness is merely the ratio between the heat transfer achieved and the maximum possible:

$$\varepsilon = q/q\text{max}$$

Note that this definition is not an efficiency and q is calculated as described in the next section.

There is a special case where the two streams have the same C (i.e. heat capacity x mass flow). This is often the case for recuperative heat exchangers in heat recovery systems. In this case the effectiveness is given by:

$$\varepsilon = NTU/(1 + NTU)$$

Temperature calculations

The exit streams of the hot and cold sides (Tho and Tco) can be calculated from the relationships:

$$q = \varepsilon^* q\text{max}$$

$$q = Ch^*(\text{Thi} - \text{Tho}) = Cc(Tci - Tco)$$

(Note that if Ch = Cc then the temperature difference between the two streams is the same).

Pressure drop losses

It can be seen from the foregoing that it is quite easy to improve the heat transfer by increasing the Reynolds number. The downside of this is that the pressure drop will increase, and an engineering judgement has to be made regarding this issue. For a given application it is usually imperative to understand this as it will influence the overall system (pump/compressor/flow) behaviour.

The pressure drop in a duct can be calculated from the following general formula:

$$\frac{\Delta P}{\Delta x} = \frac{f\rho v^2}{2d}$$

where ΔP is the pressure drop [Pa], Δx is the length [m], ρ the density [kg m^{-3}], v the velocity [m s^{-1}] and d the hydraulic diameter [m]. There is a word of caution to be applied to the friction factor, f, in the formula above. Some textbooks use the friction *coefficient* = $f/4$. Make sure that you know what is being used to avoid confusion.

The friction factor varies with Reynolds number and duct surface roughness. For laminar flow:

$$f = \frac{64}{Re}$$

For smooth pipes in turbulent flow [39]:

$$f = 0.316Re^{-0.25} \qquad Re \leqslant 20\,000$$

$$f = 0.184Re^{-0.2} \qquad Re \geqslant 20\,000$$

For rough pipes, reference needs to be made to a 'Moody' diagram (reference [39] p 491 figure 8.3) which will give the correct friction factor. As for the heat transfer correlations avoid the transition region where these correlations may be unreliable. Consult the relevant textbooks for information on abrupt transitions as these losses can be significant.

Physical properties of the fluids

There are several packages that can help with the determination of the fluid properties over a range of pressures and temperatures. Two have been used by the author: REFPROP [40]—a program from NIST that enables the easy generation of a tabulated list of fluid properties and CoolProp [41] which is open source and free. Both packages have wrappers for a variety of high-level languages and popular spreadsheets.

3.5.2 Summary and use in JT cryocoolers

A common requirement for high-capacity heat exchangers is in Joule–Thomson Cryocoolers. In this type of cryocooler, gas is expanded from a high pressure to a low pressure through an expansion nozzle. A heat exchanger is used to improve the

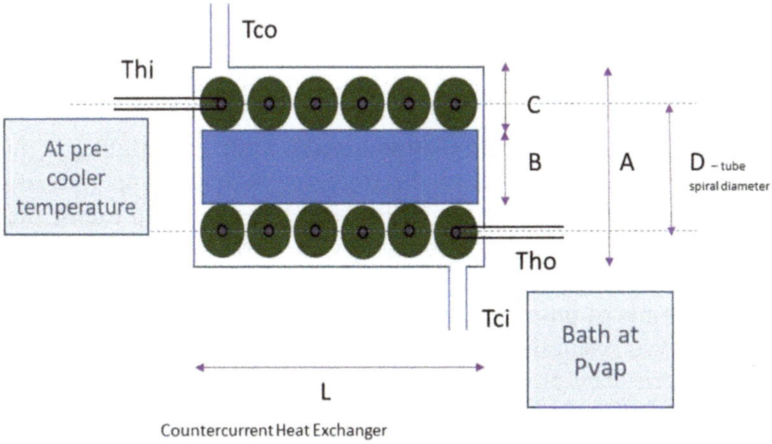

Figure 3.31. A countercurrent annular heat exchanger.

efficiency of the cycle. In this case it is usual to design the heat exchanger so that the flow on the high pressure side is turbulent as this enhances the heat transfer coefficient. On the low pressure side, the flow is usually laminar and the heat transfer poor. This is remedied by use of extended areas to reduce the thermal resistance. The temperature of the cold side of the expansion is determined by the pressure on that side so the design is usually made to minimise the pressure drop on the low pressure side.

One type of such heat exchanger is an annular type (figure 3.31).

3.6 Introduction to temperature scales

Graham Machin

3.6.1 Introduction

The reliable measurement of temperature is essential in nearly every sphere of human endeavour, for example: science, industry, meteorology, and medicine all require reliable temperature measurement. In principle, these temperature measurements could be performed directly with devices known as primary thermometers, that is, those used to perform temperature measurement according to fundamental physical principles. The most widely known primary thermometers are probably thermometers which determine the variation of pressure in a known gas, with temperature in a constant volume, the so-called 'constant volume gas thermometer'. However primary thermometers, and the temperatures derived from them, whilst being fundamental, in that they closely reflect the underlying physics, are generally not practical. They are slow, expensive, and often very large, usually requiring a laboratory full of equipment and expert staff to make accurate measurements.

In contrast there are a range of practical thermometers that can measure temperature reliably. The thermocouple, which is a deceptively simple device, is one. It is made of two dissimilar metals joined at the measurement junction and whose output, a voltage, increases with temperature. Thermocouples dominate industrial temperature measurement. Another, also of great importance, is a thermometer based on the electrical resistance of a platinum wire, known as the platinum resistance thermometer.

However, the output of such thermometers, while repeatable, is in essence arbitrary and needs to be linked to the (thermodynamic) temperatures obtained from primary thermometry methods. This need is met through use of the international temperature scales, whose temperatures are close to thermodynamic and yet simple to realise. These are established on a worldwide basis. At the heart of such scales are fixed points of pure materials, which are based on extremely reproducible and readily observed phenomena such as the change of phase. The thermodynamic temperature of such fixed points, be they the melting/freezing point of pure metals, or the triple points of water or gases, were determined using primary thermometers so that their 'true' thermodynamic temperatures were known with low uncertainties. These temperature fixed points were then used to calibrate practical thermometers, such as the platinum resistance thermometer, establishing a robust relationship between the measured quantity (e.g., electrical resistance) and the thermodynamic temperature of the fixed points.

There have been a number of scales in place throughout the 20th century of increasing sophistication. The first was the International Temperature Scale of 1927 (ITS-27) and then ITS-48, IPTS-68 (the 'P' standing for Practical) and finally the ITS-90 which is the current workhorse scale around the world. In 2000 a cryogenic scale was introduced: the Provisional Low Temperature Scale of 2000 (PLTS-2000). The main features of both these scales will be described briefly in the next section. The interested reader in the development of temperature scales is referred to the reference section [42] for more details.

3.6.2 Description of the current scales in use: the ITS-90 and the PLTS-2000

The ITS-90

The current defined scale in use around the world is the International Temperature Scale of 1990, the ITS-90 [43]. It was introduced on 1 January 1990 and superseded the previous scale (the IPTS-68). Its defined range is from 0.65 K to the highest temperatures (there is no upper limit). Formally temperatures measured using ITS-90 are designated T90, to distinguish them from those determined by thermodynamic means (designated T). However, the actual difference between ITS-90 temperatures and thermodynamic temperatures is so small that in most instances the two can be used interchangeably [44].

The purpose of the ITS-90, in line with previous scales, was 'to define procedures by which certain specified practical thermometers can be calibrated in such a way (using defined fixed points e.g., water triple point, metal freezing points) that the values of temperature obtained from them are precise and reproducible, while at the same time approximating the corresponding thermodynamic values as closely as possible'.

A typical, non-exhaustive list of fixed points used in the realisation of the ITS-90 are: the triple point of neon (24.5561 K), the triple point of argon (83.8058 K), the triple point of water (273.16 K), the melting point of gallium (302.9146 K), the freezing point of zinc (505.078 K) and the freezing point of silver (1234.93 K).

The ITS-90, for ease of realisation, is split into four parts:

- Between 0.65 K and 5.0 K T90 is realised in terms of the vapour pressure relations of 3He (0.65 K to 3.2 K) and 4He (1.25 K to 5.0 K).
- Between 3.0 K and 24.5561 K (triple point of neon) T90 is realised by means of a constant volume helium gas thermometer calibrated at three temperatures.
- Between the triple point of hydrogen (13.8033 K) and the freezing point of silver (961.78 °C): specified platinum resistance thermometers calibrated at specified sets of defining fixed points and using specified interpolation procedures.
- Above the silver freezing point (961.78 °C) the Planck radiation law in ratio form is used.

The part realised by platinum resistance thermometers is further subdivided, for the convenience of users, into 11 subranges.

Once the scale is realised, for example through calibrating a platinum resistance thermometer, over a specified temperature range, it then needs to be disseminated. This is done through calibrating thermometers against that scale—this is true whether the ITS-90 or the PLTS-2000 is realised. These thermometers are then supplied, usually to accredited calibration laboratories, who then calibrate the large numbers of practical thermometers needed by the diverse range of users in all areas of the economy.

The PLTS-2000

The lowest temperature that can be realised on the ITS-90 is 0.65 K. It became apparent that this was a serious limitation for specialised users of cryogenic equipment which routinely attained temperatures in the mK regime. To service this need the Provisional Low Temperature Scale of 2000 (the PLTS-2000) was established in October 2000.

The PLTS-2000 range is from 0.902 mK to 1.0 K and is based on the 3He melting pressure curve which has four transitions that act as fixed points for the scale. The PLTS-2000 is described in several references [45, 46].

It is impossible to give details here concerning the scales, however, much useful information about the practical realisation and dissemination of temperature scales and the practice of temperature measurement can be found on the International Bureau of Weights and Measures (BIPM) website [47].

3.6.3 Scale realisation uncertainties

Typical uncertainties in temperature for scale realisation are of the order of a few parts in 10^4. The exception to this is the water triple point, which is the best studied fixed point, which can be realised, with care, with an uncertainty of a little lower than 1 part in 10^5. Of course, when one starts using practical sensors the uncertainty is likely to be 10 or even 100 times more than this. A full discussion on uncertainties is outside the scope of this chapter. The interested reader is referred to [48], and references therein, where the concept of uncertainty and how to determine its value in practical situations is discussed.

3.6.4 The kelvin redefinition, the mise-en-pratique for the definition of the kelvin and the future of temperature measurement

On world metrology day 20 May 2019, the kelvin, along with other units of the SI, was redefined in terms of fundamental constants. The kelvin is now defined as:

'The kelvin, symbol K, is the SI unit of thermodynamic temperature. It is defined by taking the fixed numerical value of the Boltzmann constant k to be 1.380 649 × 10^{-23} when expressed in the unit J K^{-1}, which is equal to kg m^2 s^{-2} K^{-1}, where the kilogram, metre and second are defined in terms of h, c and $\Delta\nu$Cs' [47, 49]. (The defining constants of the SI relevant to the kelvin definition; h (Planck's constant) 6.626 070 15 × 10^{-34} J s, c (speed of light) 299 792 458 m s^{-1} and $\Delta\nu$Cs (hyperfine frequency of caesium 133 atom) 9192 631 770 Hz.)

The new definition is artefact-free and allows the user to use any thermometry method authorised by the Consultative Committee for Thermometry (CCT, a CC of the Committee of Weights and Measures [CIPM]) to realise temperature.

To assist temperature measurement practitioners to realise temperature according to the new definition, a document, the mise-en-pratique for the definition of the kelvin (MeP-K) [50, 51], was prepared at the same time as the new definition. The purpose of the MeP-K is to guide the metrologist from the SI unit definition to a practical, laboratory realisation of the unit. The MeP-K contains the two defined scales currently in use, but also allows for primary thermometry to be used to realise thermodynamic temperature directly if uncertainties are similar to that of the defined scales. This freedom is currently particularly relevant at high temperatures >1300 K and low temperatures <25 K where primary methods (such as Johnson Noise Thermometry at low temperatures, and indirect primary radiometry at high temperatures) may well supplant the defined scales by the mid-2020s (see, for example, selected references [49, 52]).

In this brief section I have described the need for practical scales to realise reliable temperature on a global basis. I have described the main features of the two defined scales currently in use around the world and given a brief outline of how the practice of temperature scale realisation, dissemination and ultimately measurement will change in the 2020s and beyond. The interested reader should consult the references for more information.

3.7 Practical thermometry

Beth Evans

3.7.1 Introduction

When considering thermometry, one needs to ask questions such as: what temperature range is required? What level of accuracy is needed? What stability and response time? How much money is available to spend? And is the sensor in a hostile environment? Environments such as magnetic fields, radiation, high temperatures and physical shocks can all affect the calibration and stability of sensors. It is also important to consider how and where the sensor(s) will be located, connected and monitored.

In this section we will consider the different types of sensor available and ways to obtain the optimum temperature measurement.

3.7.2 Thermometer calibration

As discussed in section 3.6, temperature scales are defined through a series of triple points, boiling points and vapour pressure scales. Interpolation between these points is achieved through use of constant volume gas thermometers and platinum resistance thermometers. Vapour pressure thermometry tables are available for the majority of fluids and can additionally be calculated using the NIST REFPROP package [40] or the free CoolProp [41]. The current temperature standard is the International Temperature Scale ITS-90. A recommended manual to the practical realisation of the scale can be found in the Guide to the Realization of the ITS-90 [53].

Standards-grade sensors are termed primary thermometers and are too sensitive to mechanical and thermal shock to use in ordinary laboratory or industrial temperature measurements. Commercially available secondary (or tertiary) thermometers, such as resistors, diodes and thermocouples, are more practical. In the past, temperature measurement was very much based around individual lab practices, each laboratory having a favoured measurement system. Calibrations were normally done on-site, usually in bath systems against vapour pressure tables. It is nowadays more common to buy pre-calibrated sensors and to use a data collection system into which the sensor calibration curves are loaded. These calibration curves can range from linear to high order polynomials. For most applications from 4 K to room temperature, suitable pre-calibrated sensors can be purchased and will be considered in this section.

3.7.3 Thermal and electrical stabilisation of temperature measurements

The precision of a temperature measurement is dependent on many factors including the suitability of the sensor chosen, the thermal contact between the sensor and the object being measured and the electrical stability of the measurement.

Thermal stabilisation

The main heat path from the outside world to the sensing element in the thermometer, particularly at low temperatures, is through the electrical leads. Thermal anchoring of the wiring at intermediate temperature stages and at the subject temperature is therefore essential. This may be achieved by winding and gluing the instrumentation wires to a bobbin or by taping them to the temperature stage. Some guidance on this is given in Hust's paper [54].

It is also important to minimise the heat conducted through the wires to the sensor as localised heating of the sensor will lead to a false temperature reading. Wire with a high electrical conductivity, but a low thermal conductivity, should be chosen such as constantan, manganin and phosphor-bronze. The diameter of the wire should be as small as is practicable. It is also important to shield the sensor from

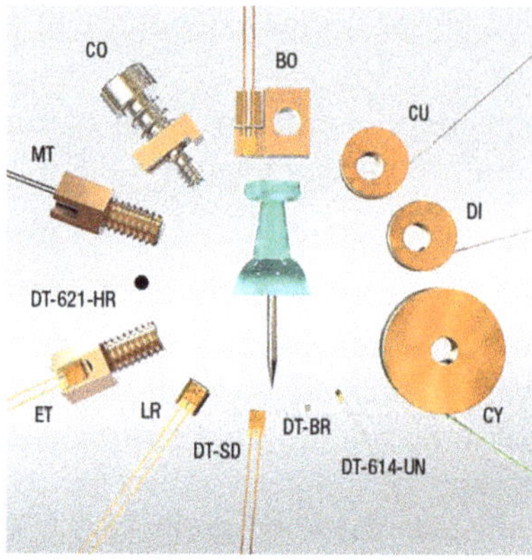

Figure 3.32. Lakeshore diode packages (Credit: with permission from Lake Shore) [55].

thermal radiation and any other heat sources to minimise the thermal gradient between the sensor and the object being measured.

Sensors are available in a range of housings depending on type. Examples of housings are shown in figure 3.32. The sensor should be mounted in close thermal contact with the object being measured. Thermal contact could be maximised using an indium gasket, thermal contact grease or an appropriate adhesive [56]. If relying solely on a mechanical fixing it should be remembered that thermal contraction may loosen fixtures.

A traditional material for thermal anchoring of sensors and wires is GE varnish, although superior materials are now available including specialist low temperature greases. These greases have high thermal conductivity and low vapour pressure. The low vapour pressure quality is important for maintaining high vacuum levels. Note that thermally conductive greases sold for heat sinking that contain a metal powder filler will not work at low temperatures. The constituent materials of these greases vary but they face a variety of problems at cryogenic temperatures due to high freezing points, a reduction in thermal conductivity at low temperatures and high vapour pressures reducing vacuum levels.

At temperatures below 1 K it is crucial to ensure that the sensor is in thermal equilibrium with the object to be measured. Good reviews of methods to achieve this are given in the works of Betts [57] and Lounasmaa [58].

Electrical stabilisation

Resistance and diode temperature sensors require a supply current to produce a voltage across the sensor which can be measured. If the leads used to carry the supply current are also the leads used for the voltage measurement, then errors are likely. This is because the voltage measurement will reflect the lead resistance as well as the effect of the sensor. In particular, small diameter wire, while reducing thermal conduction, will have high electrical resistance. For example, a typical lead resistance of 10 ohms will produce a large error in the 100 Δ resistance measurement of a standard platinum resistor at 273 K. This error will increase significantly with decreasing temperature as the measurable sensor resistance decreases. By employing

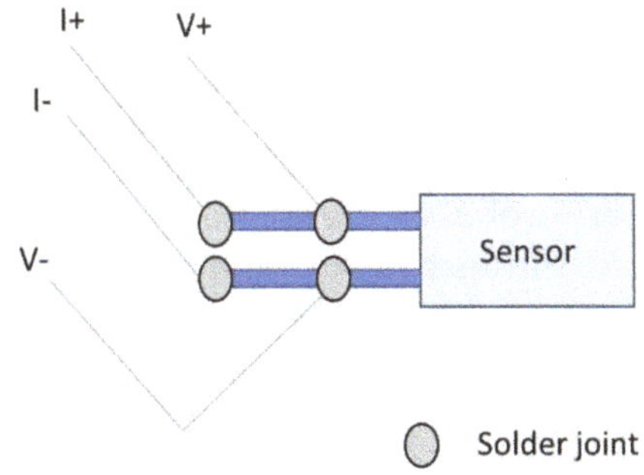

Figure 3.33. Four-wire temperature sensor wiring.

a four-wire measurement (figure 3.33) these errors can be eliminated: two leads are used to carry a larger current to power the sensor, while the other two leads are used to measure the voltage across the sensor using a very small current.

Thermal electromotive forces (EMFs) can become a significant source of error in low voltage measurements. These develop where there are dissimilar materials along the measurement chain, for example in plugs and connectors. These effects can be minimised by reversing the measurement current and then averaging the voltage readings of the forward and reverse currents. Due to their higher voltage measurements, diodes are less affected by thermal EMFs.

Other electrical errors can be caused by poor grounding of the measurement system, or by AC signal interference. Measurement leads are susceptible to AC signals in the radio frequency range. The leads act as antennae and pick up noise from instrumentation. The effect can be reduced by twisting sensor leads together in pairs. This is a particular problem for diode measurements and at very low temperatures (sub-200 mK), where it can lead to erroneous results due to overheating. A 10 μH inductance in line with a 1000 pF capacitor to ground should minimise the effects of AC interference.

3.7.4 Types of thermometers

It is not possible to buy a temperature sensor that offers a high accuracy and stable temperature readings in all environments and at all temperatures. It is a question of choosing the most suitable sensor(s), providing the best accuracy (or affordability), to cover the temperature range required [59]. In this section, sensors for the range from 300 K to 1 K are considered. At lower temperatures ($T < 500$ mK) magnetic susceptibility, Johnson noise, nuclear orientation, nuclear magnetic resonance, He_3 melting pressure curves are used for measuring temperature but these are not described in this section.

Metallic resistors

Metallic temperature sensors are positive temperature coefficient (PTC) sensors, that is, their resistance increases with temperature. A single excitation current can be used to power the sensors over a broad temperature range as power dissipation, and thus self-heating of the sensor, decreases with temperature.

Platinum resistance thermometer
The platinum resistance thermometer is the most commonly used metallic sensor. These are available in wire wound or planar configurations. The wire wound type are extremely stable and can be used as reference thermometers. The planar type is extremely cheap and very robust. These are very useful as general-purpose thermometers from liquid nitrogen temperatures up to room temperature, but they are useful to temperatures above 800 K.

Platinum cobalt resistance thermometer
The platinum cobalt resistance thermometer [60] is useful over a temperature range from 1.4 K to 300 K with reasonable accuracy.

Rhodium–iron thermometer
Rhodium–iron sensors are useful down to 1 K and can also be calibrated up to 800 K, providing a very large temperature range. They are also more resistant to ionising radiation than many other sensors.

Semiconducting sensors
Semiconducting sensors are usually negative temperature coefficient (NTC) sensors, and thus their temperature sensitivity, although low near room temperature, increases with decreasing temperature. The resistance of the sensor can change considerably with temperature and thus a variable current source and a sensitive voltmeter are required for measurements. The semiconducting sensors are generally piezoresistive and thus sensitive to stress. This means that the sensors must be mounted in a way to minimise strain, with the negative effect of reducing the thermal contact between the sensor and sample. Sensor self-heating can thus be a problem, which can also be minimised by varying the power supplied to the sensor during measurements. Calibration and subsequent monitoring when in use must be made using the same measurement method.

Common semiconducting sensors are germanium, Cernox, ruthenium oxide, carbon and carbon-glass resistance temperature detectors (RTDs).

Carbon RTDs
These are made from radio resistors and used to be commonly employed in low temperature work as they are generally cheap and easily obtainable. They are sensitive to magnetic fields and have generally been superseded in favour of more stable types. The types most commonly used in low temperature work are Allen Bradley or Speer.

Carbon-glass RTDs
These are made from carbon-impregnated glass. Although suitable for temperatures from 325 K down, they have high sensitivity below 10 K and are suitable for use in magnetic fields up to approximately 19 T [61]. Carbon-glass resistors require strain-free mounting.

Carbon ceramic TVO
TVO sensors are very robust and have long-term stability in extreme environments [62] including magnetic fields, ionising radiation and mechanical forces and vibration. They operate over the range 1.5–375 K.

Germanium RTDs
These RTDs are manufactured from a single crystal of germanium doped with arsenic, gallium or antimony to give the preferred resistance properties [63]. The sensor can thus be optimised for a particular temperature range with high accuracy over two orders of magnitude, e.g., 0.1 K to 10 K. They are useful over the range 0.05 K to 30 K. Like carbon-glass RTDs, germanium sensors are very fragile and need to be protected by a suitable housing.

Cernox RTDs
Cernox sensors are ceramic thin films sputtered onto a sapphire substrate. They have a wide temperature range, from 0.3 K to 300 K. They are resistant to ionising radiation and are suitable for use in magnetic fields. They are less bulky than other sensors and available in a range of housings.

Ruthenium oxide RTDs
These RTDs are thick film sensors. They are primarily used at temperatures below 1 K in magnetic fields, as they have very low magnetoresistance at low temperatures. They have the advantage of following a standard calibration curve.

Thermistors
Thermistors are often made from sintered metallic oxides. The advantages of this type of sensor include their small size and low sensitivity to ionising radiation and magnetic fields. Individual sensors do not have large temperature ranges and are mostly used above liquid nitrogen temperatures [64].

Semiconducting diodes
Diode temperature sensors have the advantage of being physically smaller than RTDs and this gives them a faster response time. They are also often more robust and can be used over a wide temperature range of 1.4 K to 500 K. They are, however, less accurate than some RTDs. Diodes are easy to use, requiring only a fixed current source and a voltmeter. Diodes can also be interchanged—a diode from a single series from a manufacturer can be switched with any other, using the same calibration curve, within a tolerance range. The most common type of diode is the silicon diode. These are strongly affected by magnetic fields but other diodes, such as GaAs or GaAlAs, are less sensitive.

Other types of sensors may be used at cryogenic temperatures, but their capabilities are limited. These include thermocouples and capacitance sensors.

Thermocouples
Thermocouples consist of two dissimilar metals joined together. When there is a temperature difference between the joints an EMF is produced which is a function of temperature. It is necessary to maintain one of the thermocouple joints at a known reference temperature which is usually provided by an ice bath although a cryogenic reference temperature is advantageous. The thermocouple follows a standard response curve for its type. It is important to ensure that the thermocouple leads are isothermal or the entire length of the leads will contribute to the output voltage. The thermocouples most commonly used for low temperature include copper—constantan (type T), nickel chromium—constantan (Type E) and

Table 3.6. Summary of sensor types and properties (general guide only—actual properties depend on specific sensor model).

Sensor	Temperature Range (K)	Suitable for extreme environments?	
		Magnetic field	Ionising radiation
Metallic:	–	–	–
Platinum	14–873	Some models	✓
Rhodium–iron	1.4–800	X	✓
Resistor:	–	–	–
Carbon	0.01–300	✓ (moderate fields)	X
Carbon-glass	1.4–325	✓	X
Carbon ceramic	1.5–375	✓	✓
Germanium	0.05–100	X	✓
Cernox	0.1–420	✓	✓
Ruthenium oxide	0.05–40 (<0.01–40) ULT model	✓ (moderate fields)	✓
Thermistor	77–300	✓	✓
Diode:	–	–	–
Silicon	1.4–500	X	X
GaAs/GaAlAs	1.4–475	✓ (moderate fields)	X
Other:	–	–	–
Thermocouple	Type E: 3–950 K Type K: 3–1500 K	✓ X	X
Capacitance	1.4–290	✓	X

nickel chromium—nickel-aluminium/silicon (Type K). Thermocouples can be used over the range 3–500 K, and even to above 1500 K (Type K).

Capacitors

The most common use for capacitors at cryogenic temperatures is for temperature control in high magnetic fields, rather than as a sensor. This is due to their insensitivity to magnetic fields but tendency for calibration drift with thermal cycling. Commercially available capacitors are often manufactured from $SrTiO_3$ and should be calibrated *in situ* using another type of temperature sensor. The capacitance measurement is made using an AC capacitance bridge. A summary of sensor types and properties is given in table 3.6.

3.8 Cryogenic instrumentation

Beth Evans

3.8.1 Introduction

The art of measuring temperature at cryogenic temperatures has been considered earlier in this chapter, but how are other system parameters, such as pressure, magnetic field, or cryogen flow to be determined at low temperatures? Equipment used at room temperature can rarely be employed in cryogenic systems due to the constraints of vacuum, heat loads and changes to material properties. There are many academic papers that describe successful measurement systems in a laboratory set-up, but which sensors are available commercially? This section will briefly describe sensor types available for measuring flow, pressure, magnetic fields, cryogen flow and level, along with other properties.

3.8.2 Flow

Many types of flowmeter can be used at cryogenic temperatures [65] but several factors should be taken into consideration including: the subject cryogen and its compatibility with the sensor; the range of flow measurements required; the acceptable disruption to the flow; the potential heat load caused by the flowmeter; the reliability of, and necessary maintenance required by, the sensor; its cost and availability. This section will briefly describe the flowmeters most commonly used at cryogenic temperatures.

Differential pressure flowmeters

Some of the most commonly used flowmeters are not strictly measuring mass or volume flow, but rather differential pressure. These are orifice plate and venturi type flowmeters and are shown in figure 3.34. The orifice plate flowmeter contains a sharp-edged orifice plate positioned in the centre of the tube containing the subject flow. The pressure drop across the plate is proportional to the mass flow rate. In the venturi flowmeter, a narrowing of the tube diameter is instead used to produce the measured pressure drop.

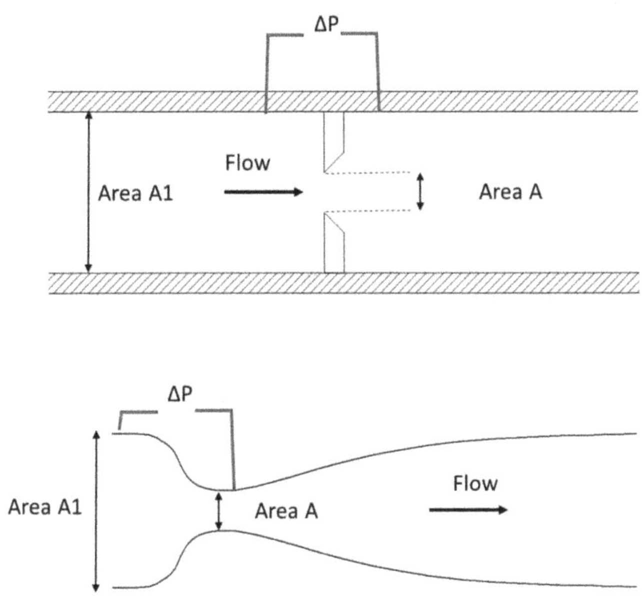

Figure 3.34. Simple diagrams of orifice plate (top) and venturi (bottom) flowmeters.

The advantage of the differential pressure type flowmeter is high reliability, as it contains no moving parts and the pressure sensor used can be located at room temperature. The orifice plate flowmeter has the disadvantage that, because the mass flow is proportional to the square root of the pressure change produced by the orifice, a large change in pressure equates to a much smaller change in flow, leading to a small operating range. Orifice plates can also produce turbulence and so either lengths of straight tube are required before and after the orifice, or flow straighteners should be used. Venturi flowmeters produce less turbulence and pressure drop and are often used for helium applications [66].

Turbine flowmeters

Turbine flowmeters provide a volumetric flow measurement. A freely spinning turbine is positioned in the centre of the fluid flow and a magnet mounted in the body of the flowmeter. Each time a turbine blade, driven by the fluid flow, passes the face of the magnet the change in magnetic permeability produces a voltage pulse. The frequency of these pulses is proportional to the speed of rotation of the turbine which is proportional to the velocity of the fluid flow. The volumetric flow can be ascertained from the fluid velocity.

A disadvantage of the turbine flowmeter is that at low flow rates the viscosity of the fluid has a significant effect and calibration is required. At higher flow rates the viscosity effect is usually negligible. Turbine flowmeters can also be damaged by large flow oscillations and thus require protection in situations such as the rapid cool-down of a transfer line.

The advantage of the turbine flowmeter is the relatively low pressure drop they produce in the fluid stream. Research has shown that the reliability of turbine flowmeters may be improved with the use of superconducting magnets and levitating bearings [67]. Commercial turbine flowmeters are available which operate down to helium temperatures [68].

Coriolis flowmeters

The Coriolis type of flowmeter provides a mass flow measurement using the Coriolis force. The subject fluid flows through two parallel tubes (inlet and outlet) which are vibrated externally. When there is no fluid flow the motion of the tubes is symmetrical but when there is fluid flow the tubes twist due to the Coriolis effect. The phase shift between the motion of the inlet and outlet tubes can be used to determine the mass flow rate.

The main advantages of the Coriolis flowmeter are the lack of cold moving parts and that no flow straighteners, or lengths of straight tube before and after the meter, are necessary. The main disadvantage is that the pressure drop can be significant and thus it is not useful for low pressure fluid measurements. Suitable commercially available flowmeters are guaranteed for temperatures down to 77 K [69].

Vortex flowmeters

Vortex flowmeters operate by producing vortices in the subject flow using a 'shedder bar'. The natural frequency of these oscillations can be measured and converted into a mass flow rate. The advantage of these flowmeters is the absence of moving parts and thus improved reliability. Vortex flowmeters for use down to liquid nitrogen temperatures are commercially available [70].

3.8.3 Pressure

There are few pressure sensors able to operate at cryogenic temperatures and all require individual calibration. A better option, where applicable, is to mount the pressure sensor at room temperature and run a capillary tube to the point of interest. There are difficulties associated with this remote measurement method, however. There is a slow response time and thus this method is most suited to static or slowly changing pressures. There is an increased risk of leaks into the system. There is also the potential for thermoacoustic oscillations [71] in the capillary tube, caused by the temperature gradient along it. These oscillations can be larger than the pressure being measured and are a particular problem in helium. Errors may also be introduced due to the pressure (thermodynamically related to temperature) being higher at the warm end of the capillary than at the cold end, with errors increasingly significant at lower pressures.

Some pressure transducers used at room temperature can also be used at low temperatures, but it is important to use a sealing method suitable for cryogenic temperatures. Common types of cryogenic pressure sensors include piezoelectric, variable reluctance and capacitance sensors. Commercially available sensors are limited and mostly available for use down to liquid nitrogen temperatures (e.g., figure 3.35) [72–74].

3.8.4 Strain

The type of strain gauge most suitable for use at cryogenic temperatures is piezoelectric transducers. These sensors generate an output voltage proportional to the force applied to the transducer. The shape of the strain gauge is altered by a change in the force across it: a tensile force stretches it, making it thinner and longer, and the electrical resistance increases; a compressive force makes it shorter and wider and the resistance decreases.

The change in resistance for a single strain gauge is very small and thus the accuracy of the measurement it provides is limited. Increasing the number of strain gauges magnifies the change and thus improves the accuracy of the measurement. Four are often combined into a Wheatstone bridge. Two gauges can be mounted in the cryogenic environment: one at the location where the strain is to be measured

Figure 3.35. Omega pressure transducer (Credit: Omega Engineering Inc.) [73].

Figure 3.36. An example of a strain gauge assembly [75]. Credit: Procter & Chester (Measurements) Ltd.

and another in an unstrained location. These two gauges then form one half of a resistance bridge, the remainder of which is at room temperature.

Temperature also affects the resistance of the strain gauge, so it is important to find a gauge material with thermal contraction properties as similar as possible to those of the sample material to reduce the effect of differential thermal contractions.

Magnetic fields may also affect some strain gauges, but thin film strain gauges have been used on superconducting magnets with some success. The adhesive (usually epoxy) used to attach the sensor is an important consideration. Figure 3.36 shows an example of a strain gauge assembly mounted on a sample.

Strain gauges are often packaged into load cells—a metal body onto which the strain gauges are secured. Load cells are commercially available for use down to helium temperatures [76].

3.8.5 Magnetic field

Hall probes and magnetoresistive sensors are the main types of sensor used for magnetic measurements at cryogenic temperatures [77], but other methods are also suitable for certain applications.

Hall probes

Hall probes are routinely used at low temperature to measure a wide range of axial and transverse fields. They contain an indium-based semiconductor crystal mounted on aluminium. The plane of the crystal is perpendicular to the probe handle and thus the probe should be held perpendicular to the magnetic field being measured. The probe will detect magnetic flux density (*B*).

Hall probes have very low temperature sensitivity, even at 4 K. They are, however, delicate and easily damaged by mechanical stresses during cool-down. They also have a finite lifetime—typically 50–100 thermal cycles. For higher accuracy they should be calibrated against a known magnetic field at the service temperature. The probes can also be used as a magnetic switch. Figure 3.37 shows examples of commercial Hall probes.

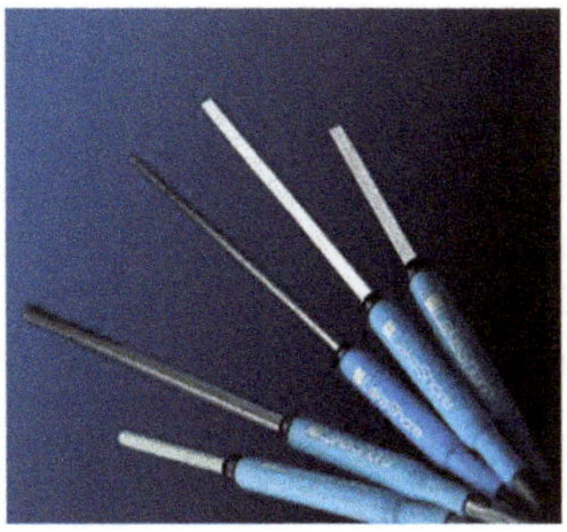

Figure 3.37. Lakeshore Hall probes (Credit: with permission Lake Shore) [78].

Magnetoresistive sensors

These are also used for magnetic measurements [79]. Magnetoresistance is the inclination of a material to change resistance when in an external magnetic field. Magnetoresistive sensors offer high accuracy and can also be used in position and proximity sensing [80]. They are best suited to measuring high magnetic fields.

Induction coils

The measuring principle of induction coils makes use of Faraday's induction law. They are the sensor used in fluxmeters and can be used to measure magnetic flux and field direction [81]. Induction coils are particularly useful for measuring pulsed or ramped fields. Rotating coils can be used for measuring the field quality of accelerator magnets. Induction coils should be calibrated at cryogenic temperatures as thermal contraction can significantly alter the calibration of coils.

SQUIDs

SQUIDs (Superconducting Quantum Interference Devices) provide the most sensitive measurement of magnetic flux of any sensor type [81, 82]. However, they are extremely noise sensitive and only suitable for very low magnetic fields.

3.8.6 Light sensors

Many light emitting diodes will operate at low temperatures and can be used for illumination and supplying fast, precise heat loads. Fibre optics, if manufactured from a suitable material, can be used to transmit laser beams and other light signals at cryogenic temperatures [83].

3.8.7 Motion sensors

Optical sensors are often used for measuring motion at cryogenic temperatures. For example, a bundle of optical fibres may be used to transmit and receive light to and from the moving subject. The intensity of the received light reduces as the subject moves away and thus can be converted into a linear distance. This method can also be used to measure the velocity of a moving subject [84].

Specific vibration sensors are also commercially available for use down to helium temperatures (e.g., PCB-(M)351B11) [85].

3.8.8 Liquid level measurement

When filling dewars, the amount of liquid cryogen in the dewar can be measured by monitoring the change in its weight. The volume of cryogen contained in cryostats or tanks is usually determined by the position of the surface of the liquid.

There are many methods to measure the level of a liquid within a container [86, 87] and the most commonly used are described below.

Capacitance level sensor
This type of sensor consists of two concentric cylindrical electrodes (tube in tube or wire in tube) placed vertically in the vessel. The cryogen liquid and vapour phases have different dielectric properties and thus the capacitance of the probe is proportional to the liquid level [88]. A disadvantage of this type of sensor is their sensitivity to ice which has a large dielectric constant.

Temperature sensor array
The presence or absence of liquid can be detected with temperature sensors which self-heat sufficiently (such as thermistors, transistors or diodes), as the rise in sensor temperature will be higher in vapour. Sensors can be positioned along a rod to provide a continuous level measurement.

Hydrostatic pressure measurement
This method uses the difference in pressure between the top and bottom of the container to estimate the liquid level. This method is mostly used with large tanks and cryogens of higher density.

Ultrasonic level sensors
These sensors emit an ultrasonic pulse directed at the liquid in the container. The return echo is used to measure the distance between the sensor and the liquid surface.

Optical level sensors
Optical sensors use infrared light to make a point-level sensing of a cryogen, rather than a continuous level measurement.

Helium dipstick
This is a very simple sensor used to measure the level of helium. It consists of a narrow tube inserted into the liquid with or without a membrane on the top end. As the dipstick is raised, thermo-acoustic oscillations in the tube change in frequency as the probe moves from liquid to vapour. This frequency difference can be detected by a thumb over the end of the tube or by listening to the sound produced [89].

Superconducting level sensor
These probes consist of a thin rod or wire of a material which is superconducting at liquid helium temperature. If heat is applied to the probe any section of wire above the liquid is heated above its critical temperature and becomes resistive. The resistance measured along the probe is used to determine the level of helium. Depending on the range of the probe resistance, some sensors may not function well just above the lambda point of helium (2.17 K) or at increased vessel pressures.

References

[1] Joel R 1971 *Basic Engineering Thermodynamics in SI Units* (London: Longman) pp 61–2
[2] Joel R 1971 *Closed cycle refrigerators, Basic Engineering Thermodynamics in SI Units* (London: Longman) ch 4
[3] Joel R 1971 *Basic Engineering Thermodynamics in SI Units* (London: Longman) pp 668–90
[4] Joel R 1971 *Convection Basic Engineering Thermodynamics in SI Units* (London: Longman)
[5] Joel R 1971 *Radiation, Basic Engineering Thermodynamics in SI Units* (London: Longman)
[6] Thipse S S 2013 *Cryogenics—A Textbook* (Oxford: Alpha Science) pp 15.1–5.24
[7] Thipse S S 2013 *Cryogenics—A Textbook* (Oxford: Alpha Science) pp 6.1–18
[8] Thipse S S 2013 Liquid bath cryostats *Cryogenics—A Textbook* (Oxford: Alpha Science)
[9] Llewellyn D T 1992 *Steels—Metallurgy and Applications* (Oxford: Butterworth-Heinemann) pp 230–31
[10] Selection of stainless steels for cryogenic applications—British Stainless Steel Association—on-line 2020
[11] Turnbull A H, Barton R S and Riviere J C 1992 *An Introduction to Vacuum Technique* (Oxford: Newnes) pp 170–84
[12] Shaughnessy B and Eccleston P 2009 Cryogenic thermal testing of the verification model mid-infrared instrument (MIRI) optics module *SAE Technical Paper* 2009-1-2410
[13] Miyakita T *et al* 2014 Development of a new multi-layer insulation blanket with non-interlayer-contact spacer for space cryogenic mission *Cryogenics* **64** 112–20
[14] Shaughnessy *et al* 2007 Thermal conductivity measurement below 40 K of the CFRP tubes for the mid-infrared instrument mounting struts *Cryogenics* **47** 348–52
[15] Shaughnessy B M 2012 Thermal design and testing of the mid-infrared instrument (MIRI) optics module. *Int. Conf. on Environmental Systems (ICES) (San Diego, CA)* pp 15–9
[16] Ade P A R *et al* 2018 Planck collaboration: Planck early results. II. The thermal performance of Planck *Astron. Astrophys.* **536** A2
[17] Ohmori T *et al* 2002 The effect of non-zero thermal radiation transmissivity of the aluminized Mylar film on the thermal performance of super-insulation *AIP Conf. Proc.* **613** 1573–84
[18] Marquardt E D, Le J P and Radebaugh R 2002 Cryogenic material properties database *Cryocoolers 11* ed R G Ross (Boston, MA: Springer) (accessed June 2020 via https://trc.nist.gov/cryogenics/)
[19] Woodcraft A L 2005 Predicting the thermal conductivity of aluminium alloys in the cryogenic to room temperature range *Cryogenics* **45** 421–31
[20] Produced using CryoComp software www.eckelsengineering.com
[21] Chuen Or *et al* 2005 Cryogenic emittance measurement and its accuracy for the James Webb Space Telescope *SPIE Proc. vol 5904, Cryogenic Optical Systems and Instruments XI;* **59040H**
[22] Petersen P 1958 Swedish Technical Research Council Report; Sartryckur TVF 29, 4,**151**
[23] Reiss H 2004 A coupled numerical analysis of shield temperatures, heat losses and residual gas pressures in an evacuated super-insulation using thermal and fluid networks: part I: stationary conditions *Cryogenics* **44** 259–71
[24] NASA TP- 1999–20 *9263 MLI Material Guidelines*
[25] ECSS-Q-ST-70-01C—Cleanliness and contamination control Nov 2008
[26] https://outgassing.nasa.gov/
[27] NASA-CR-4740 Contamination control engineering design guidelines

[28] Bergman T L, Incropera F P, DeWitt D P and Lavine A S 2011 *Fundamentals of Heat and Mass Transfer* 7th edn (Hoboken, NJ; Wiley)
[29] Ross R G 2003 Cryocooler load increase due to external contamination of low-∈ cryogenic surfaces *Cryocoolers 12* ed R G Ross (Boston, MA: Springer)
[30] Woodcraft A L and Gray A 2010 A low temperature thermal conductivity database *Proc. 13th Int. Workshop on Low Temperature Detectors (LTD13)* vol 1185 681–4
[31] Woodcraft A L 2005 Recommended values for the thermal conductivity of aluminium of different purities in the cryogenic to room temperature range, and a comparison with copper *Cryogenics* **45** 626–36
[32] https://slac.stanford.edu/pubs/slactns/tn03/slac-tn-03-023.pdf (accessed June 2020)
[33] https://trc.nist.gov/cryogenics/ (accessed June 2020)
[34] Modest M F 1993 *Radiative Heat Transfer* (New York: McGraw-Hill)
[35] NASA-TP 2005–21 *2792 Spacecraft Thermal Control Coatings References*
[36] Frolec *et al* 2019 A database of metallic materials emissivities and absorptivities for cryogenics *Cryogenics* **97** 85–99
[37] Gilmore D G (ed) 2002 *Spacecraft Thermal Control Handbook: Fundamental Technologies: vol 1: Fundamental Technologies* 2nd edn (Reston, VA: AIAA)
[38] Kays W and London A L 1964 *Compact Heat Exchangers* 2nd edn (New York: McGraw-Hill)
[39] Incropera F P and de Witt D P ed Bergman T L and Lavine A S 2007 *Fundamentals of Heat and Mass Transfer* 6th edn (New York: Wiley)
[40] https://nist.gov/srd/refprop
[41] http://coolprop.org/
[42] Cooper M and Grozier J 2017 Temperature scales; past, present and future: 1700 to 2050 *Precise Dimensions: A History of Units from 1791 to 2018* (Bristol: IOP Publishing)
[43] Preston-Thomas H 1990 The international temperature scale of 1990 *Metrologia* **27** 3–10 see also corrigendum in 1990 **27** 107
[44] Fischer J, de Podesta M, Hill K D, Moldover M, Pitre L, Rusby R, Steur P, Tamura O, White R and Wolber L 2011 Present Estimates of the Differences Between Thermodynamic Temperatures and the ITS-90 *Int. J. Thermophys.* **32** 12–25
[45] Rusby R L, Durieux M, Reesink A L, Hudson R P, Schuster G, Kühne M, Fogle W E, Soulen R J and Adams E D 2002 The provisional low temperature scale of 2000 *J. Low. Temp. Phys.* **126** 633–42
[46] http://bipm.org/c.c./CCT/Allowed/21/CCT01-40.pdf
[47] Thermometry: SI base unit kelvin, BIPM. https://bipm.org/metrology/thermometry/units.html
[48] https://npl.co.uk/resources/gpgs/guide-to-measurement-uncertainty
[49] Machin G 2018 The Kelvin redefined *Meas. Sci. Technol.* **29** 022001
[50] Fellmuth B, Fischer J, Machin G, Picard S, Steur P, Tamura O, White R and Yoon H 2016 The kelvin redefinition and its *mise en pratique Phil. Trans. R. Soc. A* **374** 20150037
[51] https://bipm.org/utils/en/pdf/si-mep/SI-App2-kelvin.pdf
[52] Machin G, Engert J, Gavioso R, Sadli M and Woolliams E 2016 Summary of achievements of the EMRP project implementing the new kelvin (InK) *Measurement* **94** 149–56
[53] https://bipm.org/en/committees/c.c./cct/guide-its90.html
[54] Hust J G 1970 Thermal anchoring of wires in cryogenic apparatus *Rev. Sci. Instrum.* **41** 622
[55] Cryogenic Temperature Sensors (lakeshore.com)

[56] Greases; Apiezon® N-Grease or Dow Corning; Adhesives—generally Stycast 2850 is used as an adhesive.
[57] Betts D S 1976 *Refrigeration and Thermometry Below One Kelvin* (Brighton: Sussex University)
[58] Lounasma O V 1974 *Experimental Principles and Methods Below 1 K* (New York: Academic)
[59] Yeager C J and Courts S S 2001 A review of cryogenic thermometry and common temperature sensors *IEEE Sens. J.* **1** 352–60
[60] Shimazaki T *et al* 2017 Characteristics of standard capsule-type PtCo resistance thermometers between 0.65 K and 25 K *Int. J. Thermophys.* **38** 96
[61] Weisend J G 1998 *Handbook of Cryogenic Engineering* (London: Taylor and Francis)
[62] Süßer M and Wüchner F 2001 Mechanical behavior of carbon ceramic TVO temperature sensors *Cryogenics* **41** 213–5
[63] Barron R F 1985 *Cryogenic Systems* 2nd edn (Oxford: Oxford University Press)
[64] Thermometrics N T C *Thermistors | Cryogenic RL Series*
[65] Konstantinov A *et al* 2000 A review of the liquid helium flow rate measurement techniques at some operating facilities with superconducting magnets *Plasma Devices Oper.* **8** 201
[66] Rivetti A *et al* 2011 Metrological performances of Venturi flowmeters in normal, supercritical and superfluid helium *Adv. Cryo. Eng.* **39** 1051
[67] Rivetti A *et al* 2004 An accurate rotor-levitated turbine flow meter for cryogenic service *AIP Conf. Proc.* **710** 420
[68] https://sino-inst.com/turbine-flow-meters/ https://flowmeters.com/turbine-flow-meter-cryogenics
[69] https://emerson.com/documents/automation/brochure-extreme-conditions-micro-motion-en-66842.pdf
[70] https://yokogawa.com/uk/solutions/products-platforms/field-instruments/flow-meters/vortex--flow-meters/
[71] Clement J R and Gaffney J 1954 Thermal oscillations in low temperature apparatus *Adv. Cryo. Eng.* **1** 302
[72] https://pcb.com/sensors-for-test-measurement/pressure-transducers/cryogenic
[73] https://omega.com/en-us/pressure-measurement/pressure-transducers/p/px1005
[74] https://kulite.com/products/product-advisor/product-catalog/cryogenic-miniature-ruggedized-pressure-transducer-ctl-375m/
[75] https://pcm-uk.com/services/strain-gauging/
[76] Custom Load Cells | Sensing Systems Corporation (sensing-systems.com)
[77] Ionescu M *et al* 2004 Cryogenic magnetic field sensor based on the magnetoresistive effect in bulk B12212+USr_2CaO_6 *Appl. Phys. Lett.* **84** 5335
[78] https://lakeshore.com/products/categories/overview/magnetic-products/hall-probes/400-series-hall-probes
[79] Schmitz B *et al* 2018 Magnetometric mapping of superconducting RF cavities *Rev. Sci. Instrum.* **89** 054706
[80] Krause O *et al* 2001 Magnetoresistive position sensor for cryogenic space applications *Proc. 9th Euro Space Mechanisms and Tribology Symp.* ESA SP-480
[81] Bottura L and Henrichsen K N 2002 *Field measurements* CERN-LHC-2002-020-MTA CAS – CERN Accelerator School: Superconductivity and Cryogenics for Accelerators and Detectors pp 118–48
[82] Fagaly R L 2006 Superconducting quantum interference device instruments and applications *Rev. Sci. Instrum.* **77** 101101

[83] https://thorlabs.com/newgrouppage9.cfm?objectgroup_id=13558
[84] https://philtec.com/industry-applications/temperature-limits/cryogenic-applications/
[85] PCB Model 351B11
[86] Barron R F 1985 *Cryogenic Systems* 2nd edn (Oxford: Oxford University Press)
[87] Weisend J G 1998 *Handbook of Cryogenic Engineering* (London: Taylor and Francis)
[88] Velichkov I V and Drobin V M 1990 Capacitive level meters for cryogenic liquids with continuous read-out *Cryogenics* **30** 538
[89] Yoshlda S *et al* 1997 Thermoacoustic Taconis oscillations in helium-4 liquid level sensors *Proc. of the 16th Int. Cryogenic Engineering Conf./Int. Cryogenic Materials Conf.* p 657

Chapter 4

Closed cycle refrigerators

Tom Bradshaw

4.1 Introduction

Conventional bath cryostats where a refrigerant (usually in liquid form) is kept cold in an insulated vacuum flask are simple but have a limited life as the refrigerant boils off. A closed cycle refrigerator (CCR) can provide refrigeration almost indefinitely and often at lower cost over the life of the installation. This chapter looks at the various types of CCRs that are commonly used to replace conventional cryogens.

The history of CCRs is a long one, and this is not the place to go into detail over what is a fascinating story [1–3].

4.2 The imperative for CCRs

Helium gas has long been recognised as a strategic commodity. It is formed underground by radioactive decay processes and commonly collects in natural gas and oil deposits from which it is purified and sold on. In liquid state it is used to cool superconducting magnets for medical applications and particle accelerators. The price of helium has fluctuated along with the fortunes of the gas and oil industry.

In the early 20th century, the United States sought to control the world supply of helium because of concerns that airships would be developed that might pose a threat to the country. This story is fascinating and related by Bo Sears in [4]. After the second world war the US stockpiled massive amounts of helium in an underground reservoir near Amarillo, Texas. In 1996 the Privatisation Act legislated to prevent the stockpiling of helium and mandated its steady release to the market, keeping the price artificially low. This is ending and the price of helium has been rising. Helium, as a liquid, is used as a refrigerant and the price is such that it has become economical, in many cases, to replace baths of liquid helium with CCRs.

Large-scale refrigeration is important for the economical storage and transport of liquid natural gas. This is comprised mostly of methane which has a high global warming potential—during transport, CCRs are used to ensure that the boil-off is minimised [5]. In hospitals, Magnetic Resonance Imaging (MRI) is a common tool

doi:10.1088/978-0-7503-2757-2ch4

for diagnostics. MRI scanners use superconducting magnets held at a temperature of 4 K, which is maintained using CCRs; sometimes in conjunction with a zero boil-off helium bath and sometimes by direct cooling.

4.2.1 Stirling engine

The origins of Stirling cycle refrigerators lie in early engine development so it is worth digressing slightly and looking at their history. The Stirling engine was an early model of the hot air engine. A hot air engine takes heat and converts that heat to motive power. An engine like this is still a refrigerator, it cools the heat source and converts that to power. If the heat source is removed and power supplied to run the mechanism it will still remove heat, causing cooling in the region where the heat was previously generated. In this configuration it is called a cryocooler. For this reason, texts on engines are still applicable to cryocoolers and vice versa. Because of their relative simplicity and ability to be miniaturised, Stirling cryocoolers have found widespread utility, for example: cooling infrared sensors in handheld night vision cameras, on tanks, fighter planes and for remote sensing from space.

The Stirling engine, figure 4.1, was developed by the Reverend Robert Stirling who filed a patent in 1816 [6]. He recognised the importance of the regenerator in a hot air engine which improved the efficiency of the device (explained in more detail below). This was a period when the quality of steam boilers was frequently poor, causing explosions. The Stirling engine used only moderate pressures and was a safer alternative. Better engineering and materials improved the safety of steam engines which eventually won the commercial battle. However, the Stirling engine has always

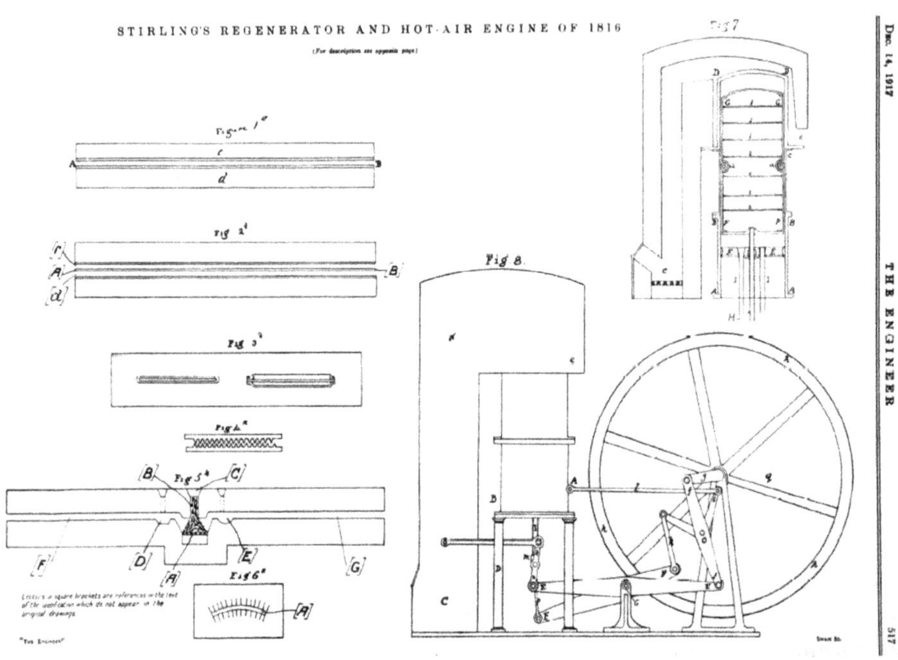

Figure 4.1. Robert Stirling's air engine (reproduced with permission from '*The Engineer*' December 1917).

had niche uses in small generators—because the combustion is external, a wide variety of fuels can be used. An example of this is the Philips 'Bungalow' generator. Only 150 of these were built. This type of generator has many advantages including near-silent operation, low maintenance and long life. These qualities stem from the inherent operation of the engine. Because power is delivered over a large part of the cycle, the stresses on bearings are low. The heat is applied externally so none of the combustion products come into contact with the lubricating oil. The near-silent operation arises from the lack of valves and continuous combustion (no explosions) [7]. However, this type of engine loses out to internal combustion engines on the grounds of efficiency and cost. Philips went on to make a name for itself producing Stirling cycle helium liquefiers. The company continues under the name Stirling Cryogenics.

Robert Stirling's air engine was turned into a refrigerator by Alexander Carnegie Kirk who was commissioned to find a substitute for an early vapour compression device that used ether. The use of ether was considered dangerous due to its highly flammable nature. In his investigations, he said in the proceedings of the Institute of Civil Engineers, 'Attention was next turned to a machine allied to Stirling's air engine, for as it was considered to be the best air engine yet produced, it was thought it ought to make a good cooling machine' [8]. The reference also details experiments made on the regenerator of the machine—this is a key component in Stirling refrigerators and anybody who has worked on this aspect of the design will be familiar with the arguments in those proceedings.

In recent years, the engine has been further developed as a refrigerator where a relatively small refrigeration capacity is required and can be found cooling infrared detectors in a variety of military and civilian applications [3].

The operation of the Stirling engine/cryocooler is relatively easy to understand. Compression of a gas using a piston in a chamber causes evolution of heat and expansion causes cooling. If another piston is introduced (a displacer) which shuttles the gas from one end of a chamber to the other, and the timing is synchronised with the compression, then the heating and cooling effects are separated. One end of the chamber is continuously heated and the other cools. What Robert Stirling understood in his engine was that there was an improvement that could be made to this cycle. Consider the gas after being expanded at the cold end of the chamber—it is displaced to the warm end where that cold is lost. Robert Stirling introduced what he called 'an economiser' (what we would now call a regenerator) [6]. This is a porous high heat capacity material that the gas flows through after the expansion phase on its way to the warm end. The regenerator is itself cooled by this action. On the return phase, after compression, the gas was pre-cooled by the regenerator on its way to the expansion space improving the efficiency of the whole process. In steady state operation the heat transferred to the regenerator around the cycle is zero (see figure 4.2).

Stirling engines and refrigerators have several mechanical configurations—these are called alpha, beta and gamma. These configurations are shown below in figures 4.3–4.5.

All the configurations shown above use mechanical linkages to phase the displacer and the compressor piston. An alternative method is to use what is termed 'free pistons'. In this arrangement the compressor and displacer are not linked but

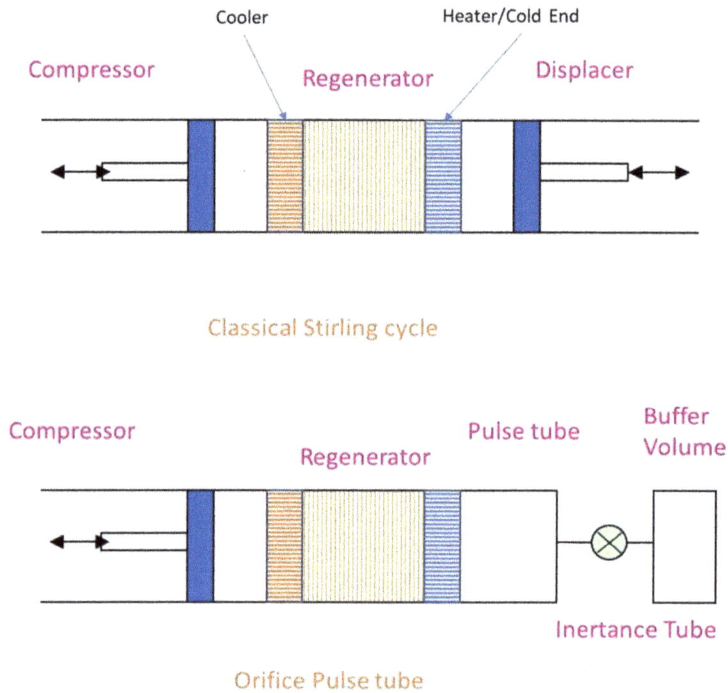

Figure 4.2. Stirling cycle and pulse tube operation.

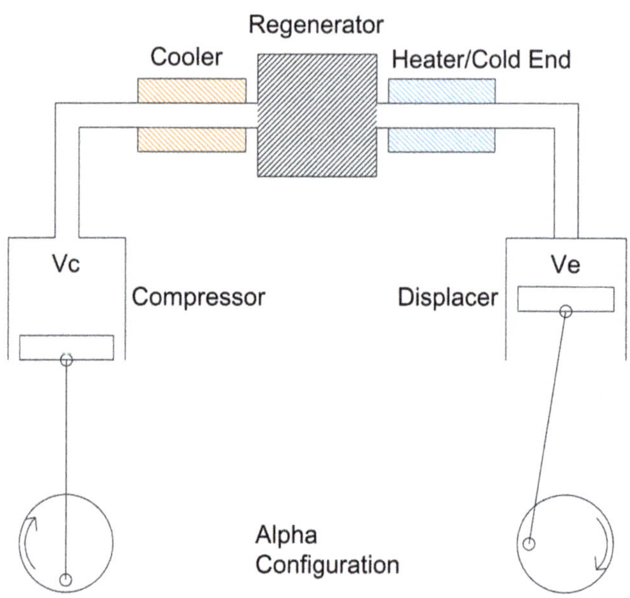

Figure 4.3. Alpha configuration is often used in automotive engines. An advantage of this type of engine is that the working gas flows through the heat exchangers allowing for improved efficiency.

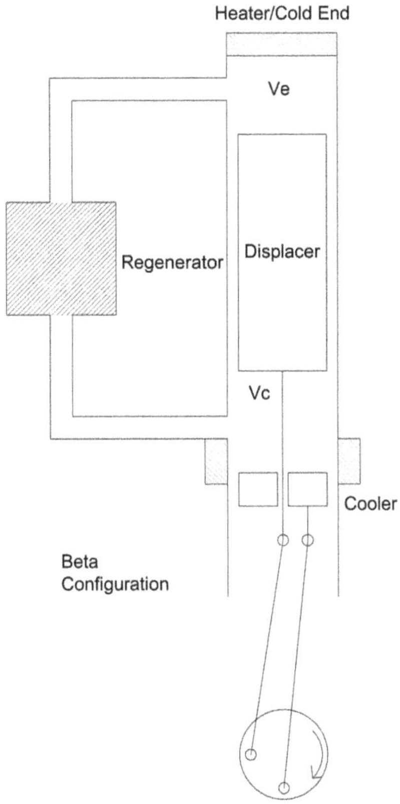

Figure 4.4. The Beta configuration has been used in both engine and cryocooler applications. The linkages between the compressor and displacer can be made using a 'rhombic drive' pioneered by Philips [9]. This gives near-vibrationless operation. A cryocooler design using this arrangement was used on several space missions [10].

driven by linear motors with the phasing determined electrically. These have some advantages as the bearings for the mechanisms are linear. Low friction and low wear can be achieved using either hydrodynamic, magnetic or diaphragm spring suspension systems. The coolers based on diaphragm spring suspension systems [11, 12] have proved to be very reliable and have been used on a wide variety of space missions for Earth observation and monitoring [13].

The optimal phase between the motion of the piston in the compressor and the displacer depends on how the engine is operating. As a thought experiment: if the temperature distribution of the engine can be maintained whilst the power piston is stopped then there will still be a pressure variation around the cycle due to the displacer shuttling gas from the heater/cold end to ambient. This contribution is in phase with the displacer motion. With the power piston in operation, the pressure variation is determined by the combination of the two pressure waveforms. The optimal phase between the compressor and the displacer is therefore less than 90° in a cryocooler and greater than 90° in an engine.

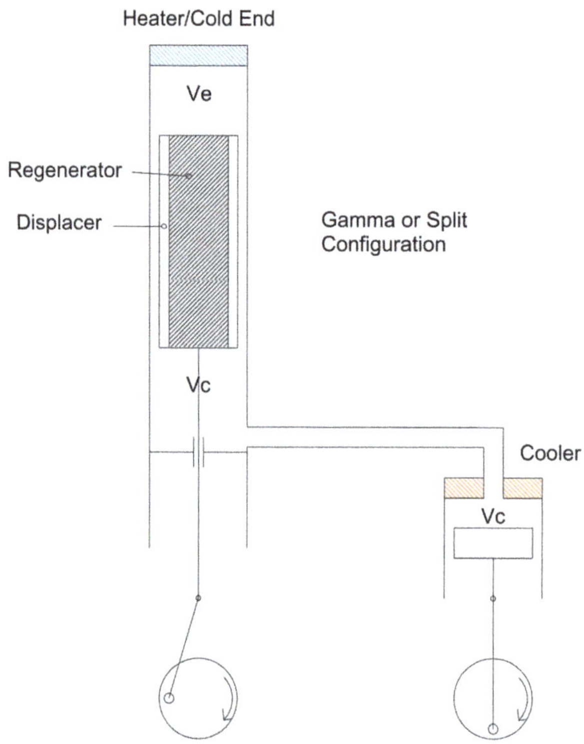

Figure 4.5. Gamma engines allow for the compressor and displacer units to be separated. This leads to a flexible system but also an inevitable reduction in specific power due to the increased dead volume. This is not always a problem and is often used in small cryocoolers, particularly free-piston ones. A free-piston compressor, running on resonance, will achieve higher efficiencies if the gas spring is linear. Note that in the diagram the regenerator is shown inside the displacer—this restricts the geometry of the regenerator but leads to a more compact system.

The regenerator is important to the action of Stirling (and similar cycles) and has to be optimised to give:

- A relatively low pressure drop;
- High heat transfer;
- Low longitudinal thermal conductivity;
- Low dead volume;
- A high heat capacity compared to the working gas.

These are conflicting requirements and present engineering challenges in optimising the form, material and the structure (engineering is the art of compromise…).

The Stirling refrigerator has several loss mechanisms driven by departures from ideal performance. A list of these losses can be given:

- Regenerator loss—this is a heat load on the cold end arising from less than perfect heat exchange in the regenerator or from a low volumetric heat capacity of the regenerator material. The gas entering the expansion chamber is at a higher temperature than the cold end temperature.

- Conduction losses—this is the heat load on the cold end down the outer containment vessel, the displacer material and along the regenerator.
- Pressure drop losses—this is a reduction in the cooling power due to the reduced pressure swing in the cold end. This can arise from pressure drop in the transfer lines, ducts and regenerator as well as expansion and contraction losses.
- Shuttle heat transfer—this arises from the radial heat transfer from the displacer to the outer cold finger when the displacer is away from its mean position. An alternative description is 'bucket brigade loss'.
- Imperfect heat transfer in the heat exchangers—the engineering of the heat exchangers is a balance between heat transfer and increasing the dead volume which limits the pressure swing.
- Thermal load from the external environment—this is a heat load on the cold end from thermal radiation from the surrounds. There can be a heat load from air being present around the cold finger—this is commonly reduced to zero by ensuring that there is a vacuum in that space.
- Compression losses—these arise from imperfect irreversible heat transfer in the compression space. There has been a lot of work done on this—it essentially boils down to minimising the surface area of the compression space, although frequency of operation is also important.

At low temperatures the material used for the regenerator becomes important. The heat capacity of the working gas (usually helium) is increasing while that of the regenerator is falling as T^3. Improvements to the regenerator performance can be made by choosing materials that have low temperature phase transitions. A high heat capacity is associated with these transitions that can improve the regenerator effect. Figure 4.6 shows the volumetric heat capacity of helium and a range of materials with low temperature heat capacity anomalies.

Modelling the thermodynamic cycle and the associated losses can be quite a task. Simple isothermal models can easily be set up on a spreadsheet. Treating the volumes as adiabatic was first tackled by Finkelstein [14]. This method is quite a reasonable representation of the cycle and, combined with an appreciation of the various loss mechanisms, is a good basis for engine optimisation. There are several textbooks and web-based resources that can help with this [15]. Proprietary programs are also available to aid the design processes. Although not for the newcomer, and now quite an old text, the *Stirling Engine Design Manual* is a useful read [16].

At small cooling powers and moderate temperatures ($T > 50$ K) Stirling refrigerators can be made in a compact and highly efficient form.

4.3 Gifford McMahon cryocoolers

These are probably the most widespread of all the cryocooler types. The two-stage units can offer temperatures of around 4 K with additional cooling to around 40 K. These replace the traditional helium bath plus liquid nitrogen shield arrangements.

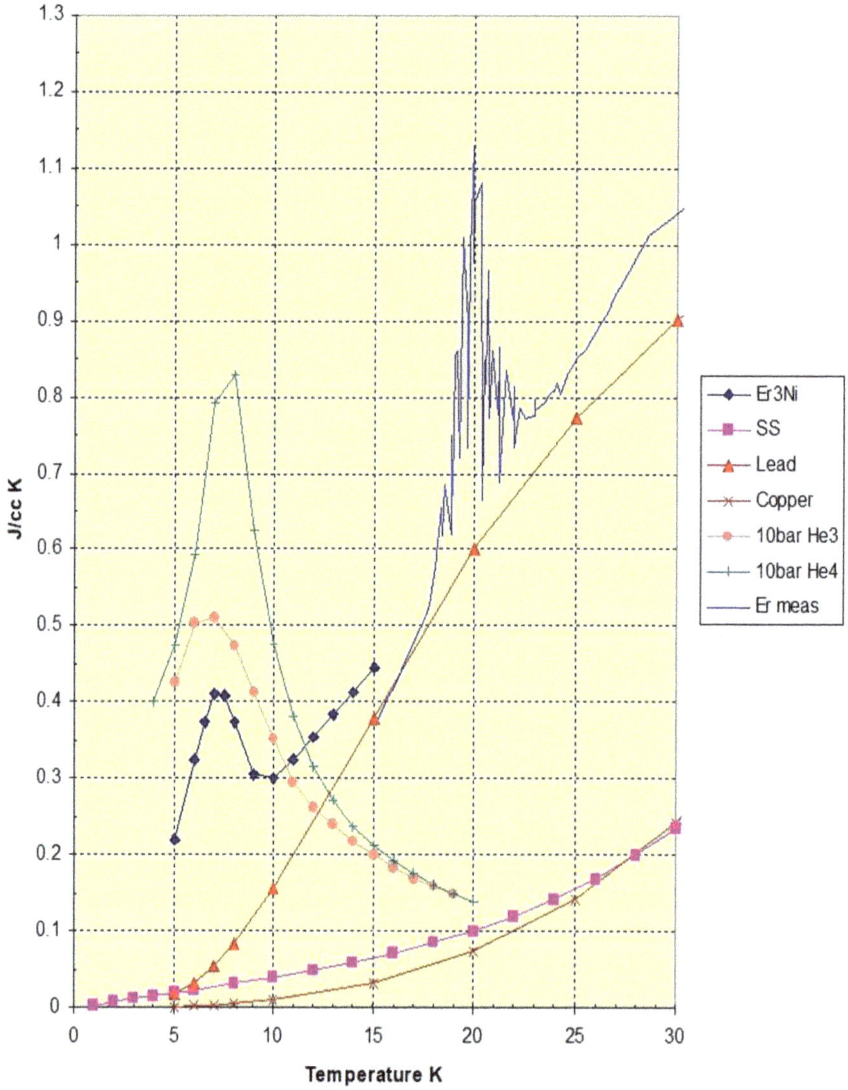

Figure 4.6. Volumetric heat capacity of helium compared to a range of materials.

The operation of these coolers is similar in principle to the Stirling refrigerator described previously, except that the pressure is provided by a 'DC' compressor (usually a scroll-type) and the pressure variation by a switching valve. This switches the pressure in the displacer unit between the low and high pressure as shown in figure 4.7. The rotary valve is synchronised with the displacer motion so that expansion takes place when most of the gas is in the expansion space (the displacer is 'down' in the figure). The connection to the high pressure line is made when the displacer is near the cold end and most of the gas is at the room temperature part. The regenerator (orange hatched part) performs the same function as in the Stirling.

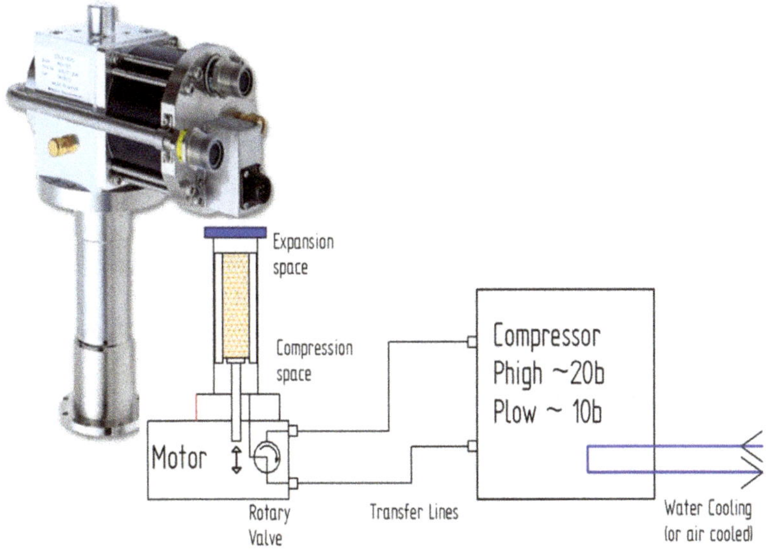

Figure 4.7. Gifford McMahon cryocooler inset is a Sumitomo single stage cold head (photograph permission of SHI Cryogenics Group RD-125D 77K Head shown).

Because of the high cooling powers available with these devices a challenge is often thermal contact with the cold stages of the refrigerator. Implementation in a cryostat needs to take into account thermal contraction issues and any possible issues arising from vibrations. The compressors take a significant amount of power and may require water cooling. The transfer lines can be quite long as this is where the helium gas reservoir is stored—any excess heat of compression is removed in these lines.

4.3.1 Use with superconducting magnets

These devices can be multi-staged with a stepped displacer providing an intermediate stage. Commonly a two-stage unit can provide ~40 K at the intermediate stage and 4 K at the cold end. The intermediate stage is useful for cooling a radiation shield (which reduces the 4 K load) and is appropriate for use cooling the current leads in superconducting magnets.

The current leads to a helium temperature superconducting magnet generally are constructed in two stages. The first stage is made from brass or copper with the cross-section optimised so that the Joule heating and conduction losses are minimised. The second stage to helium temperatures, is made from an appropriate high temperature superconductor (HTS).

Care must be taken in the in the placement and orientation of the HTS leads with respect to the magnetic field [17].

The stray magnetic field from the magnet also affects the cryocooler operation from its influence on the drive motor and the cryocooler regenerator. There is some degradation of the regenerator operation in magnetic fields as the design often takes advantage of the high heat capacities occurring with magnetic phase changes at low temperatures. Depending on the level of the field the motor may also require shielding. Some variants offer remote switching valves which can be located away from high field regions.

4.4 Pulse tube refrigerators

This is an interesting cryocooler that relies on the phasing between the pressure wave and the gas motion by using an inertance tube and a buffer volume shown in figure 4.8. This combination obviates the need for a displacer so there are no moving parts at low temperatures.

In the early days of the development of this type of refrigerator the inertance tube was variously replaced by an orifice or even a hand valve. The importance of using an inertance tube was realised later on, so the literature should be read with this in mind. A useful introduction to pulse tube refrigerator design has been written by Pfotenhauer *et al* [18].

There are several variants of pulse tube refrigerators which are defined by the frequency of operation. At high frequencies these are known as acoustic pulse tubes (described below [19]). At lower frequencies, the operation can be visualised by comparison with phase shifting in an electrical RC circuit.

Smaller refrigerators have been developed for cooling infrared detectors and optics in outer space for remote sensing applications. Because of the cold stage being in the centre of a linear design (figure 4.8) the cooler is more often configured in a 'U' shape. Designs have also been made with the pulse tube in the centre of an annular regenerator. An advantage over the Stirling model is that the gas is moving at the cold end which enhances the heat transfer, a disadvantage is that the flow through the regenerator is higher for a given capacity. Both effects need consideration in the design.

Large devices with capacities of order 1 W at 4 K (and below) have been developed. As with the larger Gifford McMahon cryocoolers the challenge is thermal contact with the cold stages. The compressor requirements are similar, however.

Although there are no moving parts at the cold end of a pulse tube, vibrations are transmitted through the transfer lines from the compressor. The pressure wave also

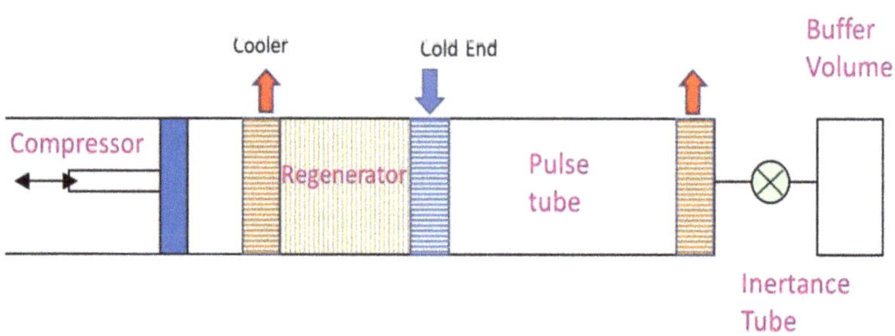

Figure 4.8. Pulse tube cryocooler.

causes an expansion and contraction of the thin-walled cold finger which modulates the position of the cold end. In operation, the orientation of the cold finger is important due to convective effects in the pulse tube. An orientation with the cold finger at an angle other than with the cold finger down results in reduced performance.

As with the Gifford McMahon cryocooler, operation in a high field region will require the switching valve motor to be shielded.

4.5 Thermoacoustic refrigerators

These devices use a standing wave in a closed tube. The components located in this tube (figure 4.9) are a hot heat exchanger, a cold heat exchanger and a device called a 'stack'. The stack conventionally consists of a strip of metal wound into a spiral. The coils on the spiral are separated from each other by a distance corresponding to a small multiple of the thermal penetration depth of the working gas. When used as a refrigerator, acoustic power is introduced into the tube by means of a loudspeaker-type device.

A good review of these devices is given by Swift [19]. The advantages of these devices are that they are simple, have few parts and are cheap to manufacture.

The disadvantages are that they tend to have low efficiencies and the position of the heat exchangers makes them difficult to integrate into systems. Despite the drawbacks these devices have been deployed successfully in large-scale energy recovery systems and have the potential in niche applications.

4.6 Joule–Thomson refrigerators

The principle of operation of a Joule–Thomson refrigerator is cooling obtained by adiabatic (i.e. no heat exchange) expansion of a gas through a nozzle. Strictly, the Joule–Thomson process concerns just the expansion of the gas through a nozzle.

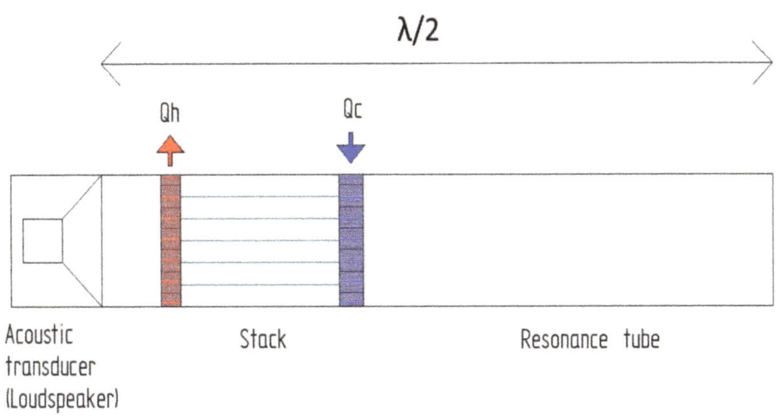

Figure 4.9. A basic thermoacoustic cooler.

A closed cycle refrigerator consisting of:

- compressor;
- heat exchanger;
- expansion nozzle;

… is termed a Linde Hampson Cryocooler. In this configuration the cooler can be used both as a refrigerator and a liquefier.

An excellent overview of this type of refrigerator is given by Maytal and Pfotenhauer [20]. The expansion needs to take place at the correct conditions where maximum cooling power will be obtained if the start point is on the inversion curve and the end point inside.

For the initial design of a Joule–Thomson cooler a useful plot is the 'inversion curve'. The inversion curve for some common refrigerants is given in figure 4.10. The inversion curve is a locus of points where the Joule–Thomson coefficient is zero. The Joule–Thomson coefficient is defined as:

$$\left(\frac{\delta T}{\delta P}\right)_H = -\frac{1}{C_p}\left(\frac{\delta H}{\delta P}\right)_T$$

where T is the temperature, P the pressure, C_p the specific heat and H the enthalpy.

Outside of this curve expansion from a high pressure to a low pressure will result in heating. Inside the curve cooling will result, with the maximum cooling power obtained from expansion from a point on the inversion curve to a low pressure.

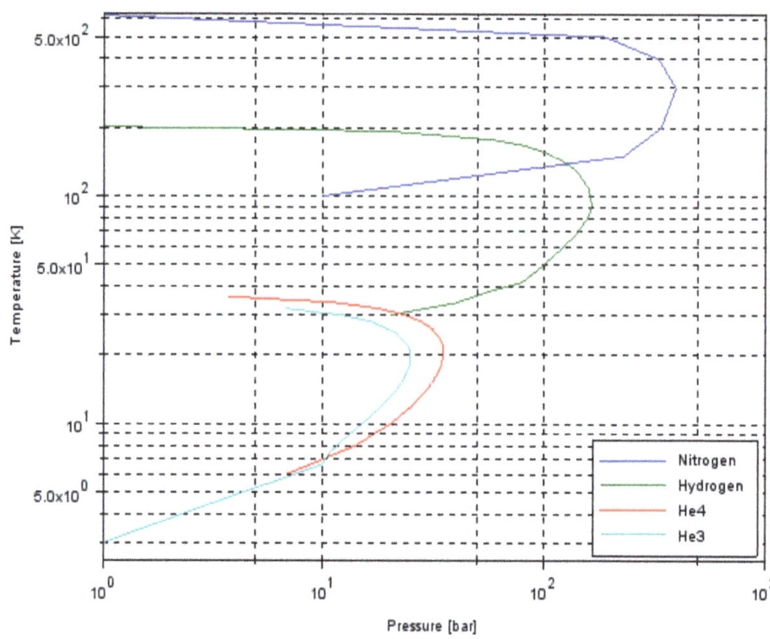

Figure 4.10. The inversion curve for several gases.

Some gases will produce cooling on expansion at room temperature, for others, pre-cooling stage(s) will be required. These two types of cooler are illustrated in figure 4.11. As a rule of thumb, it can be shown that the inversion curve meets the temperature axis at a temperature of approximately 6.8× boiling point at atmospheric pressure. To liquefy helium with a boiling point of 4.2 K, for example, pre-cooling to below 28 K will be required.

The counterflow heat exchanger is a key component in Joule–Thomson systems [21]. In general, the high pressure stream requires less surface area than the low pressure stream due to higher Reynolds numbers (and subsequent heat transfer coefficients) being generated for a given mass flow. In systems where liquefaction occurs the mass flow in the return stream is lower than that for the high pressure stream, adding an even greater requirement on the heat exchanger effectiveness.

Very small heat exchangers have been made using glass lithography [22]. For larger refrigerators tube-in-tube or finned tubes coiled in an annular space are used.

The cooling power is simply given by the mass flow times the enthalpy difference at the entrance to the Joule–Thomson heat exchanger:

$$Q = m(H(p, T)) - (H(p, T))$$

The two occurrences of H(P,T) refer to the enthalpy being a function of pressure and temperature).

To calculate this, an analysis of the heat exchanger is required.

The advantages with these refrigerators are that they can be made simply in a variety of sizes. They can also be made closed cycle with a compressor or open cycle off a high-pressure bottle. The disadvantages are low efficiency and the need to use a gas tailored to the temperature of interest. Much work has been done on gas

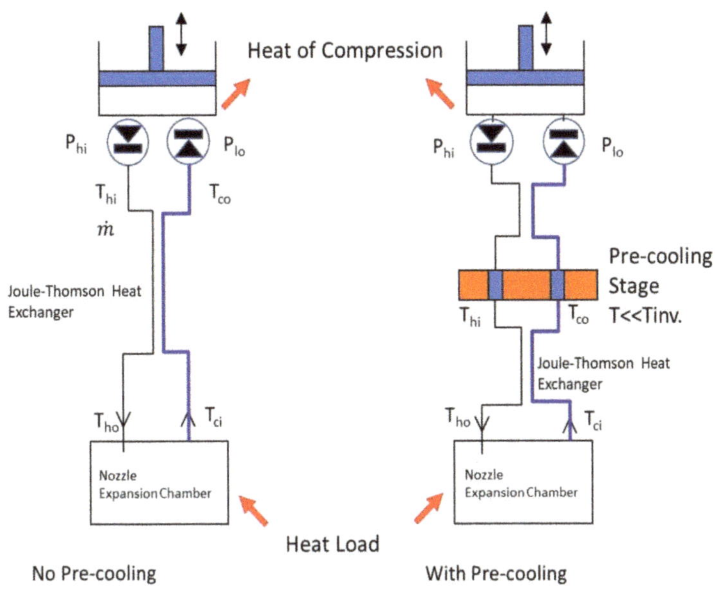

Figure 4.11. Joule–Thomson coolers—with and without pre-cooling.

mixtures, for example [23], which offer improved cooling power and a tailored temperature.

4.7 Vapour compression refrigerators

This cycle is used in the food industry, car air-conditioning systems, heat pumps and many other applications. It may seem strange to include this cycle in a textbook on cryogenics, however, low temperatures can be achieved through the use of cascade systems which are described later in this section. Due to the components being produced in huge numbers these are generally cheap, reliable systems.

4.7.1 Standard vapour compression refrigerator

The standard vapour compression refrigerator is the bedrock of most domestic refrigerators. The refrigerant is compressed (steps 1–2 in figure 4.12) and the high pressure gas condenses in the condenser (2b–3). The condenser is the system of coils and fins at the back of the refrigerator. The gas liquefies and exits as a high pressure liquid. The next stage (3–4) is expansion through a metering valve—this reduces the

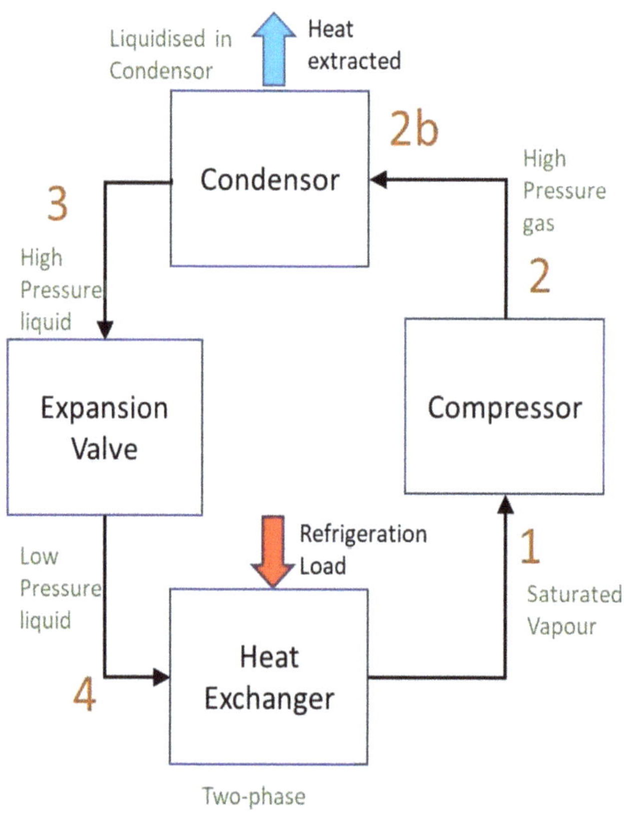

Figure 4.12. Standard vapour compression refrigerator.

pressure of the liquid and cools it to low temperatures. The cold gas then enters the heat exchanger (4–1) which is inside the refrigerator box. This is sometimes called the evaporator. The contents of the box are refrigerated and the low pressure liquid turns into a two phase gas/liquid mixture. The gas then enters the compressor where the circulation continues.

(As an aside, if this cycle is used to refrigerate the outside air then the heat of compression can be used for domestic heating. This is the basis of a heat pump.)

The vapour compression cycle shown in figure 4.13 is a cycle using R134a—commonly used as a refrigerant in domestic applications as it has a low global warming potential. The power requirement on each stage of the cycle can be easily calculated with a knowledge of the mass flow [kg s^{-1}]. The refrigeration is the difference in enthalpy at position 1 (H1) minus the enthalpy at position 4 (H4) multiplied by the mass flow (m).

$$Q_{\text{ref}} = \dot{m}(H_1 - H_4)$$

The work done is similarly the difference in enthalpy between positions 2 and 3 multiplied by the mass flow.

4.7.2 Cascade refrigerators

The standard vapour compression cycle is commonly used to refrigerate food to around −18 °C. A cascade can be set up to produce lower temperatures that in the cryogenic region. The cascade refrigerator operates by using two almost independent

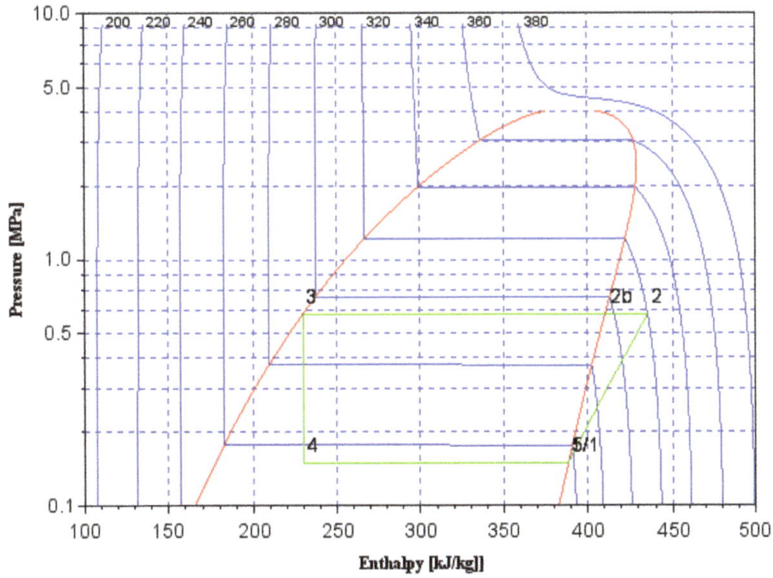

Figure 4.13. The vapour compression cycle.

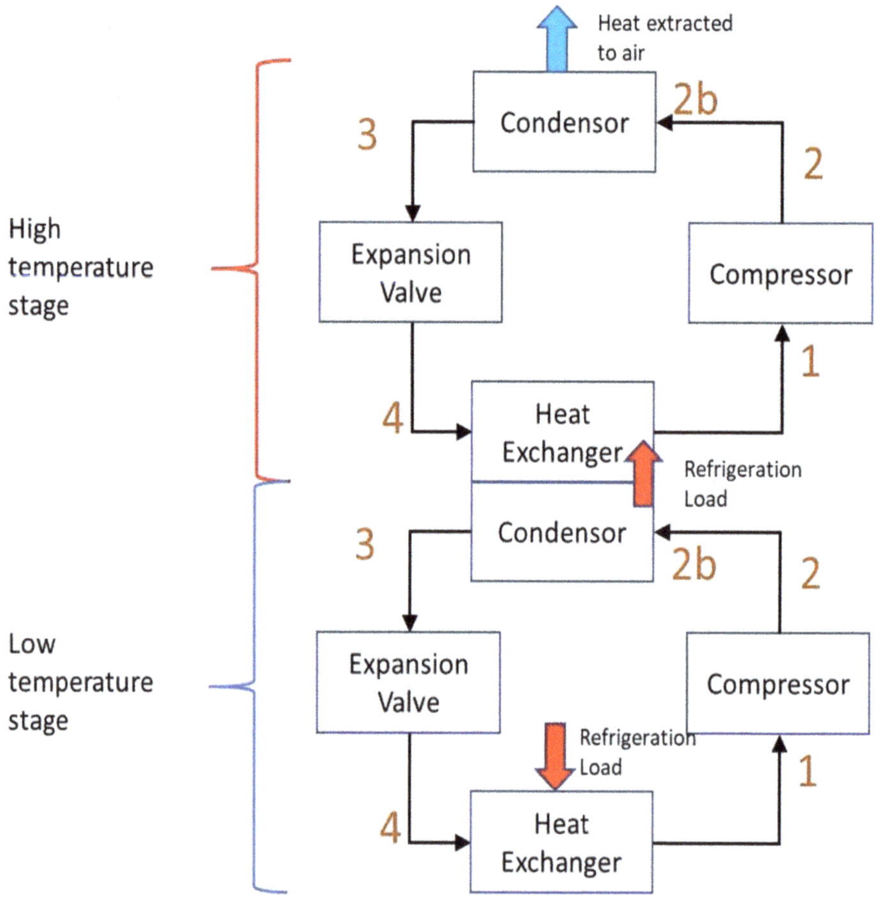

Figure 4.14. Cascade refrigerator.

refrigeration cycles with a high temperature stage being used to cool the condenser of a low temperature stage. This is illustrated in figure 4.14.

These types of refrigerators can achieve quite low temperatures. As an example, the refrigerant R23 when used in a cascade system can achieve 193 K.

4.7.3 Refrigerants

Refrigerants are designated by their ASHRAE number. This nomenclature is a way of naming refrigerants and classing them in terms of flammability and toxicity. Some refrigerants (in particular, chlorofluorocarbons) have been phased out. Others are being phased out in favour of those that have a low impact on the ozone layer and low global warming potential. The list of allowable refrigerants changes so it is worth consulting the ASHRAE website [24] for up-to-date information. Many modern refrigerants are mixtures that are selected for various properties (e.g. low flammability, toxicity).

4.7.4 Refrigerator design

The design of a refrigerator starts with the boundary conditions—required temperature, reject temperature and cooling power. From this the refrigerant(s) can be selected and the required mass flow. Once that is established the compressor and heat exchangers can be specified. There are many good textbooks that can help with the design of vapour compression refrigerators. Complete skids incorporating the compressor and heat exchangers can be purchased to aid with construction.

Properties of refrigerants can be determined from available libraries such as the US National Institute of Standards and Technology (NIST) REFPROP [25] (they can also supply vapour compression cycle analysis programmes). A free open source version is called Coolprop [26]. Both versions have high level interfaces. Both are not databases—fluid properties are calculated from first principles and both can deal with mixtures.

4.8 Turbo Brayton refrigerators

Turbo Brayton refrigerators are found in many industrial applications where large amounts of cooling power are required, although small units have also been developed. In particular, they are used in marine applications to re-condense liquefied natural gas during transport.

Turbo machinery provides a low pressure ratio and high volumetric throughput. The components are as follows and illustrated in figure 4.15.

- A compressor;
- Heat exchanger for the removal of the heat of compression;
- Heat exchanger for heat recovery;
- An expander;
- A heat exchanger for the thermal load.

In many applications the power dissipated in the expander can be recovered to reduce the compressor load. With clever design the compressor and the expander turbines can be mounted on the same spindle.

In cryogenic applications work is put into the cycle and heat is taken out—this is normally referred to as the reverse Brayton cycle.

For most applications the turbines for the compression and expansion stages are available commercially. The rotor design is usually made to limit aerodynamic losses and eddies at the tips of the turbine blades. A useful introduction to the design of hydro kinetic machines is given in Walshaw and Jobson [27].

These coolers typically find application in aircraft for air-conditioning systems. They are also used in the liquefied natural gas industry to liquefy the gas and during bulk marine transport to re-liquefy the boil-off gas. Small units have been used on spacecraft in a variety of applications, including one on the Hubble NICMOS instrument [28]. Because of the high power capability there is interest in using them for superconducting power applications such as power lines and fault current limiters [29].

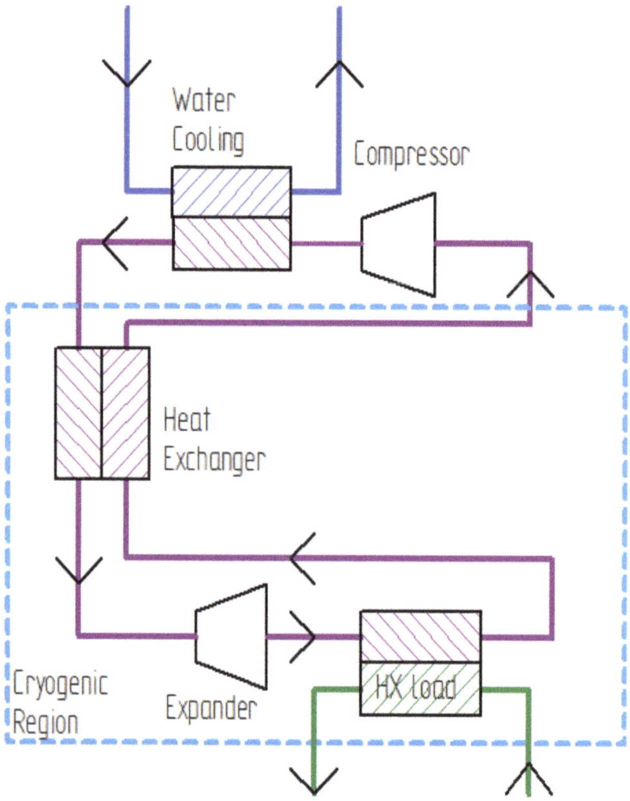

Figure 4.15. Brayton refrigerator.

4.9 Thermoelectric coolers

Thermoelectric coolers (sometimes called Peltier coolers) work by what is called the Seebeck effect. Two junctions made from dissimilar materials will create a voltage when a temperature difference is applied between them. This is the basis of a thermocouple temperature sensor. Similarly, if an electrical current is applied through the two junctions a temperature difference is formed.

These coolers are found in a variety of applications—most office water coolers employ them. They are attractive because of the silent operation and relative economy. The disadvantages are the low efficiency and the relatively low temperature difference that can be achieved. As the current through the junctions is increased, resistive heating increases and balances the cooling effect. This limits the temperature difference and cooling power that can be achieved. For low temperatures, the device can be cascaded (figure 4.16).

The material for the thermoelectric cooler is usually a bismuth telluride alloy. The cooling power is proportional to the number of elements × current. Note that when a current is applied there is simultaneous heating on the hot side and cooling on the cold side. Also note that a temperature difference across the elements is established. If the heat is not removed effectively from the hot side, then the effectiveness of the

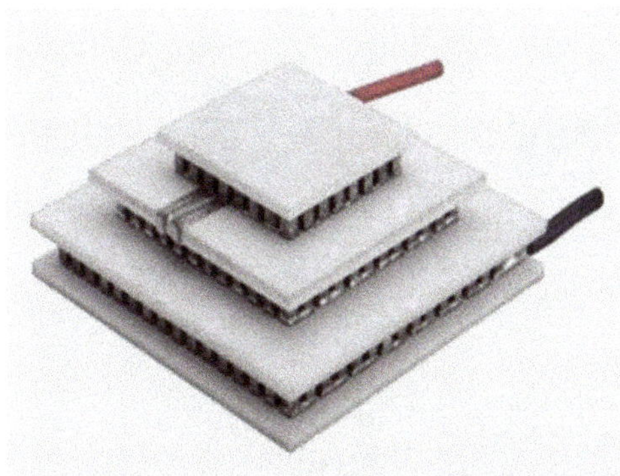

Figure 4.16. A cascade thermoelectric cooler [30] (picture with permission from Marlow Industries).

cooler is much reduced. This hot side heat removal issue is the main design issue in this type of cooler and needs to be treated carefully.

4.10 The Carnot cycle

The Carnot cycle is a reference cycle and is the 'best that you could ever hope to do thermodynamically'. Nicholas Leonard Sadi Carnot (1796–1832) analysed the perfect cycle in his '*Reflections on the Motive Power of Fire*' (1824) [31]. He was a French military engineer who died aged 36 of Cholera. His work is around the time of other developments, notably Stirling 1816 and Kirk 1874. As we will see, it is not a practical cycle as certain stages assume perfect material behaviour.

This cycle is often put forward as a demonstration of the thermodynamics of heat engines. It is better understood if practical cycles are understood first.

The cycle is shown diagrammatically in figure 4.17. The piston positions a–d are directly above the graph. There are two isothermal stages b–c and d–a where the hot or cold sources are placed above the piston. There are two adiabatic stages a–b and c–d, where insulators are placed above the piston.

It is often used as an exemplar so you will see cryocooler performance compared as a percentage of the equivalent Carnot cycle. Irreversible engines will always have a lower efficiency than this.

There are certain conclusions that can be made from this:

(a) To have an engine you need a temperature difference—any temperature difference can produce work.
(b) An engine takes in heat at the high temperature and gives out heat at the low temperature so trying to restore the thermal equilibrium.
(c) It follows that to produce a temperature difference work must be applied.

All heat transfer processes in the Carnot cycle are isothermal and reversible. During the adiabatic parts of the cycle perfect insulation is assumed. It doesn't

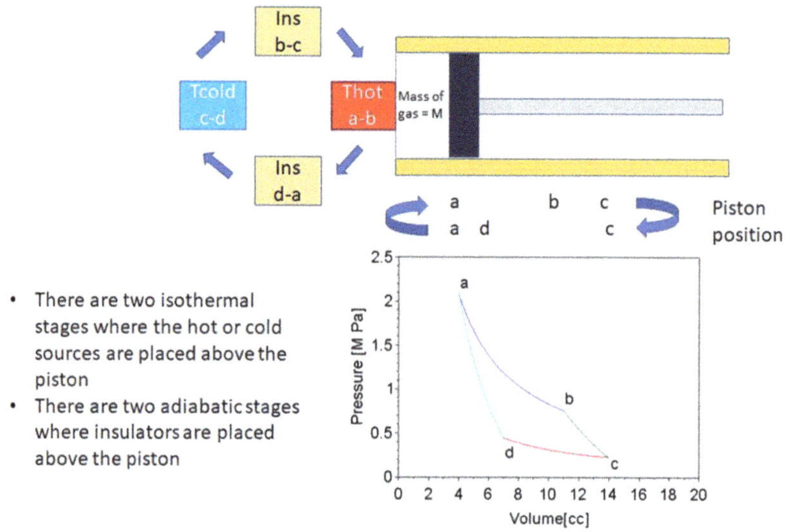

Figure 4.17. A representation of a Carnot cycle.

matter which gas is used (in practical engines it does, because of losses). It is assumed that the gas is ideal, i.e.—not close to a phase change. These considerations give rise to an important, yet simple expression for the efficiency:

$$e = \frac{W}{Q_a} = \frac{T_{\mathrm{hot}} - T_{\mathrm{cold}}}{T_{\mathrm{hot}}}$$

where:

e = efficiency;

Q_a = heat added;

W = work done (PV loop area);

T_{hot} = temperature of hot stage;

T_{cold} = temperature of cold stage.

The reference for *Reflections on the Motive Power of Fire* [31] is a modern reprint as the original is very rare and extremely expensive. This is an easily obtainable volume that contains a foreword by E Mendoza, the 'Reflections' by Sadi Carnot and two other papers; 'Memoir on the Motive Power of Heat' by Emile Clapeyron and 'On the Motive Power of Heat, and on the Laws which can be deduced from it for the Theory of Heat' by Rudolf Clausius. The two latter papers extend the work of Carnot.

Carnot introduced a thermodynamic cycle that is the best that can be achieved and in doing so debunked the idea of perpetual motion machines. He realised that any engine is driven by a temperature difference ('... wherever there exists a difference of temperature, motive power can be produced') and the engine is agnostic of the working gas. He also introduced the concept of a thermodynamic cycle that could be repeated without decay. He also noted that putting work into a cycle will produce a temperature difference—important for the action of closed cycle refrigerators.

This is an excellent and, perhaps surprisingly, a very readable volume. The introduction by Mendoza puts the work into context both in a historical and scientific sense. The papers by Clapeyron and Clausius turn the arguments into a more analytical form.

4.11 Summary

This chapter can only be an introduction to an extraordinarily diverse branch of cryogenics. The imperative for the adoption of CCRs stems from the fluctuating and often high price of helium. Most equipment is designed to operate around the clock, for example to reduce the amount of boil-off from helium baths or liquid natural gas storage containers. A plethora of equipment that operates in the infrared requires cooling to low temperatures ranging from satellites in space [32] to night vision goggles. Over the past few decades, several sea and land surface temperature radiometers [33] have been flown in space which are used to understand global warming, climate change and associated weather issues. This set of instruments has provided global measurements dating back to 1991. Every weather report that you see is based on data collected by these instruments which are only possible because of the developments in closed cycle refrigeration.

CCRs operating at cryogenic temperatures have entered mainstream activities and are seen as a cost-effective, reliable means of refrigeration.

References

[1] Andersen S A 1959 *Automatic Refrigeration* (London: MacLaren & Sons Ltd)

[2] Scurlock R G (ed) 1992 *History and Origins of Cryogenics* (Oxford: Clarendon)

[3] Walker G 1989 *Miniature Refrigerators for Cryogenic Sensors and Cold Electronics.* (Oxford: Oxford University Press)

[4] Nuttall W J, Clarke R H and Glowacki B A (ed) 2012 *The Future of Helium as a Natural Resource* (Milton Park: Routledge)

[5] Das N K, Bhandari R K and Mallik S C 2012 India: harnessing helium from the Earth's interior *The Future of Helium as a Natural Resource* ed W J Nuttall, R H Clarke and B A Glowacki (Milton Park: Routledge) ch 7

[6] Patent number is #4081. From the research that I have been able to do it was probably filed at the Chancery in Edinburgh in 1816. I believe that it should be in either NRS reference C19 or C20

[7] Air-Engine Generator MP 1002 CA Philips Instruction Manual

[8] Kirk A 1874 On the mechanical production of cold *Minutes of Proc.—Institution of Civil Engineers* **37** 1

[9] Storace A 1971 A miniature, vibration-free, rhombic-drive, Stirling-cycle cooler *AdvCryoEng.* **16** 185–94

[10] Lindale E and Lehrfeld D 1983 Life test performance of a Philips rhombic-drive refrigerator with bellows seals, NASA document 19840007273 *Proc. Refrigeration for Cryogenic Sensors NASA CP-2287*

[11] Werrett S T, Peskett G D, Davey G, Bradshaw T W and Delderfield J 1986 Development of a small Stirling cycle cooler for spaceflight applications *AdvCryoEng.* **31** 791–99

[12] Bradshaw T W, Delderfield J, Werrett S T and Davey G 1986 Performance of the Oxford miniature Stirling cycle refrigerator *AdvCryoEng.* **31** 801–9

[13] Gibson A S, Reed J, Bradshaw T W and Linder M 2008 Heritage overview: 20 years of commercial production of cryocoolers for space *AIP Conf. Proc.* **985** 493

[14] Finkelstein T 1960 Generalised thermodynamic analysis of Stirling engines *SAE Annual General Meeting* **118B**

[15] Urieli I and Berchowitz D M 1984 *Stirling Cycle Engine Analysis* (Bristol: Hilger)

[16] Martini W R 1983 *Stirling Engine Design Manual* 2nd Edn DOE/NASA194-1 NASA CR-168088

[17] Green M A *et al* 2008 The effect of magnetic field on the position of HTS leads and the cooler in the services tower of the MICE focusing magnet *IEEE Trans. Appl. Supercond.* **18** 1447–50

[18] Pfotenhauer J M, Gan Z H and Radebaugh R 2008 Approximate design method for single stage pulse tube refrigerators *Adv. Cryog. Eng.* **53** 1437–44

[19] Swift G 1995 Thermoacoustic engines and refrigerators *Phys. Today* **48** 22–8

[20] Maytal B-Z and Pfotenhauer J M 2013 *Miniature Joule–Thomson Cryocooling: Principles and Practice, Int. Cryogenics Monograph Series* (New York: Springer)

[21] Maytal B-Z and Pfotenhauer J M 2013 Heat exchangers *Miniature Joule–Thomson Cryocooling: Principles and PracticeInternational Cryogenics Monograph Series* (New York: Springer) ch 3

[22] Little W A 1984 Microminiature refrigeration *Rev. Sci. Instrum.* **55** 661

[23] Piotrowska A *et al* 2017 Performance analysis of Joule-Thomson cooler supplied with gas mixtures *IOP Conf. Ser.: Mater. Sci. Eng.* **171** 012080

[24] https://ashrae.org/technical-resources/standards-and-guidelines/ashrae-refrigerant-designations (accessed July 2020)

[25] https://nist.gov/srd/refprop

[26] http://coolprop.org/

[27] Walshaw A C and Jobson D A 1972 *Mechanics of Fluids* 2nd Edn (London: Longman)

[28] Swift W L, Dolan F X and Zagarola M V 2008 Advances in cryogenic engineering: Trans. of the Cryogenic Engineering Conference CEC *AIP Conf. Proc.* **985** 799–806

[29] Ozaki S, Hirai H, Hirokawa M and Kobayashi H 2018 Development of 10 kW turbo-Brayton refrigerator for HTS power applications *IOP Conf. Series: Materials Science and Engineering, vol 502, 27th Int. Cryogenics Engineering Conf. and Int. Cryogenic Materials Conf. 2018 (ICEC-ICMC 2018) (Oxford, 3–7 September)*

[30] https://marlow.com/resources/thermoelectric-technology-guide/ (accessed May 2021)

[31] Sadi Carnot E M (ed) 2005 *Reflections on the Motive Power of Fire* (New York: Dover) (this is a modern reprint).

[32] Duband L 2015 Space cryocooler developments *Phys. Proc.* **67** 1–10

[33] https://sentinel.esa.int/web/sentinel/user-guides/Sentinel-3-slstr/overview/heritage

IOP Publishing

Cryogenics

Fundamentals, foundations and applications

Tom Bradshaw, Beth Evans and John Vandore

Chapter 5

Very low temperature techniques

Chris Lawson, Viktor Tsepelin, Lionel Duband and Tristan Valenzuela

This chapter draws on a team of experts to look at how low temperatures are achieved and maintained—these techniques almost always require multiple stages of cooling, so a knowledge of the previous chapters is highly recommended.

The applications for very low temperatures (<1 K) used to be limited to research into the more esoteric properties of helium. These days very low temperatures are required in a range of real-world activities such as quantum computing and cryptography. Astronomy in the far infrared relies on bolometers and superconducting detectors which themselves rely on low temperatures for improved sensitivity. Sorption coolers offer probably the lightest solution for cooling in the 300 mK range and are therefore important for space applications. Laser cooling is a fascinating technique that uses different techniques to those usually encountered in conventional cryogenics but offers extremely low temperatures, albeit on a limited number of atoms.

5.1 Dilution refrigerators

Chris Lawson

5.1.1 Introduction

Before the development of the dilution refrigerator, low temperature physics found itself stuck in the temperature range of around 0.2–0.3 K achievable using adiabatic demagnetisation of paramagnetic salts [1], or by reducing the vapour pressure above liquid ^{3}He [2]. It is important to note that these are, typically, one-shot techniques, setting an absolute limit on the possible experimental time.

Interestingly, the theoretical groundwork for a continuous process which utilised as-yet-undiscovered phase separation effects in ^{3}He/^{4}He mixtures was laid out some years before the phenomenon was observed [3, 4].

It was a further 10 years, in 1965, before a continuous dilution refrigerator was experimentally realised, [5] (Edwards, and others, discovered the finite solubility of

doi:10.1088/978-0-7503-2757-2ch5

^3He in ^4He at the absolute zero in 1965), [6] (P Das, R de Bruyn Ouboter, K W Taconis, one-shot dilution refrigerator, 1965), [7] (H E Hall, P J Ford, K Thomson, continuous dilution refrigerator, 1966).

The dilution refrigerator has played (and continues to play) a critical role in advancing the science of low temperature physics. It is unique in its ability to provide a continuous cooling power at temperature below 300 mK, and as such serves as a launch pad for much lower temperature cooling systems such as nuclear demagnetisation stages [8].

The successful commercialisation of dilution refrigerator systems has truly led to a quantum leap in capability across many science disciplines, from quantum turbulence to gravitational wave detection.

5.1.2 Theoretical description

The basic physical principles leading to the creation of a dilution refrigerator are as follows: ^4He, and its lighter isotope ^3He, have long played their part in cryogenics. By reducing the vapour pressure above the liquid surface, it is possible to reach very low temperatures, below 1 K. This is in part due to helium's ability to remain liquid even down to the absolute zero of temperature. Van der Waals interactions between individual helium atoms are so weak that they may be overcome by the inherent vibrational energy granted to them by quantum mechanics (so-called zero point energy).

It is due to these quantum mechanical effects that a mixture of ^4He and ^3He together possess the fantastic properties required for dilution refrigeration to work.

If a $^3He/^4He$ mixture is cooled below approximately 870 mK, a spontaneous phase separation occurs. That is to say that the fluid ceases to be a miscible combination of both isotopes, instead we enter a situation where there are two discrete fluids. The first component of which is a ^3He rich phase, consisting of pure ^3He. This floats on top of the second component; a *dilute* phase that is mostly ^4He, but which also contains 6.6% ^3He. This seemingly arbitrary percentage is a result of the same zero point energy effects which give helium its useful low temperature characteristics, coupled with the fact that ^3He atoms are fermions and must therefore obey the Pauli exclusion principle (this leads to the case where it is energetically favourable for ^3He atoms to reside in the ^4He). This can be seen in the phase diagram for helium mixtures, shown in figure 5.1. In designing a dilution refrigerator, one would aim to introduce the correct volumes of ^3He and ^4He such that this phase separation will occur in the **mixing chamber** of the fridge.

Once this situation is established, we can take advantage of the very different vapour pressures of ^3He and ^4He to preferentially extract ^3He atoms from the dilute (mostly ^4He) phase. This occurs in the **still** of the dilution fridge, an apt name, as this process is the same as used when one wishes to create one's own spirits (a subject for the next British Cryogenics Council eBook perhaps!)

The removal of ^3He from the dilute phase produces an osmotic pressure; the 6.6% ^3He concentration must be preserved. The only remaining source for replenishment of ^3He is the concentrated phase, therefore ^3He atoms will cross the phase boundary

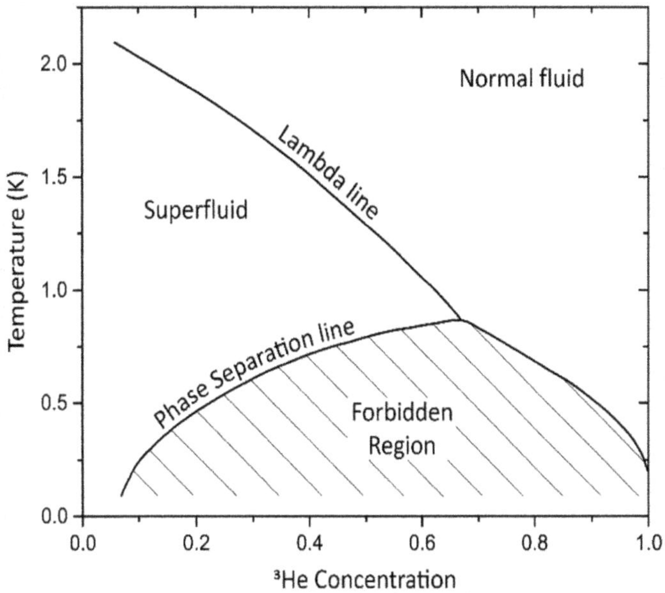

Figure 5.1. Helium mixture phase diagram.

between concentrated and dilute phase. It is known that the enthalpy of ^{3}He in the dilute phase is larger than that of the concentrated phases, leading to a heat of mixing. Energy must be removed from the surroundings to support this process.

Transferring ^{3}He atoms across the phase boundary at rate $\dot{n}_3$ will give a cooling power Q according to the difference in enthalpy of the dilute and concentrated phases, H_d and H_c respectively.

$$\dot{Q} = \dot{n}_3[H_d(T) - H_c(T)]$$

Running through this calculation yields a description of the cooling power for the dilution process as:

$$\dot{Q}(T) = 84\dot{n}_3T^2$$

Note that this indicates the cooling power of a dilution fridge should go as T^2. This process is shown schematically in figure 5.2.

In reality, the creation of a real dilution refrigerator involves a large number of specialist engineering techniques (some developed purely for their manufacture), several of which will be mentioned in the next section.

5.1.3 Practical description

The modern dilution refrigerator (now widely available commercially, with fantastic performance and ease of use) is a culmination of not only 50 years of dilution refrigerator powered research, but also the development of modern methods in manufacturing techniques, materials research and industrial control. Considerations

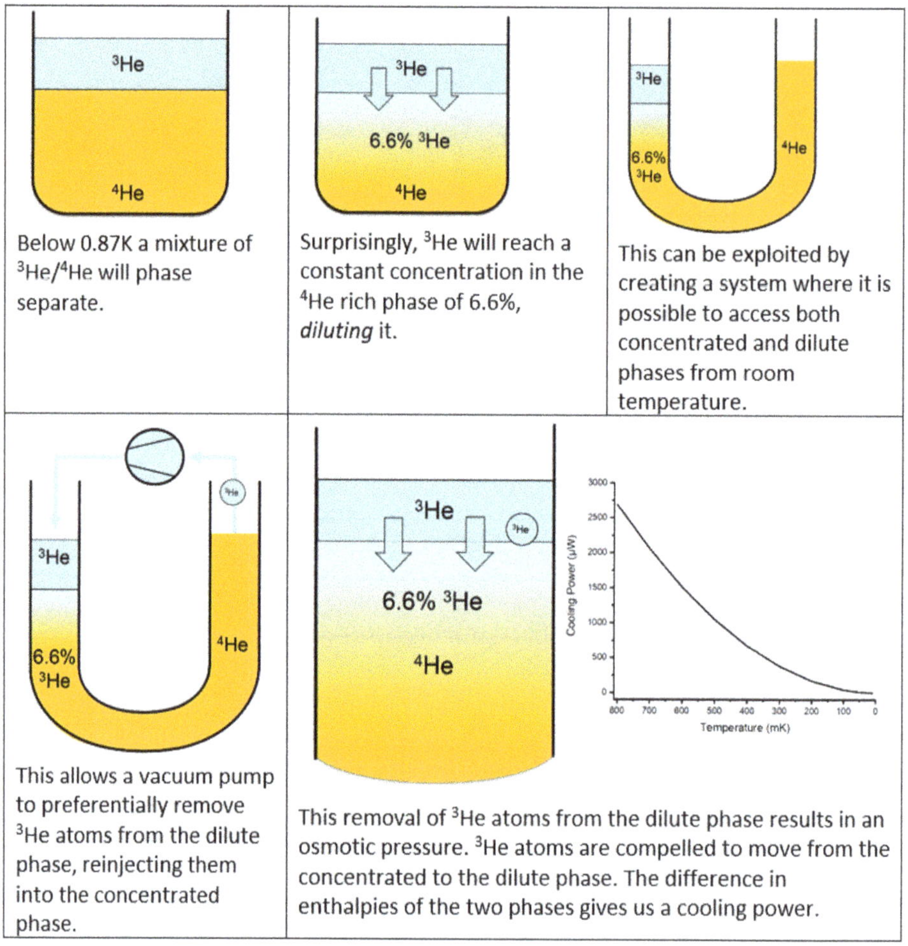

Figure 5.2. Operation of the dilution refrigerator.

in thermal design (mentioned previously in the design of cryostats) become critically important at dilution fridge temperatures. Even the highest performance modern fridge will have cooling power of not much more than 1 mW at 100 mK. An overall dilution refrigerator design is shown in figure 5.3.

5.1.4 Gas handling

Beginning at room temperature, the contemporary dilution fridge will almost certainly be operated by a computer-controlled gas handling system. To inject, recover and store the $^3He/^4He$ mixture (simply 'mixture' from here on) the gas handling system will feature leak-tight connections to the still and condenser side of the dilution fridge. Care must be taken in the connection and testing of these fittings, as there is potential to lose or otherwise pollute the carefully calculated mixture.

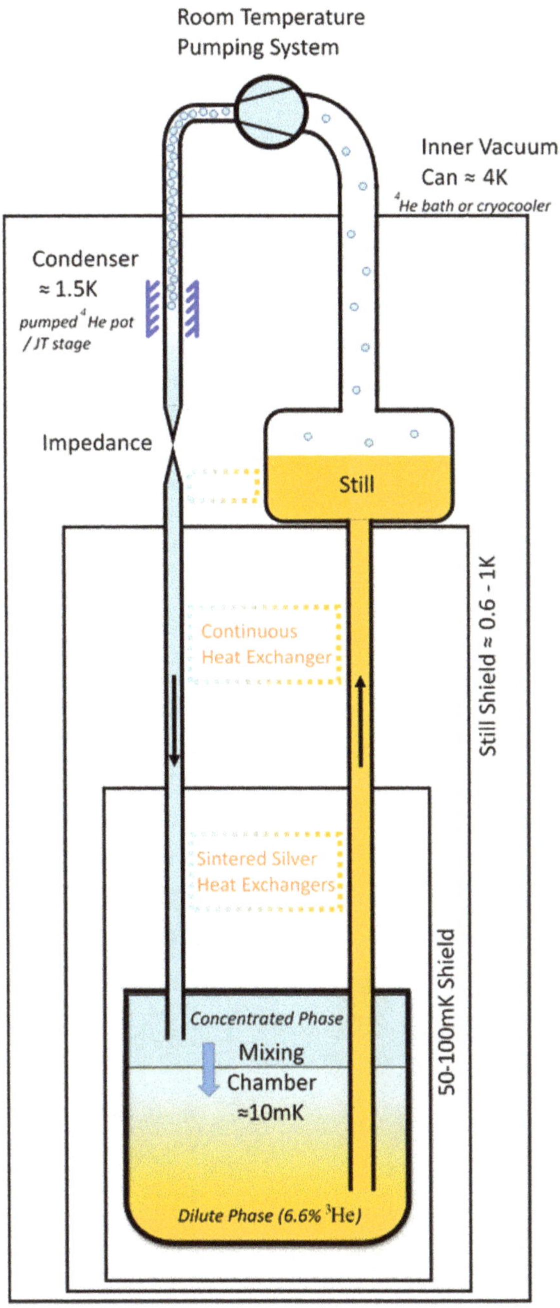

Figure 5.3. Dilution refrigerator components.

A large volume will be used to store the mixture when the fridge is not in use, preferably at pressures below 1 atm to prevent losses in case of a leak. How large this is will depend on individual refrigerator specification.

Mixture will be circulated through the gas handling system and the fridge using vacuum pumps and compressors. Typically, dry pumps (i.e. oil-free) are preferred for this purpose. The pump throughput will directly affect the fridge's ultimate performance, and so should be matched to a particular refrigerator specification. If a compressor is used, for example to provide high pressure for a Joule–Thomson cooling stage (necessary in fridges which do not feature a liquid ^{4}He bath), then it should be of a double membrane design to prevent possible mixture leakage. All pumps should be maintained according to the manufacturer's maintenance schedule.

Before the mixture enters a dilution refrigerator it should pass through a liquid nitrogen trap at 77 K. This will freeze out almost all possible contaminants, thus protecting the fridge's fine impedance (described later) from blockage. The trap itself takes the form of a metal cylinder filled with activated charcoal, giving a very large surface area and acting as a cryopump. Under normal circumstances the trap will not capture a great deal of non-mixture gases, however, it is best practice to periodically check its contents via a warm-up (whilst isolated and connected to a pressure gauge). Often the first sign of a problem can be 'unusual' gases appearing in the trap. Hydrogen contamination will not be captured by a nitrogen trap, if you are unlucky enough to experience this then special steps need to be taken, such as the use of a chemical trap in-line with the condensing circuit. An example gas handling system is shown in figure 5.4.

It is worth noting that there have also been successful dilution refrigerator designs that are 'closed cycle', using systems such as alternating cold sorption pumps to circulate ^{3}He without the need for a room temperature mechanical pump [9]. This is advantageous as it removes a major source of vibrational heat leak [10].

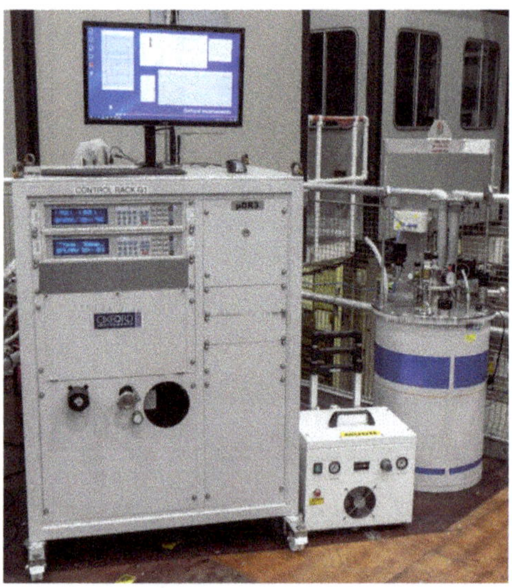

Figure 5.4. An example gas handling system (credit: ISIS Neutron and Muon Source, Science and Technology Facility Council).

5.1.5 Low temperature environment and mixture condensation

The low temperature environment surrounding the active dilution fridge circuit serves two purposes. Firstly, it facilitates the condensation of ^{3}He/^{4}He mixture into the liquid state. Secondly, it acts as a thermal shield for the colder parts of the refrigerator.

For decades, this environment was provided by a liquid ^{4}He cryostat at 4.2 K, with a pumped '1 K pot' of helium fed continuously from the bath. Support of a wet helium system requires significant effort, one may either recapture evaporated helium and re-liquefy (if your institution is lucky enough to have a helium liquefier) or continuously purchase helium from an external supplier to make up what has been lost through evaporation and pumping. As helium became a more expensive commodity (and very difficult to get in some parts of the world) it was natural that a 'dry' solution would be extremely popular.

Pioneering work using Gifford McMahon (GM) cryocoolers and pulse tube refrigerators (PTRs) showed that this was indeed possible [11, 12].

In the modern era, the vast majority of dilution refrigerators that are manufactured utilise PTRs, avoiding the vibrational issues which occur with a GM cooler and removing the need for a liquid helium cryostat. In these designs a two-stage PTR is used, a warmer stage that serves as a thermal shield around 50–60 K, and a colder stage at 4 K or below. Once mixture has passed below the 4 K stage, it will meet the main flow impedance, this exists to generate enough pressure (of order 100 mb) to allow the incoming ^{3}He to condense. Additionally, to aid in the condensation of fridge mixture, a Joule–Thomson cooling stage is used in most systems, avoiding the need for a pumped 1 K pot.

5.1.6 Heat exchangers

Descending into the dilution fridge, we encounter heat exchangers. Well-designed and expertly made heat exchangers are essential if one is to reach the ultimate performance of a dilution fridge system. Their purpose is to make maximum use of the cooling power of extracted mixture from the mixing chamber, to reduce the temperature of incoming reinjected ^{3}He. Heat exchangers are *the* critical component in any dilution fridge, essentially determining the minimum achievable temperature.

The challenge here is that at low temperatures there exists a thermal boundary resistance, known as the Kapitza resistance R_K, which obstructs the flow of heat Q between two surfaces leading to a step in temperature T thus:

$$\Delta T = R_K \dot{Q}$$

Critically, the Kapitza resistance is inversely proportional to temperature to the third power:

$$R_K \propto \frac{1}{A\,T^3}$$

As we are working at extremely low temperature this proves to be quite a problem in the design of heat exchangers. Ultimately the only option is to try and maximise the

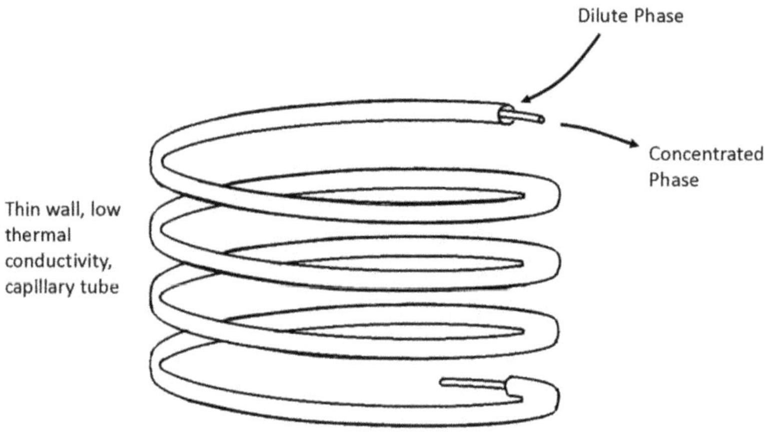

Figure 5.5. Example concentrated heat exchanger (credit: Tom Bradshaw).

contact area A. Accurate modelling of the heat exchanger is complex; temperature gradients (and indeed mixture concentration) will change along the length of the heat exchanger. Two examples are shown in the following paragraphs.

Most fridge designs will use a comparatively simple 'continuous heat exchanger' at higher temperatures (e.g. just below the still). This design is relatively cheap and simply to make, whilst still giving a sufficiently large surface area for heat exchange. As shown in figure 5.5 this heat exchanger consists of a tube-in-tube design using different sized capillaries. Heat will be conducted away from the injected concentrated phase to the extracted dilute phase through the capillary walls.

To achieve even lower temperatures than are achievable with continuous heat exchangers alone, a more advanced design is required. So-called 'step' or 'sintered' heat exchangers overcome the tough requirement for large surface area by using sintered silver powder as a surface for heat exchange. Further considerations include large channels, which reduce viscous heating effects due to the flow of liquid helium. The viscosity of helium increases greatly as temperature drops and puts a limit on the lowest base temperature achievable by a reasonably sized fridge. For example, designs by Frossati [13, 14], pictured in figure 5.6, have been used in some of the most high-performance refrigerators.

5.1.7 Mixing chamber

Similar ideas involving sintering to maximise surface area may also be found in the mixing chamber of the refrigerator. Here the phase boundary is maintained and provides continuous cooling to the experiment. Most often, the experiment is tightly connected to the mixing chamber via some metal-to-metal contact. But in some designs the mixing chamber may be disassembled, allowing the experiment to sit directly in the ^{3}He/^{4}He mixture and take advantage of the fantastic thermal conductivity of liquid helium.

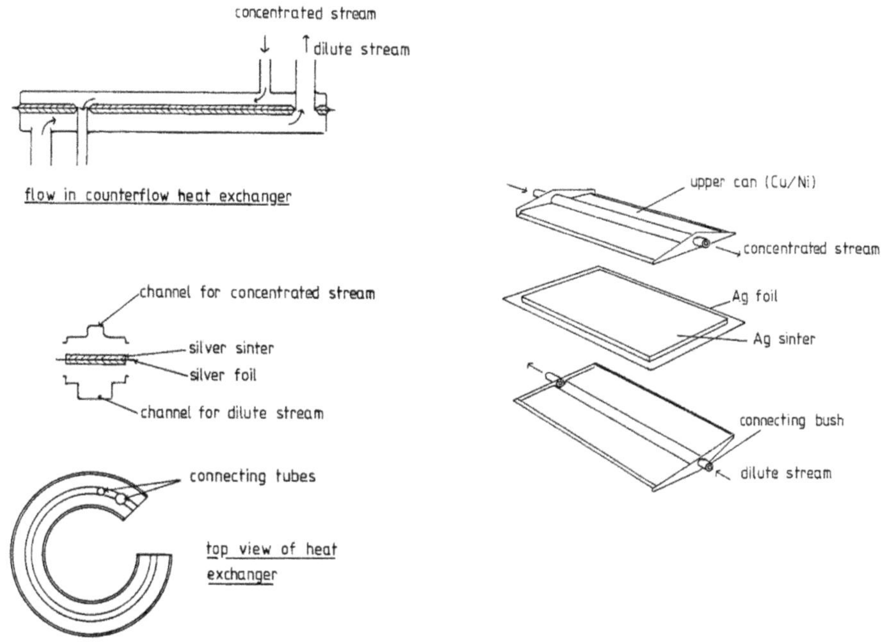

Figure 5.6. Sintered heat exchanger example (credit: Tom Bradshaw).

5.1.8 Thermometry, wiring and thermal anchoring

Thermometry and readout at these temperatures is a large subject in itself and is covered elsewhere in this book. Commercial dilution refrigerators typically come equipped with resistance thermometry, usually read using a low amplitude AC resistance bridge, such as Lakeshore 372. As resistance measurement is a secondary thermometer, it will have been calibrated against some primary thermometer at the factory. Examples of primary thermometry used at these temperatures include ^{3}He melting curve [15], nuclear orientation [16], pulsed nuclear magnetic resonance [17] and noise thermometry [18–20].

5.1.9 Fixing incorrect fridge mixture

The most traumatic event possible for most dilution fridge users is a loss of precious ^{3}He/^{4}He mixture. This can occur in some singular dramatic event or may happen over time if a small leak is present. In the latter case, it may only be noticed once the loss is enough to affect fridge performance, usually causing strange effects as phase boundaries shift out from the mixing chamber or still.

Once the sinking feeling of terror has subsided, one is left with several options to try to correct the issue. The first step is obvious; the mixture will have to be topped up to the full value. However, the ratio of ^{3}He:^{4}He is important and will have to be determined.

If the fridge is still able to remain operational then a 'single-shot' can be performed; this involves halting the re-injection of mixture into the fridge and

Figure 5.7. Acoustic gas analysis of dilution fridge mixture using an SRS BGA [R Price and C Lawson, Internal technical note, ISIS Neutron and Muon Source].

extracting virtually pure ^{3}He from the still. One will see a small drop in base temperature (we have removed the heat leak caused by incoming mixture) and a slow drop in still pressure. Once all ^{3}He is removed, the mixing chamber temperature will warm and the still pressure will drop suddenly. If cooling the fridge is not possible, one must analyse at room temperature. This can be done via mass spectrometry. Recent trends involve the use of acoustic gas analysis to determine dilution fridge mixture (an example is shown in figure 5.7). This technique has the advantage of being very robust and non-destructive.

Further information on diagnosis and resolution of common dilution refrigerator problems can be found in references [21, 23].

5.1.10 Modern applications

The dilution refrigerator allows researchers to remove thermal noise in a system, such that previously hidden low energy or quantum behaviour may emerge. A secondary application is to act as a launch pad for the creation of even lower temperature environments. The most highly performing commercial dilution fridges reach temperatures of around 3 mK, with high cooling powers sometimes exceeding 1 mW at 100 mK. The lowest temperatures possible with dilution refrigeration are achieved only with custom-made research machines that exist in specially-built shielded rooms (shown in figure 5.8), for example 1.75 mK in [24] and 1.9 mK in [25].

Dilution fridges have always been a core component of modern condensed matter research, such as spin systems [26], superconductivity [27] and atomic level microscopy [28].

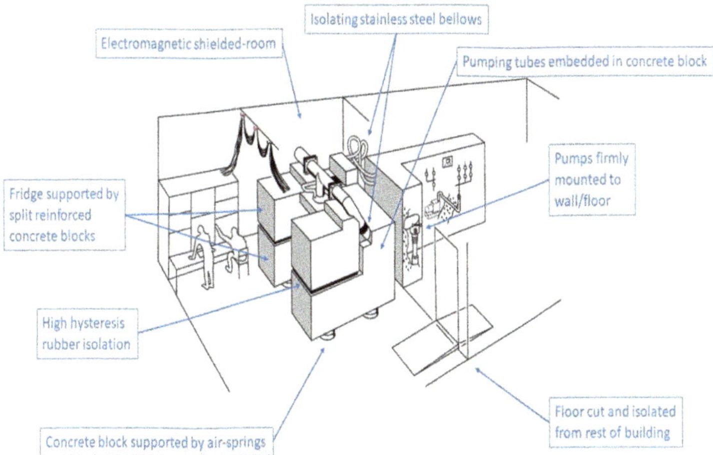

Figure 5.8. The lengths to which one must go to achieve the ultimate dilution refrigerator performance. Source: G Pickett, Lancaster University.

Some of the most adventurous applications involve the creation of entirely plastic dilution fridges for use in high magnetic fields [29] and ultra-low temperature thermal conductivity measurements of LEGO bricks! [30].

The commercialisation of dilution fridge technology, especially the introduction of a closed-cycle cryocooler as a substitute for a liquid helium bath, has led to a massive increase in the number of dilution fridges in use today. This age of 'milliKelvin for the masses' has been largely driven by the enormous demand for quantum computation research, with many types of qubits requiring very low temperatures for their quantum behaviour to manifest.

5.1.11 Quantum computation

There is a vast amount of resource worldwide being dedicated to the creation of a quantum computer, and the humble dilution refrigerator is at the heart of many of these efforts. Dilution refrigerators feature in high profile machines from Google [31], IBM [IBM Q] and D-Wave [32], their gold plated flanges and shields taking centre stage in publicity photos, as shown in figure 5.9.

The qubits studied in these systems are typically superconducting and so the fridge is needed firstly to keep the system below the superconducting transition T_C, but also to reduce the thermal noise in the system which will compromise the coherence time of said qubits if left unchecked [33].

One major challenge in this work is in wiring and electronic readout. As it is difficult to multiplex qubits, individual input/output is usually required for each one, leading to a huge thermal load on other stages of the dilution fridge. Additionally, the use of cryogenic amplifiers and electronics dumps even more power on the fridge's high temperature stages. This has driven the creation of monstrously large machines that feature multiple cryocoolers, whose main role is to reduce the thermal

Figure 5.9. Artist's rendition of the Google Sycamore processor mounted in the cryostat. (Credit: Forest Stearns, Google AI Quantum Artist in Residence. From: https://ai.googleblog.com/2019/10/quantum-supremacy-using-programmable.html.)

load on the colder parts of the fridge by supplying excess cooling power at 4 K. Further details may be found in [34] and references therein.

A related area of study has been the pursuit of very low electron temperatures using dilution fridges. It is known that there exist thermal boundary resistances not only between different metals, but also at an atomic scale between the lattice, electrons and nuclei. As quantum computation requires reduction of thermal noise, this includes the electrons present in electrical connections to the qubits [35, 36]. Later attempts make use of demagnetisation techniques, described later [37].

5.1.12 Quantum fluids and solids

Some of the earliest uses for dilution refrigeration were in quantum fluids and solids research. Here researchers aim to investigate phenomena such as superfluidity, quantum turbulence [38–40], thin films [41], and quantum plasticity [42].

5.1.13 Cosmological phenomena

The ability to remove background thermal noise is critical when attempting to measure the extremely small signals prevalent in some cosmological phenomena. One impressive example of this is the creation of a gravitational wave detector using a CuAl sphere weighing over 1 tonne cooled to 65 mK [43]. At these temperatures, SQUID magnetometers may be used to measure the sphere's deflection, at magnitudes small enough to theoretically detect gravitational waves [44].

Figure 5.10. The CUORE dilution refrigerator, capable of producing the coldest cubic metre in the known universe. Image from the CUORE collaboration website: https://cuore.lngs.infn.it/en/images/cryoconstruction.

Dark matter detection also benefits greatly from the ultra-low temperatures provided by dilution refrigeration. The CUORE experiment, based underground in the Laboratori Nazionali del Gran Sasso, uses a custom-made dilution fridge (shown in figure 5.10) to cool 741 kg of tellurium oxide crystal detectors and support structure to 10 mK. This makes CUORE the coldest cubic metre in the known universe. By searching for neutrinoless double beta decay, researchers hope to investigate whether neutrinos are their own anti-particles [45].

Astronomical detectors, such as transition edge sensors, require very low temperatures to operate. Compact, fast-cooling, dilution systems have been developed to aid in their testing and use in terrestrial observatories [46–48].

It is also interesting to note that dilution refrigerators have even been used beyond this planet! The Planck Satellite featured a single-shot dilution refrigerator (with the pumped ^{3}He ejected directly into space!) which ran for three years. This was used to cool detectors that allowed the satellite to produce a complete and detailed map of the cosmic microwave background [49].

5.2 Adiabatic demagnetisation—electronic and nuclear

Viktor Tsepelin

5.2.1 Introduction

Adiabatic demagnetisation employs magnetic fields as a control parameter for temperature and is the only technique that can reach temperatures into the submilliKelvins in condensed matter physics. Sometimes it is referred to as the magnetocaloric effect. Adiabatic demagnetisation is also the oldest technique used to reach significant sub-Kelvin temperatures. Historically it originated nearly a century ago with demagnetisation of electronic magnetic moments in paramagnetic salts [50, 51]. The state-of-the-art apparatus with paramagnetic salts was able to reach temperatures down to 2 mK. However, adiabatic demagnetisation is a single-shot technique and despite huge advances it was eventually superseded by ^{3}He–^{4}He mixture dilution refrigerators continuously delivering temperatures down to 2–8 milliKelvin.

Application of adiabatic demagnetisation to nuclear magnetic moments instead of electronic moments allowed temperatures down to microKelvins to be reached. The first practical realisation of the nuclear demagnetisation method was achieved in the 1960s by Kurti *et al* [52], who succeeded in reducing the temperature of nuclear spins in copper to about 1 microKelvin. It took some time and serious development to achieve significant refrigeration of the lattice and conduction electrons essential to cool down samples [21, 51, 53].

As mentioned already, in the temperature range from several hundred to a few milliKelvin, the ^{3}He–^{4}He mixture dilution refrigerators dominate the mass market, as typical adiabatic demagnetisation refrigerators based on paramagnetic salts reach approximately 50–100 mK and offer only limited power. As a result, adiabatic demagnetisation refrigerators are used or developed when liquid cryogens are undesirable or auxiliary parts of their systems, for example pumps, are impractical, e.g. space applications [54].

Recently, a number of sealed, self-contained systems working on sorption of ^{4}He, ^{3}He and ^{3}He–^{4}He mixture have become available commercially, where the user does not have to worry about helium liquids and external pumps. These resulted in the development of hybrid systems combining ^{3}He–^{4}He mixture dilution refrigerators with adiabatic demagnetisation for space application [55].

Nuclear adiabatic demagnetisation requires high magnetic fields and starting temperatures of a few tens of milliKelvin, and employs ^{3}He–^{4}He mixture dilution refrigerators with moderate cooling power as a precooling stage [21, 53, 56]. Hence, the commercial systems that produce sub-milliKelvin temperatures are rare due to level of customisation required and high development cost. If temperatures below 2–5 mK are required for your project, consider developing and building an adiabatic demagnetisation refrigerator or contact existing low temperature laboratories for collaboration and provision of a cooling platform. There may even be opportunities for fully funded access, for example in 2019 the European Union funded European MicroKelvin Platform [57] to provide an ultra-low temperature regime for experiments by external users.

5.2.2 Principle of operation

The procedure for adiabatic demagnetisation cooling is identical regardless of whether electronic or nuclear magnetic moments are used. The term adiabatic implies isolated as well as reversible. The entropy of such ideal systems stays constant, and for adiabatic demagnetisation the entropy of a system is a function of the ratio between magnetic field and temperature. Thus, to maintain the system entropy constant the reduction of magnetic field will decrease the temperature of the system. In a nutshell, the technique requires a suitable working body (demag stage), a constant temperature heat reservoir (precooling stage), a magnet with ability to control magnetic field at the demag stage and a heat switch determining a degree of thermal isolation (connection) to the precooler. Figure 5.11 shows the schematics of the essential parts.

The adiabatic demagnetisation technique offers a one-shot cooling and the general sequence is shown in figure 5.12; the first stage is the so-called isothermal magnetisation of the working body and usually carried out at the lowest reachable temperature by the precooler. The figure depicts this stage as a red vertical arrow on the entropy–temperature diagram, and two inset cartoons of the system on the right.

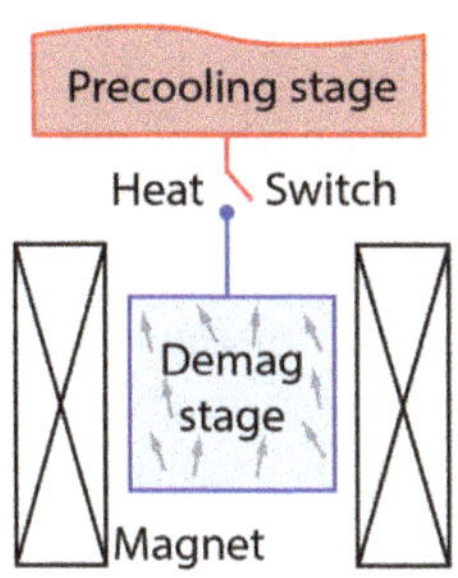

Figure 5.11. The schematics of the essential parts of adiabatic demagnetisation setup.

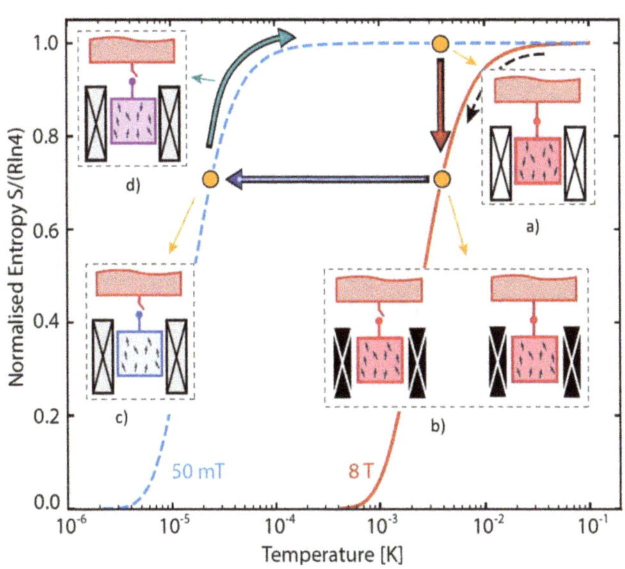

Figure 5.12. Principal sequence of the adiabatic demagnetisation technique.

The top cartoon image (figures 5.12(a) and (b)) shows that the demagnetisation stage in the absence of magnetic field is thermally linked to the precooling stage via the heat switch. While the magnet is energised and generated magnetic field orders magnetic moments of the demag stage, the entropy is reduced. During the magnetisation the produced heat is absorbed by the thermal reservoir maintained at a constant temperature. It is worth mentioning that the heat required to be removed in the isothermal magnetisation is half that compared to the process (black dashed arrow) where the temperature of the bath is allowed to increase during magnetisation [21, 51]. However, for dilution refrigerators temperature is usually allowed to increase since the cooling power of the mixing chamber increases as a square of the temperature. At the end of the isothermal magnetisation a desired magnetic field is reached and the temperature of the demag stage and thermal bath are equal (figure 5.12(c)).

The second stage is an adiabatic demagnetisation of the magnetic moments and is shown as a dark blue horizontal arrow. Here the heat switch is opened and isolates the demag stage from the precooling stage (bottom centre inset (b)). The subsequent reduction of magnetic field happens without entropy change of the working body and results in cooling and lower temperature of the demag stage. The demagnetisation is stopped at a desired magnetic field which is shown as figure 5.12(c). The last stage corresponds to warming up of the system due to non-ideal isolation and resulting thermal heat load. The green curved arrow depicts this as the top left cartoon. The whole process can be repeated once the temperature exceeds a desired working range.

It is appropriate to differentiate adiabatic demagnetisation cooling and refrigeration. In the cooling only the magnetic moments of the system have to reach low temperatures, while for refrigeration the goal is to cool something else, for example a sample or another adiabatic demagnetisation stage. Hence, for the refrigeration it is important to consider the thermal transfer of the low temperature of the magnetic moments to the rest of the system.

Figure 5.13 shows an example of a refrigeration setup that has a sample and two thermometers. All components are equally critical to the success of the project. The design should start around the sample and desired programme of measurements. The amount of heat released during measurements should be negligible compared to the sample heat capacity. The sample heat capacity and thermal conductivity will determine the self-cooling time constant of the sample and how large the cooling stage should be. Thermal links must provide good connection between the sample,

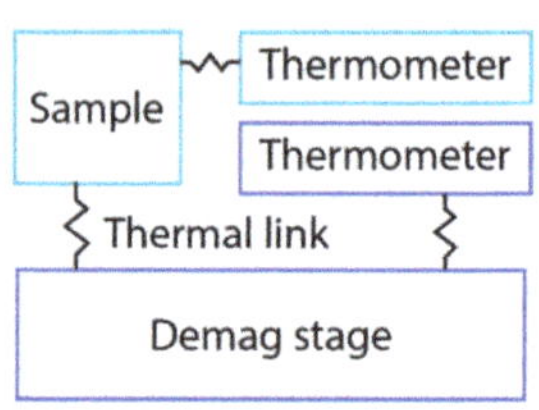

Figure 5.13. An example of refrigeration setup that has a sample and two thermometers.

demag stage and thermometry. Thermometry choice is non-trivial but should be accurate and fast [21, 53, 58].

After reaching the desired temperature we need to maintain it for a long enough time to measure sample properties. The entropy of the magnetic moments is often large compared to the lattice and conduction electron entropies of the stage and therefore neglected in the cooling. For refrigeration, the situation is not as straightforward. Philosophically there are two approaches to design of demagnetisation stages: the largest possible working stage or a minimalistic approach required to cool the sample. The first approach uses bulk demagnetisation stages and examples are given in the references [60]. The Lancaster demagnetisation refrigerator for research on superfluid ^{3}He is an example of a minimalistic approach to the nuclear stage [56, 61, 62]. The latter approach is also used in on-chip demagnetisation of nano-sized objects [58].

5.2.3 Theoretical description

The adiabatic demagnetisation technique is an elegant example of direct application of statistical mechanics of distinguishable particles. The majority of statistical mechanics textbooks describe the behaviour of spin ½ solid in a magnetic field and have excellent introductions to the physics underpinning demagnetisation techniques [63–65]. For more specific applications and in-depth discussion, the interested reader is referred to more general books on low temperature physics [51, 53, 54]. All these books have a proper mathematical description of underlying processes and detailed mechanisms of heat leaks. Here we concentrate on concepts and some of the crucial equations.

Let us consider the behaviour of a spin ½ solid in a magnetic field B, where each spin can be aligned up or down. In a strong enough magnetic field, all spins will be aligned, while in a small field their orientation will be more random. According to the statistical definition of entropy the perfectly aligned system of spins has entropy equal to zero. The randomly oriented spins correspond to a maximum entropy, which for a two-level system equals $R\ln(2)$ per mole, where R is the universal gas constant. More generally, a system with the total angular momentum quantum number I has $(2I + 1)$ different orientations of spin and maximum entropy per mole of $R\ln(2I + 1)$. Temperature T also plays a substantial role, since thermal energy shared between spins can change orientation of spin in the magnetic field and increase disorder and consequently the entropy.

The interplay between temperature and magnetic field is determined by the ratio of magnetic $\mu_B B$ and thermal $K_B T$ energies. Here μ_B is a value of the magnetic moment and the magnetic energy $\mu_B B$ is governed by a so-called Zeeman effect, which lifts degeneracy of the energy levels. Figure 5.14 demonstrates splitting of the energy levels due to applied magnetic field B and their relative populations for a spin ½ solid at temperature T. The population of a level with the energy E is determined by the Boltzman distribution that has a functional form $e^{-E/K_B T}$.

It is appropriate to describe what happens to a system of spins in zero applied magnetic field when temperature tends towards the absolute zero. On one hand the spins should be randomly oriented as the external magnetic field is not present and

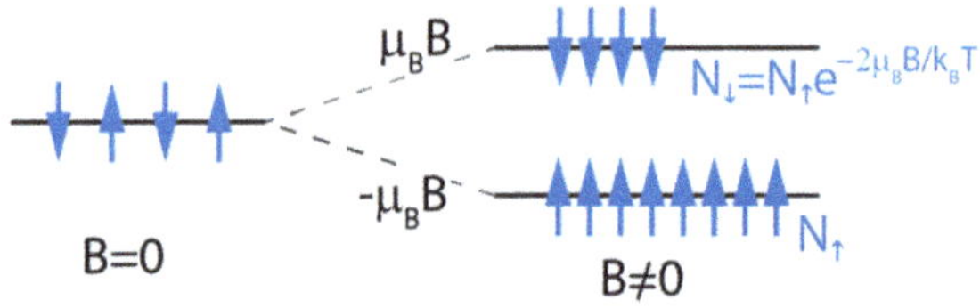

Figure 5.14. Energy levels of two-level system with and without magnetic field.

yields a finite entropy. On the other hand, the third law of thermodynamics states that entropy must tend to zero as temperature approaches the absolute zero. This suggests that the spins must be ordered. The paradox is resolved by existence of interaction between the spins. This interaction can be perceived as an internal magnetic b, which governs a system behaviour below a certain temperature and leads to magnetic ordering. The method of adiabatic demagnetisation only works at temperatures above the ordering transition for a given material.

In the limit where the interaction energy between spins is negligible compared to the thermal energy $K_B T$ the entropy of a system of spins with the total angular momentum quantum number I is given by:

$$S = R\ln(2I+1) - \frac{\lambda}{2\mu_0}\frac{(B^2+b^2)}{T^2}$$

where λ is the molar Curie constant of a given material and $\mu_0 = 4\pi \times 10^{-7}\text{NA}^{-2}$ is the vacuum permeability. Figure 5.12 plots the entropy dependence for copper nuclear spins ($I = 3/2$ and $= 3.22 \times 10^{-6}\,\mu_0 J\text{KT}^{-2}\,\text{mol}^{-1}$) as a function of temperature for the magnetic field of 50 mT and 8 T.

In the limit of small internal field b the equation shows that entropy depends only on a ratio of magnetic field and temperature $S\left(\frac{B}{T}\right)$. Hence the reduction of magnetic field 100 times should yield the same temperature decrease. During the adiabatic demagnetisation the value of the magnetic moment stays constant and the distances between the energy levels remain proportional to the magnetic field. Since the entropy of the system is preserved, the Boltzmann distribution of the spins over the energy levels remains the same, and hence the temperature falls proportionally to the field. The final temperature of demagnetisation is:

$$T_f = \sqrt{\frac{B_f^2 + b^2}{B_i^2 + b^2}}\,T_i$$

where T is the initial (precooling stage) temperature and B_i and B_f are the initial and final magnetic fields, respectively. The internal magnetic field of copper b equals 0.36 mT, and demagnetisation from 8 T to 50 mT with a starting temperature of 5 mK will reach the final temperature of 0.03 mK.

As with any cooling technique there is a compromise between the final temperature and an available cooling power. The heat capacity of magnetic moments is given by:

$$C_n = \frac{\lambda}{\mu_0} \frac{(B^2 + b^2)}{T^2}$$

After demagnetisation to the final magnetic field, heat capacity is inversely proportional to T^2, which leads to accelerating warm-up of the demagnetisation stage under a constant heat leak. Quite often the final magnetic field is slowly decreased to compensate for heat load and to maintain the required temperature.

5.2.4 Practical description and limitations

The nuclear magnetic moment is on the order of the nuclear magneton $\mu_{\rm N} = 5.05 \times 10^{-27}\ {\rm JT}^{-1}$, while the electronic magnetic moment is comparable to the Bohr magneton:

$$\mu_{\rm B} = \frac{e\hbar}{2m_e} = 9.27 \times 10^{-24}\ {\rm JT}^{-1}$$

The difference of 2000 between the electronic and nuclear moments results in strikingly different temperature ranges, magnetic fields and environment for adiabatic demagnetisation of nuclear and electronic moments.

Adiabatic demagnetisation of electronic moments is realised using paramagnetic salts. The refrigerators based on this principle are commonly referred to as adiabatic demagnetisation refrigerators (ADRs). The temperature range of an ADR is determined by a choice of salts and their ordering temperature. Commonly used salts are ferric ammonium alum (FAA) with $I = 5/2$, manganous ammonium alum (MAS) with $I = 5/2$, chromic potassium alum (CPA) with $I = 3/2$, and cerium magnesium nitrate (CMN) with $I = 1/2$. Their minimum working temperatures are approximately 35 mK, 180 mK, 12 mK and 2 mK, respectively [51, 53, 66]. The paramagnetic salts have very poor thermal conductivity and are grown directly onto a metallic matrix made from a few thousands of copper or gold wires. Figure 5.15(a) shows an example of a metal matrix made from 1600 gold wires used in Astro-E satellite for cooling an x-ray detector. The material of the wires needs to be chosen carefully as paramagnetic salts can chemically react with the underlying matrix.

Figure 5.15. (a) A matrix of metal wires for growing paramagnetic salt used in Astro-E satellite for cooling an x-ray detector (Credit: Space Science and Engineering Center (SSEC), University of Wisconsin-Madison). (b) Schematics of a gas-gap heat switch.

Furthermore, the low temperature salts have to be sealed after the growth to prevent dehydration. The weblink (https://www.ssec.wisc.edu/adr/) of Space Science and Engineering Center (SSEC), University of Wisconsin-Madison shows a series of images for step-by-step assembly of the adiabatic demagnetisation refrigerator for Astro-E satellite.

To isolate the demagnetisation stage from the precooling stage in ADRs a gas-gap heat switch is often used [51, 54, 67–69]. Figure 5.15(b) shows a schematic of a gas-gap heat switch. At temperatures below 1 K any surface acts as a sorption pump, and charcoal or a Zeolite pill/pellet that has a large surface area per gramme are commonly utilised to absorb large quantities of gases. In normal conditions a gas-gap heat switch has a vacuum and hence is not thermally conducting. To change the state of a switch a charcoal or Zeolite pellet is warmed up by an electrical heater and deabsorbed gas fills the gap and thermally connects stages. Once a heater is switched off the gas gets reabsorbed into the pellet. There are also carefully designed passive switches that change its state automatically with the temperature of a precooling stage [67, 69]. Expensive ^{3}He gas is typically used as a conducting medium since ^{4}He forms a superfluid film on the surfaces and affects the thermal performance.

For temperatures significantly below 1 K, superconducting heat switches are used [51, 53, 54, 69]. Figure 5.16(a) shows a sketch of a superconducting heat switch. The principle is to use an external magnetic field to change the phase of a metal from the superconducting state to normal. Free conduction electrons in a normal metal are responsible for thermal conductivity while in a superconducting state they are paired and only the lattice phonons conduct heat. The operating temperature of a superconducting switch is determined by the transition temperature of the metal and as a rule of thumb should be an order of magnitude lower than that. For example, an aluminium heat switch effectively operates below 100 mK since its superconducting transition temperature is at 1.2 K. In a normal operating mode, the ratio of thermal conductivities between normal and superconducting state usually is in the order of 10 000. At high temperatures a heat switch starts to thermally conduct (leak), which causes its performance to deteriorate.

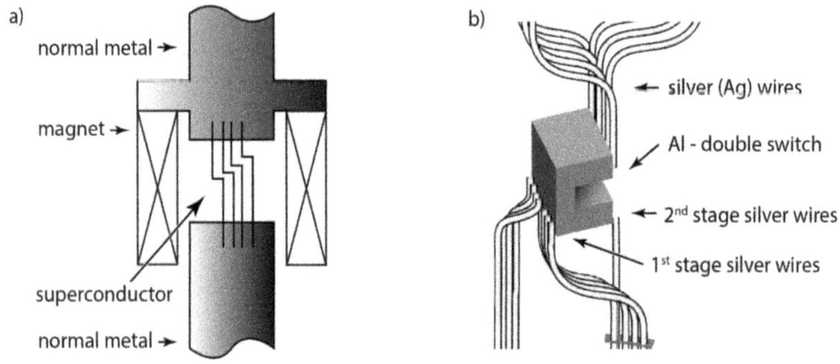

Figure 5.16. (a). Schematics of a superconducting heat switch. (b). An artistic impression of double heat switch used in Lancaster nuclear demagnetisation refrigerator (G R Pickett, Lancaster University).

Superconductive heat switches are the current low temperature community standard to thermally isolate the nuclear adiabatic stages from the precooling stage. Copper is often used as the normal metal part of a heat switch due to its availability, purity, price and how easy it is to work with. Superconducting material must form an almost seamless joint to the normal metal, otherwise thermal conductivity in normal state will be poor and precooling will be inefficient. Figure 5.16(a) demonstrates that a part of the superconductor is placed perpendicular to the external magnetic field in order to mitigate possible magnetic flux trapping and thermal leaking in superconducting state. Figure 5.16(b) shows an example of a double superconducting heat switch that can isolate two nuclear demagnetisation stages in parallel. This particular design uses aluminium for the superconducting part and silver wires as a normal metal conductor [61].

Due to the smallness of the nuclear moment the nuclear demagnetisation technique has typical starting temperatures below 10 mK and requires magnetic fields in excess of 6 T [51, 53, 54, 62]. Even in these conditions the realisable polarisation of the nuclear moments and entropy reduction is much smaller than for electronic magnetic moments in paramagnetic salts. Nevertheless, the polarisation is good enough for successful experiments and reaching temperatures in the order of microKelvin. The choice of material for nuclear demagnetisation refrigeration requires careful consideration as heat absorbed by nuclear magnetic moments is not directly linked to a sample [62]. Hence, metals are essential for nuclear demagnetisation due to their high thermal conductivity. To cool the sample the cold nuclear magnetic moments (spins) interact with the conduction electrons, and weakly with the lattice vibrations (phonons), which in turn provide a link to the sample. Figure 5.17 illustrates an example of thermal path between the sample and nuclear moments. Heat leaks into each subsystem need to be evaluated as they result in temperatures differences.

Let us consider the thermal path via the subsystems. Spin–spin interaction time is short and is in the order of 1 ms, hence after demagnetisation nuclear moments reach equilibrium temperature fast. The thermal path between nuclear moments and electrons is described via so-called Korringa law. The time constant of spin-electron interactions is approximated by $\tau_1 = K/T_e$, where T_e is the conduction electrons temperature and K is the Korringa constant of the material determined by strength of hyperfine coupling between electrons and spins [51, 53, 54, 62]. For example, copper has a Korringa constant of 1.27 Ks, while in indium it equals 0.09 Ks [54]. This implies that at 100 microKelvin their spin-electron interaction time constants are 3.5 h and 0.25 h, respectively.

The long-time constants show that the electron subsystem can become overheated if the heat load is too high during measurements. Depending on the system, the

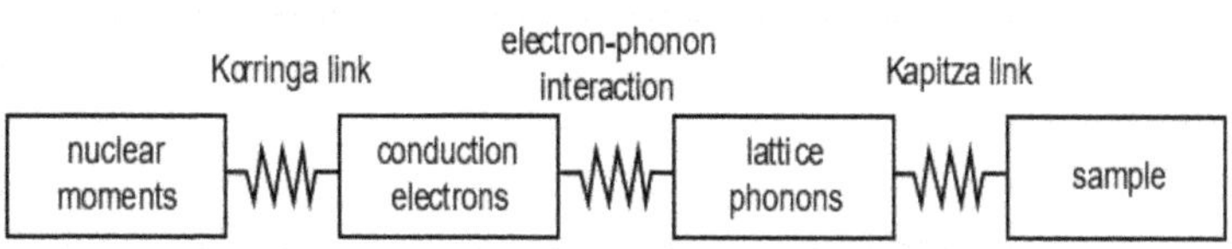

Figure 5.17. Subsystems for nuclear demagnetisation.

permitted heat leaks should be picoWatts or even smaller. Often it can be assumed that phonons and conduction electrons are at the same temperature. However, it is essential to remember that the heat flow between the electron and phonon systems has a very strong functional dependence on temperature $\dot{Q} = \Sigma V\left(T_{\mathrm{ph}}^5 - T_{\mathrm{e}}^5\right)$, (where V is the volume of a stage and Σ is the material dependent coefficient [59]). The thermal link between sample and lattice is equally crucial and has to take into account, among other things, what subsystem of the sample has to be cooled down: lattice phonons, conduction electrons, or both? This question becomes increasingly relevant as there is a drive to cool nano-sized samples to sub-milliKelvin temperatures.

5.2.5 State-of-the-art

Adiabatic demagnetisation refrigerators for space missions are optimised for weight reduction and have strict requirements for heat rejection at higher temperatures since each mission has a defined lifetime with limited resources. Space mission refrigerators are designed to cool microcalorimeters or bolometers in infrared, millimeter-wave and x-ray astronomy [54–56, 67]. They reach a typical minimum temperature of 30 mK to 50 mK and have a cooling power of about 1 μW at this temperature.

An example of a five-stage adiabatic demagnetisation refrigeration for the Constellation-X astronomy mission is shown in figure 5.18. This multi-satellite mission employs arrays of microcalorimeters to detect x-rays with the detectors cooled to 50 mK [67]. The prototype of the refrigerator aimed to deliver 50 mK with

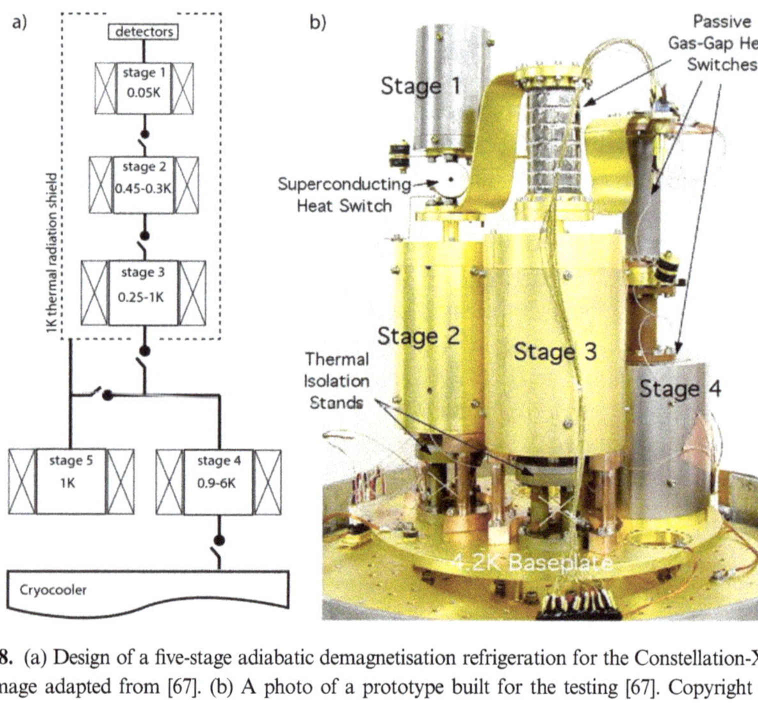

Figure 5.18. (a) Design of a five-stage adiabatic demagnetisation refrigeration for the Constellation-X astronomy mission. Image adapted from [67]. (b) A photo of a prototype built for the testing [67]. Copyright (2004), with permission from Elsevier

the cooling power of 6 microWatts by using cascade demagnetisation. The technology behind cascade demagnetisation has been available since the 1950s [50] but has substantially changed since its introduction. Multi-stage refrigerators are now used for quasi-continuous cooling, where a sample stage stays at a constant temperature. To achieve a constant temperature, series or triad designs can be used. The series design is shown in figure 5.18 with the lowest stage delivering a stable temperature, while other stages are demagnetised sequentially and thus periodically transfer heat from the low temperature and dump it to the higher temperature stage. In the triad design three adiabatic refrigerators run in parallel. One stage is absorbing heat at low temperature, the second stage is rejecting heat at high temperature, and the remaining stage is transiting between low and high points. The example where both designs are used is a seven-stage refrigerator that is being developed for LiteBIRD, which is a JAXA-led L-Class mission to study the cosmic microwave background [55] and aims to continuously deliver 100 mK with a cooling power of 2.2 microWatts.

While the development of continuous adiabatic demagnetisation refrigerators (CADR) has already improved cooling power-to-weight ratio compared to single-stage refrigerators, extending the temperature range with each stage of the adiabatic refrigerator adds weight. To reduce the mass further the SAFARI system (the infrared instrument developed by a European based consortium for SPICA, a Japanese led mission) plans to use hybrid architecture based on the combination of a 300 mK ^{3}He sorption cooler and one-shot adiabatic demagnetisation stage [56]. The SAFARI detectors are transition edge sensors and require temperatures down to 50 mK and 0.4 microWatt heat load.

Cascade demagnetisation is also quite common in state-of-the-art nuclear demagnetisation refrigerators. The YKI-group from Aalto University achieved a world record in 2000 by cooling nuclear moments in a rhodium sample to 100 picoKelvin (https://ltl.tkk.fi/wiki/LTL/World_record_in_low_temperatures). The researchers used a chain of two nuclear demagnetisation stages precooled by a powerful ^{3}He–^{4}He mixture dilution refrigerator in their quest to reach the lowest temperatures [70]. The first nuclear stage was made from copper cooled to a temperature of 50 μK, and was used to precool the second nuclear stage (the rhodium sample) in a magnetic field of 7.5 T. The rhodium spins after demagnetisation have demonstrated a temperature of 100 pK. The idea behind the measurements was to study magnetic properties and superconductivity of rhodium [71]. Figure 5.19 shows the lowest temperature components of the apparatus used by the YKI-group.

The Lancaster low temperature group uses a two-stage demagnetisation refrigerator for cooling metals and studying superfluid ^{3}He at the lowest temperatures [61, 62]. The amount of copper in the Lancaster nuclear stages is carefully calculated and follows a minimalistic approach that uses just enough material to cool the required sample. The group has cooled superfluid ^{3}He to approximately 80 microKelvin by using a combination of a powerful ^{3}He–^{4}He mixture dilution refrigerator [57] and two nuclear stages that are demagnetised simultaneously using the same external magnetic field. Both stages cool liquid ^{3}He, and the first stage surrounds the second and acts as a thermal shield for the second stage. Helium liquid

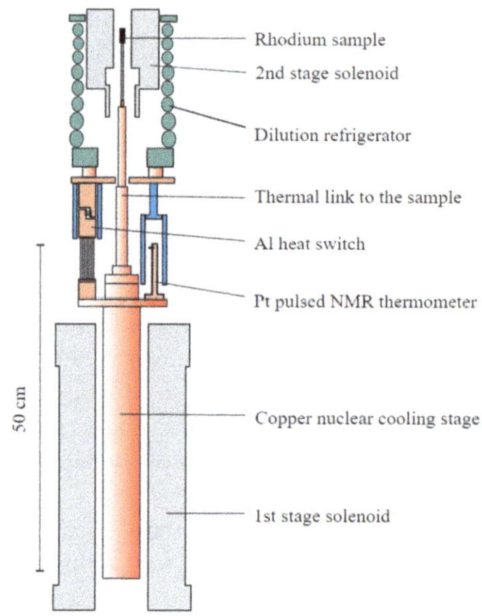

Figure 5.19. Schematics of Aalto University double demagnetisation refrigerator. Sketch is adapted from https://ltl.tkk.fi/wiki/images/1/18/YKI.pdf and Yao *et al* [70], copyright (2000), Plenum Publishing Corporation, with permission of Springer.

cooled by both stages is weakly linked via small capillaries (figure 5.20). Superfluid ^3He at extremely low temperatures can be used to study cosmological analogues of the early Universe [72, 73] and quantum turbulence [74].

5.2.6 Modern trends

The majority of nuclear demagnetisation cryostats have been precooled using so-called 'wet' dilution refrigerators, which have liquid ^4He cryoliquid at 4.2 K as a starting temperature (the previous section in this chapter is dedicated to the dilution refrigeration). Nearly all manufacturers of commercial dilution refrigerators are selling cryogen-free ('dry') systems based on pulse-tube precoolers, as they are much easier to operate. Hence, there is a growing appetite to build a nuclear demagnetisation refrigerator on such 'dry' systems. Oxford Instruments in collaboration with Royal Holloway have built a demagnetisation refrigerator using the pulse tube dilution fridge with a $PrNi_5$ nuclear stage that has reached 600 microKelvin [75]. BlueFors Cryogenics in collaboration with Aalto University have cooled down superfluid ^3He to 160 microKelvin using a copper nuclear stage on a cryogen-free experimental platform [76].

Following the recent developments of continuous refrigeration for space applications there is a desire to design and build similar continuously running systems using nuclear demagnetisation techniques [77, 78]. One of the drivers behind the development is to cool samples to sub-milliKelvin temperatures that have large spin–lattice relaxation times, e.g., a-Si films [78]. For such samples typical single-shot demagnetisation times are insufficient to achieve proper thermalisation.

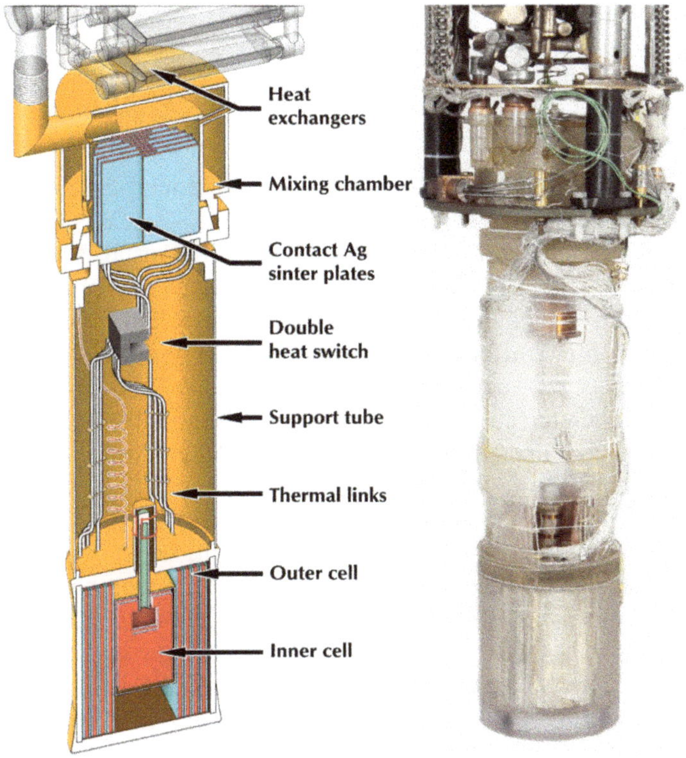

Figure 5.20. Lancaster demagnetisation refrigerator for studying ^3He. Copyright G Pickett and V Tsepelin, Lancaster University.

The past few years have shown the emergence of a significant effort to produce sub-milliKelvin electron temperatures in nano-sized devices. The development is driven by expected improvements for quantum technologies, quantum sensors and metrological applications. Lower temperatures provide a quieter noise environment and often can result in a discovery of new phenomena otherwise hidden. While it is straightforward to cool electrons in bulk samples, it is difficult in nano-scaled structures. The challenge of cooling on-chip electrons is underpinned by the steep temperature dependence of phonon–electron coupling and its proportionality to the volume of the material. The latest developments have broken a 4 milliKelvin barrier that stood for more than 15 years [59, 79–81]. The first success came by using a powerful dilution refrigerator and achieved a low electron temperature by immersing the incoming leads in the ^3He–^4He mixture in the mixing chamber [79]. Simultaneous adiabatic demagnetisation of on-chip and off-chip nuclear copper stages reached an electron temperature of 2.8 milliKelvin [80]. The latest attempt of an on-and-off-chip demagnetisation of indium nuclear spins has yielded electron temperature below 500 microKelvin [81]. The paper by Jones *et al* [59] contains a comprehensive review of the most recent advances in refrigeration of nano-sized devices and describes techniques used for their cooling and demonstrates the importance of electrical filtering.

5.3 Helium evaporative sorption coolers

Lionel Duband

5.3.1 Introduction

A current trend in low temperatures, whether for ground-based experiments or space-borne systems, is the suppression of cryogenic liquid tanks in favor of mechanical or self-contained coolers. Although liquid cryostat is a straightforward technique, widely used and offering multiple advantages, the advent of reliable autonomous coolers has changed the way the cryogenic chains are now designed. This technological breakthrough saves space, mass, offers ease of operation, fast turn-around and does not require consumable cryogenic fluids. This is of particular interest at a time when the cost of helium gas has substantially increased over recent years. This approach comes with a number of challenges which are not discussed here (interfaces, temperature stability, induced vibration, limited cooling power, etc) and extends all the way to the ultra-low temperature. As a matter of fact, one of the biggest dilution refrigerator manufacturers uses multi-staged pulse tubes for the precooling. Indeed, in this case several systems must be combined in cascade to cover the entire temperature range. This section is limited to the sub-Kelvin domain, i.e. systems capable of cooling an object to temperatures below 1 K and, more precisely, to the helium sorption coolers. Sorption cooler is a generic term that encompasses many types of coolers sharing a common feature: the use of either a sorption pump or a sorption compressor. This technique (not limited to, but mostly used with, nitrogen, hydrogen and helium) can be used to drive Joule–Thomson loops, operate self-contained dilution refrigerators or pump on liquid baths. We focus here on the helium evaporative cooler.

5.3.2 Principle

Evaporative cooling involves the same physical mechanisms as perspiration, which is used to regulate body temperature. For water, the change from the liquid to the gaseous state requires energy; however, in the absence of any external source, this energy is taken directly from the thermal agitation of the liquid's atoms leading to a cooling effect. This property holds for nearly all fluids and, when used with helium, it allows sub-Kelvin temperatures to be reached.

The physics involved is linked to the states of the fluid, which in its liquid phase only exist for well-defined pressure-temperature couples, the so-called phase diagram and saturation vapor pressure curve. Any deviation from this curve results in an increase or decrease in temperature (or in pressure). Note that it is possible to obtain out-of-equilibrium states (supercritical or subcritical helium for example) but they require an external action (cold source, increase in pressure). Obviously, to improve the efficiency of this cooling process, the rate of evaporation is increased by pumping.

Evaporation leads to a loss of matter, and consequently part of the bath will be consumed in this process. The ultimate temperature that can subsequently be

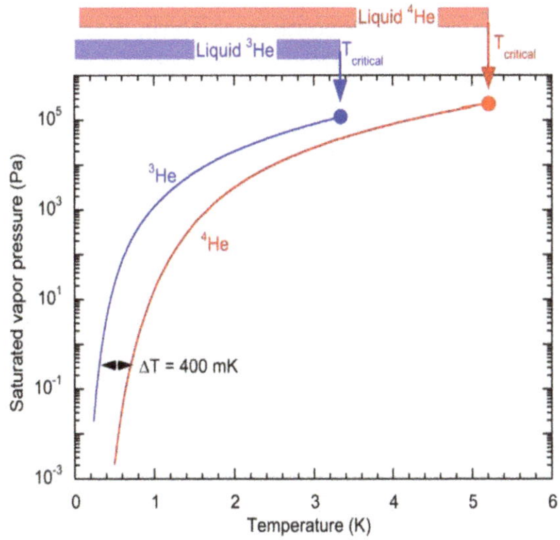

Figure 5.21. Saturated vapor curves for ^{4}He and ^{3}He.

obtained depends on the final pressure conditions above the bath. The saturation vapor pressure curves of the two helium isotopes (figure 5.21) are used to determine these equilibrium states. The two isotopes have different properties, and, in particular at equivalent pressure, liquid ^{3}He is significantly colder than ^{4}He. Therefore, its use makes it possible to obtain much lower temperatures. In practice, these temperatures are limited to approximately 750 mK and 240 mK, respectively for ^{4}He and ^{3}He.

Some useful numerical values are given in the table below.

Properties	^{4}He	^{3}He
Boiling T (T_B) STP	4.2	3.2
Critical point (K/10^{-5} Pa)	5.2/2.23	3.32/1.16
Enthalpy of vaporization (J g^{-1})	21 @ T_B	8.44@ T_B
	18.7 at 750 mK	8.3 at 240 mK
Density at T_B (g cm^{-3})	0.125	0.058
Gas viscosity at T_B (μPa s)	1.3	1.2

5.3.3 Basic sizing

Amount of helium and fraction lost during cooldown

The initial state of the system is a mass (m) of liquid helium at a temperature (T_i), with a target of a temperature drop to 'T_f'. The initial temperature T_i must be lower than the critical temperature for the liquid phase to exist, i.e. 3.32 and 5.2 K, respectively, for helium 3 and 4 (see figure 5.21). However, we can briefly mention that it is possible to produce liquid from helium gas under pressure starting from a temperature above the critical temperature using an isentropic expansion (the so-called Simon expansion). We come back to the liquid bath and its evolution during pumping. The associated equations are simple: each quantity of matter (dm) which evaporates requires an energy (Ldm), where L in J kg^{-1} is the enthalpy of evaporation at the considered temperature ($dm \ll m$). This energy is taken from the bath and leads to a decrease of its temperature according to mC_PdT, where C_P in J·kg^{-1}·K^{-1} is the specific heat of liquid helium at the considered temperature. Thus, we can write:

$Ldm = mC_PdT$ which can be written $\frac{dm}{m} = \frac{C_P}{L}dT$ leading to $R_{T_i}(T_f) = \frac{mT_f}{mT_i} = \exp\left[\int_{T_i}^{T_f} \frac{C_P}{L}dT\right]$ where $R_{T_i}(T_f)$ is the residual amount of liquid in percent.

It is assumed here that the other specific heat (mC_P) involved (cryostat materials, experiment to be cooled, etc) are neglected compared to that of helium, and moreover that the cooldown time (pumping) and/or the system's own losses, are small enough to be neglected for this calculation. These two hypotheses are generally correct at first order. A mathematical description of the enthalpies of evaporation and heat capacities of the two isotopes can be made, and thus the previous equation can be integrated between the initial temperature T_i and the final temperature T_f (figure 5.22).

It is thus possible to determine the residual quantity of liquid at any temperature (T) for a cooldown from an initial temperature corresponding to the boiling temperature (T_B) under atmospheric pressure (figure 5.23). For a lower initial temperature (T_L), it is easy from these curves to determine the residual quantity at a temperature T by simply using $R_{T_\mathrm{L}}(T) = \frac{R_{T_\mathrm{B}}(T)}{R_{T_\mathrm{B}}(T_\mathrm{L})}$.

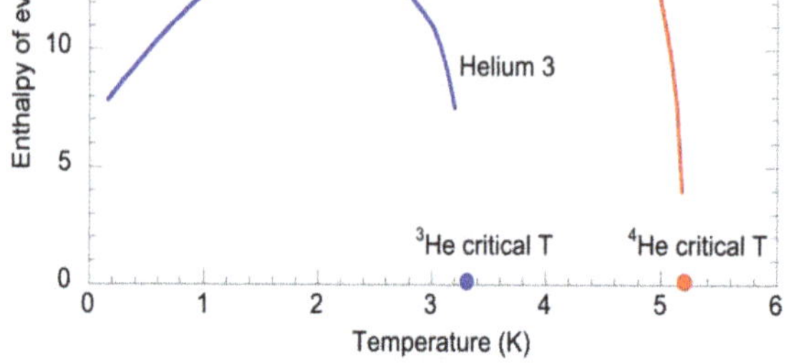

Figure 5.22. Enthalpy of evaporation for ^{4}He and ^{3}He.

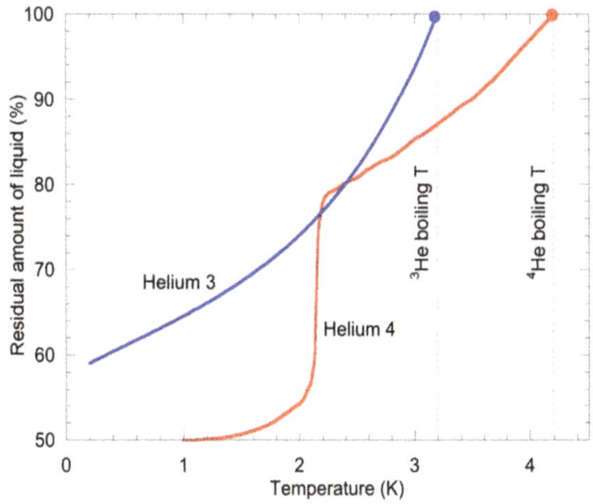

Figure 5.23. Residual amount of liquid after pumped down from the boiling temperature of a ^{4}He or ^{3}He bath.

The ^{4}He curve features a particular shape with a sudden loss of liquid associated with the heat capacity anomaly at the superfluid transition at 2.17 K. In practice, obtaining temperatures below 2 K from of an atmospheric pressure bath (4.2 K) consumes almost half of this bath.

The previous calculation gives the mass of liquid after pumping and therefore the amount of energy available for the experiments to be carried out. Knowing the heat input to the bath Q (parasitic and applied loads) one can easily compute the autonomy (t) following:

$$t = \frac{mL}{\dot{Q}}$$

Once again, it is necessary to consider the particular case of ^{4}He for which, below 2.17 K, a superfluid phase with peculiar properties appears, in particular the superfluid film known as Rollin film. This film can impact the design of pumped baths since it will rise along the walls of the cryostat and the pumping line, to then evaporate when the temperature becomes greater than 2.17 K. Obviously, once evaporated, the helium is evacuated through the pumping line and therefore leads to a continuous mass loss from the bath. This phenomenon should not be neglected and can contribute dramatically to the losses, since the associated enthalpy of evaporation is lost. Fortunately, technical solutions are available to limit this effect. The most common is to integrate a diaphragm into the pumping line, close to the evaporator. Indeed, the thickness of the Rollin film is more or less constant and therefore the mass flow rate is proportional to the wetted perimeter. Typically, a 200 μm diaphragm reduces it while limiting the pressure drop in the line which is linked to the ultimate temperature of the bath.

The ultimate temperature is determined by the pressure above the bath (P_{bath}), and therefore by the performance of the pumping line and the pump, and by the thermal resistance—the so-called Kapitza resistance—between the liquid and the walls of the container (generally copper).

These coolers are energy devices, much like the adiabatic demagnetisation refrigerator (ADR): they provide a number of 'cold joules' at low temperature, and the autonomy of the cooler at its operating temperature will depend on the load to the evaporator. Once the evaporator runs out of liquid the system must be recycled again. The duty cycle efficiency (ratio of cold duration to total cycle duration) depends on the available resources at the upper heat sinks. In the case of a pumped helium bath, for instance, the duty cycle can approach 98% or even 99%. However, for mechanical-based cryogenic chains, the full chain must be seen as a whole and optimised accordingly. For continuous operation, two systems can be combined in parallel, or in series with one system acting as a thermal buffer while the other is recycled.

Sizing of the pumping line

We will not detail the fluid mechanics and pressure drop equations for this process, but we can establish some simplified formulae to obtain an approximate value of the achievable temperatures. If we consider a pumping line of length L and diameter Ø, the thermal load on the bath generates a mass flow rate m and the pressure drop ΔP induced by the pumping line is given by:

$$\Delta P\ (Pa) = \frac{\dot{m}P}{\rho C}$$

with C ($m^3\ s^{-1}$) the conductance of the line and ρ ($kg\ m^{-3}$) the density of the gas.

The mean free path λ of the gas molecules is a key parameter and defines three regimes, molecular ($\lambda \gg Ø$), viscous ($\lambda \ll Ø$) and intermediate between the two. Assuming an ideal gas (reasonable at low pressures) and taking an average temperature for the line, the pressure on the bath can be written (in SI units):

—Molecular regime: $P_{\text{bath}} = \frac{7\dot{m}L}{Ø^3}\sqrt{\frac{T}{M}} + P_P$

—Viscous regime: $P_{\text{bath}} = \sqrt{\frac{0.4115\dot{m}L\eta T}{Ø^4 M} + P_P^2}$

where P_P is the pressure at the end of the line (pump) (in Pa), η is the gas viscosity ($Pa\ s^{-1}$) and M the molecular mass ($kg\ mol^{-1}$).

In most cases, the regime is intermediate or viscous. It can be seen that the pressure drop increases with temperature, and consequently for systems with pumping extending over a large temperature range (room T for instance), a succession of sections of increasing diameters will be used.

These rather simple equations allow us to obtain predicted temperatures relatively close to the results from more advanced models that consider thermal gradients and integration over successive line sections. One of the reasons also comes from the fact that the saturated vapour pressure varies exponentially with temperature: an error on the value of the pressure leads to an error on the temperature. Conversely, as the temperature decreases, an additional gain ΔT on the ultimate temperature requires a significant gain on the pressure.

One should also pay attention to the Kapitza resistance. Different theories have been proposed with mixed success for this resistance but, from an engineering point of view, particularly at low temperature, the fact is that there is a temperature difference between the liquid helium and the wall of the container subjected to heat transfer, i.e. the practical temperature accessible to the user. In most cases this ΔT is inversely proportional to T^3 or T^4 and the surface area. To limit this effect, it is thus important to maximise the exchange area (S).

In practice, the minimum temperatures accessible on an optimised system are 750 mK and 240 mK, respectively, for ^{4}He and ^{3}He. It can be shown that for ^{3}He, the ultimate temperature could be around 160 mK to 180 mK; at this temperature level, adiabatic demagnetisation or dilution are more suitable solutions and in practice the lowest temperatures obtained experimentally with evaporation systems are 237 mK [82] and 204 mK [83]. Reaching this level of performance requires having an extremely efficient pump, which can be achieved with adsorption pumps using activated carbon (see the references for further information) [84]. We then benefit from two factors, an excellent pumping capacity (low P_P in the previous equation) and operation of the pump at cryogenic temperature, typically 5 K for helium, which makes it possible to considerably reduce the pressure losses (see previous equations). The geometry of the cooler is reduced to its simplest expression, a pump, a condenser, an evaporator and a line connecting these elements (figure 5.24). In this arrangement, operation of the cooler is gravity based. No external piping is required and the system is simply controlled by an electrical resistor which heats the adsorption pump to desorb the helium and condense it once the saturated vapour pressure at the condenser temperature is reached (figure 5.24, point A). The helium falls by gravity into the evaporator and then can be pumped by simply cooling the adsorption pump thanks to the thermal link which connects it to

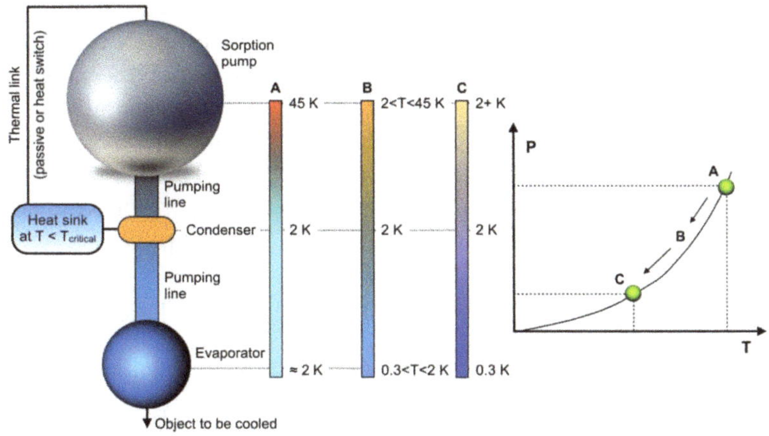

Figure 5.24. Principle of an evaporative helium sorption cooler.

the cold source (figure 5.24, points B and C). This type of system can therefore be extremely compact and easy to integrate. At first order the heat lift (related to *m*) is independent of the size of the cooler. For illustrative purposes, a 2 l and a 20 l STP ^{3}He unit may both provide 100 μW heat lift at 350 mK, but one will last at least 10 times longer.

5.3.4 Sorption pumping

Adsorption is the physical mechanism by which a gas can be trapped onto a material surface: if a gas is brought into contact with a solid surface, some of the molecules striking the surface will be retained for a finite period of time, resulting in a significantly higher molecular concentration at the surface than in the gas phase. Depending on the magnitude and nature of the attractive forces the effect is described as chemisorption or physisorption. In the former a chemical bond is formed and the process involves a transfer of electric charges; in the latter the process relies on the relatively weak van der Walls forces. In the following discussion we focus on physisorption. Evidently materials with high specific area, such as activated charcoal (typically 1200 $m^2\ g^{-1}$), are required. The amount of gas adsorbed is then a strong function of pressure and temperature; it increases as the temperature decreases and the pressure increases. Thus, by varying the temperature it is possible to provide either a compression or a pumping effect. The former can be used, for instance, to drive a Joule–Thomson loop, the latter can be used to first create the thermodynamic conditions necessary to produce the liquid phase and then perform evaporative pumping on this liquid bath to reduce its temperature. This fully static process is highly effective in terms of pressure amplitude, and in effect, a ratio of six orders of magnitude (10^6) can be achieved, a performance unmatched by mechanical systems. In addition, physisorption, or the ability to vary the pressure by varying the temperature, can also be used to design very efficient gas gap heat switches as discussed below.

With a few exceptions, the range of interest of sorption cooler remains below 10 K. This is due in part to the fact that to alternatively heat up or cool down the element containing the porous material requires an amount of power/energy that becomes prohibitively large as the temperature increases (increase of the associated heat capacities). The adsorption capacity of activated carbon is generally described by adsorption isotherms: for different temperatures, the amount of gas adsorbed is reported per gramme of carbon as a function of pressure. There are many predictive approaches but, in practice, and for helium, we use the work of Roubeau [85] on adsorption isotherms with the following mathematical description:

$$\log(P) = [153\log(C + 112) - 450]\left(\frac{1}{T} - 0.016\right) + 1.9 + \log(C)$$

where *P* is the pressure in torr (mmHg) and *C*, the quantity adsorbed, in cm^3 NTP of helium per gramme of activated carbon. The cycle described in the previous paragraph is shown in figure 5.25 (green dashed line).

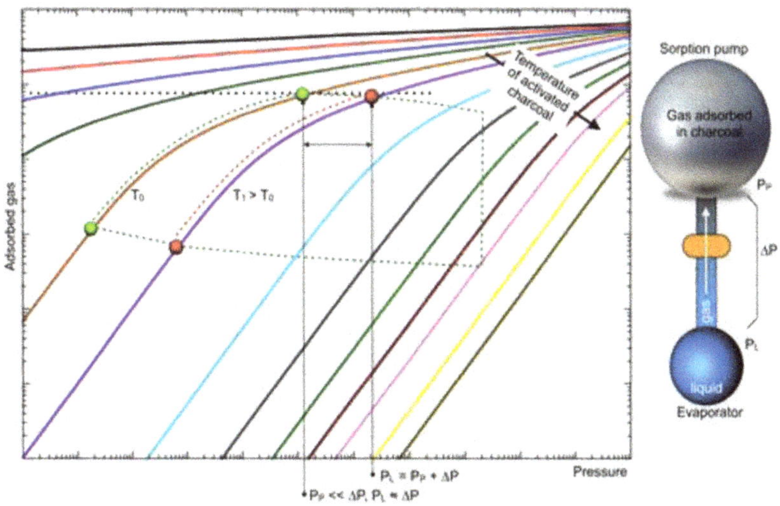

Figure 5.25. Sorption cooler cycle.

The sizing of the adsorption pump must be such that when the pump is cold (operation at low T of the bath), its internal pressure P_P can be considered zero, or in any case much lower than the pressure drop ΔP due to the line, whatever the filling rate of activated carbon. This is, moreover, a criterion for correct sizing of the system, which guarantees temperature stability (figure 5.25).

In an ideal cycle, when the last drop of liquid has evaporated, the pressure in the pump remains much lower than ΔP and thus the pressure above the liquid (P_L) is only driven by ΔP. When the amount of gas versus the mass of charcoal is not adequate, the pressure in the pump at the end of the cycle becomes of the same order of magnitude as ΔP, and consequently the pressure P_L within the evaporator increases and the temperature drifts upward (figure 5.25, red dashed line). This effect can be used to regulate the temperature of the evaporator. Indeed, this regulation can be achieved in two ways: by applying power directly to the evaporator (over a limited temperature range beyond which the required powers become prohibitive for the operation of the system, an approach which, of course, is detrimental to the autonomy of the system), or by increasing the pressure in the pump by simply heating it. The benefits of the latter approach are twofold since, in addition to any impact on autonomy—which in reality increases as the enthalpy of evaporation of helium increases—this method allows one to regulate over a wide range of temperature.

An important characteristic of adsorption pumping is its exothermic nature (loss of degrees of freedom of the trapped molecules). Each quantity of gas evaporated and then adsorbed within the activated carbon will generate heat and therefore heat transfer between the pump and its cold source. The associated energy actually represents one of the major contributions to the thermal balance and therefore to the performance calculation of these machines. This heat of adsorption L_A can be extracted from the adsorption isotherms [84] using a Clausius–Clapeyron type equation (SI units):

$$L_A = -R\left[\frac{\partial(\ln P)}{\partial\left(\frac{1}{T}\right)}\right]_C$$

where L_A (J mol^{-1}) is the molar heat of isosteric adsorption (constant C), and R (J mol^{-1} K^{-1}) the molar constant of the gases.

As the most energetic sites within activated carbon become populated, the heat of adsorption decreases. The agreement between the heat calculated from the adsorption isotherms and the experimental measurements is relatively good. From a practical point of view and, provided that we are in the first third of the cycle, we can use the following experimental result: each μW dissipated on the ^{3}He evaporator (300 mK) generates approximately 45 μW at the pump level. In the case of a ^{4}He system, this factor is about 14. This characteristic can be used to obtain an estimate of the system's internal thermal losses. Indeed, by measuring the heat dissipated at the pump, one can directly deduce the heat dissipated on the evaporator. To obtain a more approximate value of the internal losses, several autonomy measurements can be carried out under different applied loads, under the same recycling conditions. The amount condensed and the losses can be easily extracted from the plot $1/t$ versus the applied loads.

Finally, the sorption pump is also the main driver for the charging pressure, i.e. the internal pressure at ambient temperature (self-contained unit). Its volume must be large enough to accommodate the right amount of activated charcoal, but should not be too large either, so that the internal pressure at room temperature is above 5 MPa and even preferably around 7–8 MPa for ^{4}He. Indeed, it can be shown that this pressure has a strong impact on the condensation efficiency and thus on the autonomy of the cooler. The previous numbers have been calculated for condensation temperatures of 1.5 K and 4.2 K, respectively, for ^{3}He and ^{4}He.

5.3.5 Multi-stage systems

For a ^{3}He system, the need to have a cold source at a temperature below the critical temperature (3.2 K) can be a limiting factor. This temperature is at the limit of the capabilities of current mechanical cryocoolers and can be essentially obtained from a pumped ^{4}He bath or a ^{3}He Joule–Thomson loop. However, one can use a simple thermal architecture by associating a first evaporative ^{4}He stage, which can be operated from 5 K, and which in turn provides temperature lower than 2 K for a second ^{3}He stage (figure 5.26).

It is possible to cascade several stages such as ^{4}He/^{3}He/^{3}He (figure 5.26) to provide net heat lifts at various sub-Kelvin temperatures, and have a last stage where thermal losses by conduction are shunted by the upper stages and thus allowing temperature close to 200 mK to be obtained [86].

Figure 5.26. Showing left: Artemis two-stage cooler ($^4He/^3He$) mounted on the APEX telescope, Chile; right: BICEP Array three-stage cooler ($^4He/^3He/^3He$) deployed in Antarctica. Source: Laurent Clerc DSBT Univ. Grenoble Alpes.

5.3.6 Space applications

Cryogenics for the space environment implies specific constraints, the main ones being the mechanical strength required to survive take-off, the microgravity conditions, and the need for reliable systems able to operate without failure for 5 to 10 years. The solution to this last requirement is to develop systems without any friction or even better without moving parts. In that respect, evaporative sorption coolers are perfect candidates. They have no moving parts, are vibrationless and can be designed to be self-contained and compact with a high duty cycle efficiency. These features and their consequent reliability make them very attractive for space applications.

5.3.7 Space-borne evaporative helium cooler

As described previously, in a laboratory configuration, gravity is used to move the liquid from the condenser to the evaporator and to confine the liquid to the evaporator. For operation in microgravity, another mechanism is needed. The idea is first to condense the liquid directly into the evaporator by establishing the necessary temperature gradients, and second to use a porous material to hold the liquid by capillary attraction. Both the surface tension and the vapour pressure provide forces that drive and hold the liquid at the coldest point (i.e. evaporator). As a consequence, a heat switch is required to make and break thermal contact between this evaporator–condenser and the cold source, respectively, during the condensation and low temperature phases. Another switch is usually also used on the pump to improve the thermodynamic efficiency of the cycle (figure 5.27). This thermal architecture was used on the European Space Agency's HERSCHEL satellite. This space observatory, intended to study the birth of stars and the evolution of galaxies, carried three scientific instruments, two of which, PACS and SPIRE, featured a ^3He adsorption system for cooling bolometric detectors to 300 mK (figure 5.28) [87]. These two systems operated reproducibly and without failure over the lifetime of the mission, i.e. almost four years in orbit.

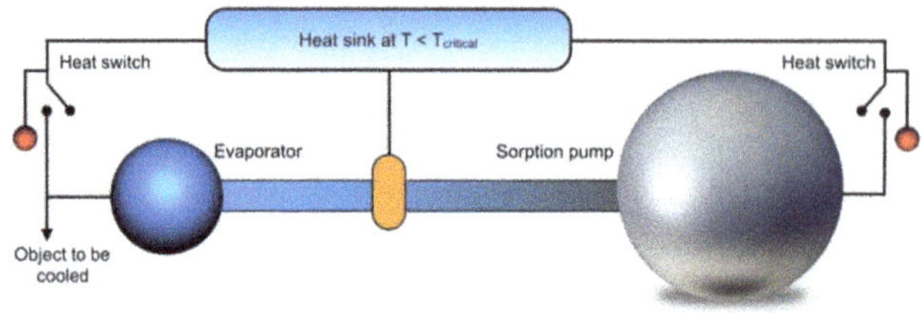

Figure 5.27. Evaporative helium sorption cooler—space version.

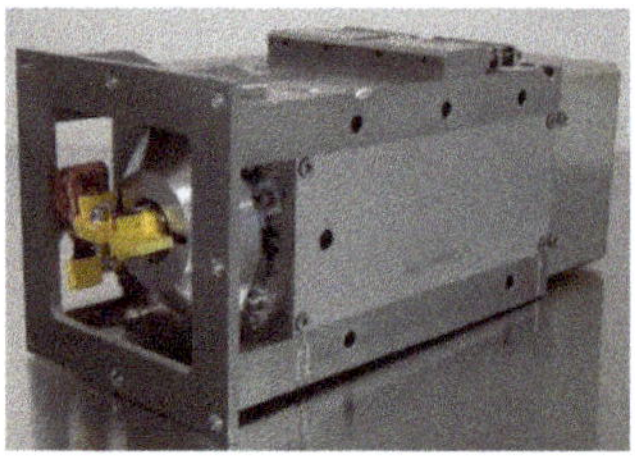

Figure 5.28. HERSCHEL 300 mK. Source: Lionel Duband Univ. Grenoble Alpes, CEA INAC-SBT.

5.3.8 Suspension system

The suspension system must ensure mechanical strength during lift-off and guarantee excellent thermal isolation of the suspended elements once in operation at low temperature. The relevant physical properties are a low thermal conductivity (k), a high breaking strength (σ) and a reasonably high Young's modulus (Y) to avoid low resonant frequencies. Depending on whether the instrument is launched at room temperature or already cold, the mechanical criteria must be considered accordingly. Most future cryogenic missions will feature a warm launch and thus 20 °C is assumed in the following discussion. For most geometries, the resonant frequencies are proportional to the square root of the Young's modulus and we can therefore define a material quality ratio following $\frac{\sigma\sqrt{Y}}{k}$.

The higher the ratio, the better the material. Of course, other criteria must be considered such as the ease of use, the compatibility with welding and/or bonding, the resistance over time, etc. The ratio defined above is plotted in figure 5.29 as a function of temperature for different materials.

Kevlar outperforms all other materials and therefore justifies its use in many suspension systems.

5.3.9 Gas gap heat switches

Cryogenic heat switches based upon different physical mechanisms have been developed, but the gas gap heat switch using a miniature adsorption pump remains the most commonly used due to its simplicity and absence of any moving parts. It consists of two copper pieces separated by a small gap, which is filled with or emptied of helium gas to achieve the switching action. The two copper pieces, for instance

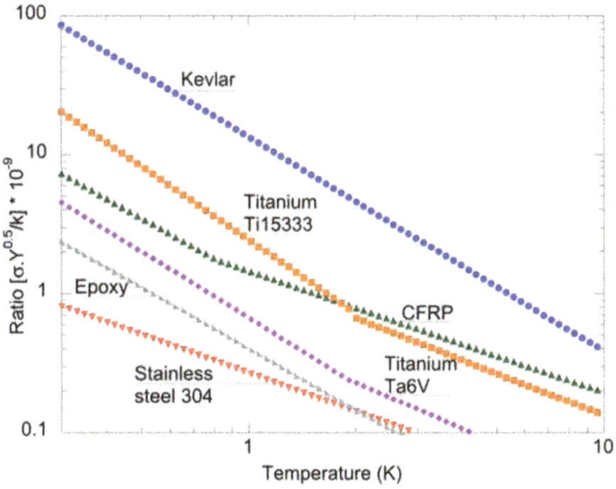

Figure 5.29. Performance of several materials for the suspension system.

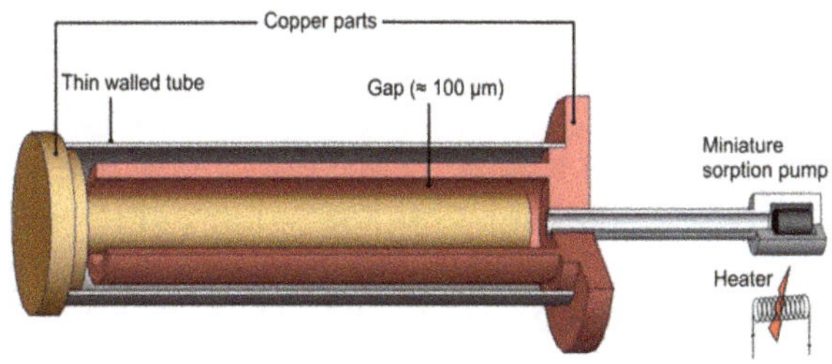

Figure 5.30. Schematic of a gas gap heat switch.

concentric cylinders, are held together by a thin walled tube (stainless steel or low thermal conductivity titanium alloy such as Titanium 6AL4V), which also provides the mechanical support as well as the confinement of the gas. A miniature sorption pump allows the control of the presence of gas within the switch (figure 5.30). One of the disadvantages of this solution is to have a finite off conductance, acceptable in most cases. In the ON state, the heat transfer through the exchange gas is proportional to the exchange area and inversely proportional to the gap, which then should be minimised. The typical gap of 100 μm between the two copper parts requires a carefully thought-out design and some tooling to guarantee the alignment.

ON to OFF ratios of over four decades can be obtained. As an example, a switch with a diameter of 8 mm and a length of 60 mm features ON/OFF conductance (closed/open) of typically 40 mW K^{-1} and 5 μW K^{-1} at 2 K, respectively. ^3He can be used as the exchange gas, since it exhibits a thermal conductivity much higher than that of ^4He in this range.

5.3.10 Hybrid cooler

Several future space missions call for much lower temperatures in the vicinity of 50 mK. These temperatures are outside the range of helium evaporative cooler but

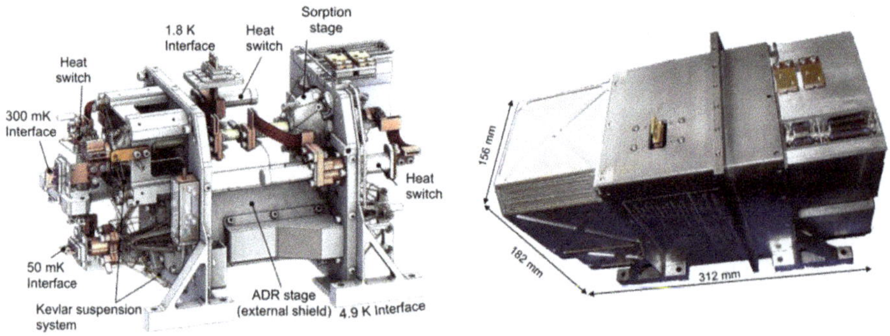

Figure 5.31. SPICA SAFARI EM hybrid cooler. Source: Florian Bancel (3D drawing) and Jean Louis Durand DSBT Univ. Grenoble Alpes.

this technology can still be used. Indeed, a clear objective of these missions is to maximise the amount of onboard science (instruments) within the overall mass allocation, and thus one of the mottos for cryogenics is the lighter the better. In that respect a 300 mK sorption cooler is the lightest possible solution to address the 0.25–2.5 K range (typically less than 300 grammes). It can then be coupled to a miniature magnetic stage (see section 5.2) to provide cooling down to few tens of mK. Because the ADR interfaces with a sub-Kelvin temperature, only a limited magnetic field ($\approx$1 T) is used to force the electronic spins of the atoms of the paramagnetic salt to align, thereby transferring magnetic entropy to thermal entropy. Heat is dissipated to the 300 mK interface. Then once the salt pill has been thermally isolated by means of an efficient gas gap heat switch, the field is reduced producing a cooling effect. Both stages are one-shot systems, the operation being a succession of a recycling phase followed by a cold operating phase. They can be recycled indefinitely once the stage runs out of liquid or field. Duty cycles in excess of 75%–80% can be obtained depending on the available resources at the upper heat sinks. This so-called hybrid cooler [88], providing cooling at 300 mK and 50 mK, has already been qualified to a high technology readiness level (TRL 6) (figure 5.31).

This section is partly based on a previous work published in the 'Techniques de l'ingénieur' (BE 9817).

5.4 Laser cooling and low temperatures in atomic physics

Tristan Valenzuela

5.4.1 Introduction

During the first half of the 20th century, as modern physics was blooming, experimentalists embarked on the quest for precise measurements that would allow a better understanding of the quantum world. During these early years, experiments were mainly based on vapours in spectroscopic cells. One of the main limitations of this kind of experiment is that due to the intrinsic distribution of velocities of the individual particles, the spectroscopic signals suffered from Doppler broadening and shifts contributing to substantial systematic and statistical uncertainties.

The first steps towards narrowing the broadened lines was achieved by use of collimated atomic, and molecular, beams interrogated in the transversal direction thus minimising the velocity spread of the particles in the relevant direction. This, however, required a very accurate alignment of the spectroscopic probe to avoid any possible systematic shift due to the projection of the longitudinal velocity of the beam on the interrogation direction. In order to overcome those limitations, during the second half of the century there was a big push towards achieving better control of the motion of the atoms that would allow increased precision in the measurements. This led to the development of several trapping and cooling techniques that are now state of the art in atomic physics. In the next sections, a brief description of these techniques and an overview of the applications will be presented.

5.4.2 Ion trapping

For trapping charged particles the most common approach is to create a potential minimum using electromagnetic fields. The electromagnetic force is required to satisfy the Laplace equation $\vec{\nabla}\vec{F} = 0$ and therefore it is not possible to create a potential minimum in three dimensions in free space using static fields alone. The two approaches, the Penning and Paul traps, to solve this issue were developed by H G Dehmelt and W Paul [90], respectively, who shared the Nobel Prize in Physics in 1989 for that work.

In a Paul trap, a pseudo-potential minimum is created by switching the polarity of two pairs of electrodes that generate a saddle-shaped potential that rotates at the switching frequency ω. Therefore, by choosing the appropriate value for the intensity of the electric field and of ω in accordance with the charge-to-mass ratio of the desired ions a stable trapping pseudo-potential can be achieved. See mechanical analogue in figure 5.32.

In a Penning trap, the potential is achieved by applying a DC electric field across three electrodes (see figure 5.33) and superimposing a DC magnetic field such that the magnetic field is in the axial direction. Then, charged particles will be trapped in harmonic potential with proper frequencies given by the charge-to-mass ratio and the intensity of the electric and magnetic fields.

5.4.3 Inductive cooling

In both types of trap, charged particles will oscillate moving towards and away from the electrodes. The presence of the ions will induce a mirror charge on the electrodes and as the particles move, the mirror charge will also move, effectively creating a mirror oscillatory current. This effect can be used to dissipate energy from the system by forcing the current to flow through a resistive element. To optimise the system, a resonant LC circuit can be coupled to it (figure 5.34). Furthermore, as the charged particles reach thermal equilibrium with the dissipative circuit, by inserting the electronic circuit in a cryostat the charged particles can be cooled down to the temperature given by the macroscopic cooling system. Using this method it has been demonstrated [92] that a single electron can be cooled down to

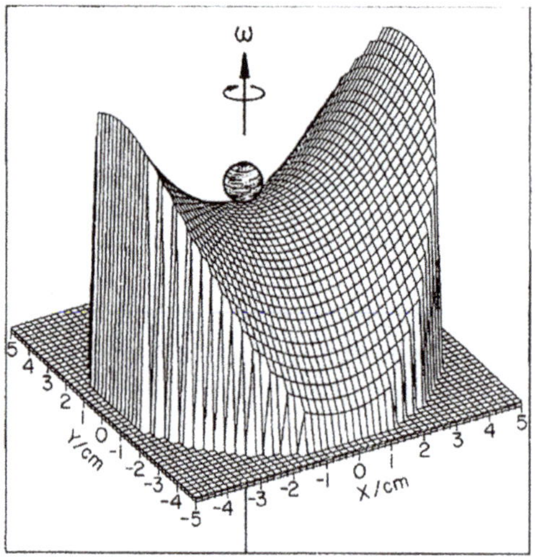

Figure 5.32. Illustration of the rotating saddle potential where the ball at the centre represents the trapped particle. Reprinted with permission from [89], figure 8, copyright (1990) by the American Physical Society.

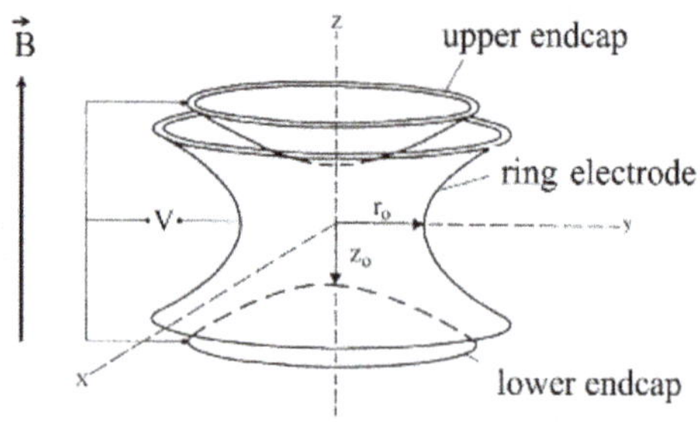

Figure 5.33. Typical electrode arrangement for a hyperbolic Penning trap [91], copyright (1969), EDP Sciences, Springer-Verlag, Società Italiana di Fisica, with permission of Springer.

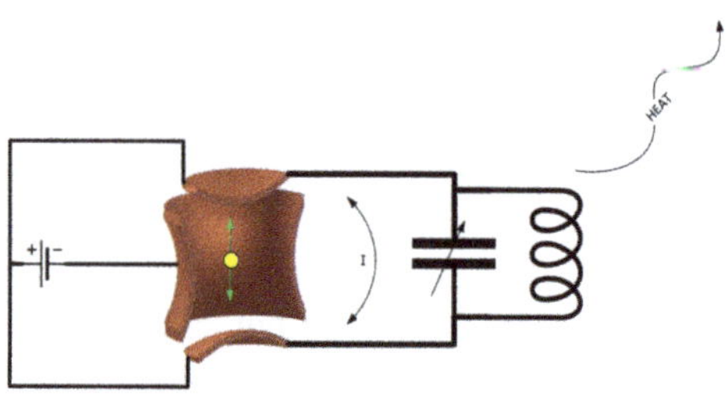

Figure 5.34. Schematic of the electrical connections for inductively cool ions contained in a Penning trap. As the charged particles (yellow circle) oscillate up and down, an image charge is created on the top and bottom electrodes. This image charge travels, as a current (I), through the external circuit leading to heat dissipation through Joule effect.

temperatures that allow observing the quantisation of the energy levels of the quantum harmonic oscillator, and the ground level of the oscillator has been observed.

5.4.4 Sideband cooling

When an atom or molecule is oscillating in a harmonic potential, spectroscopic signals are modulated, due to the Doppler Effect, at the oscillation frequency. In cases where the linewidth of the atomic (or molecular) transition from the ground state $|g\rangle$ to an excited state $|e\rangle$, is narrower than the oscillation frequency it is possible to observe this modulation as a sideband of the transition line. Therefore, if a red detuned radiation is applied with energy matching the gap between the ground state and the excited state minus the energy gap between two levels of the harmonic oscillator a transition $|g,n\rangle \rightarrow |e,n-1\rangle$ can be driven (figure 5.35). Then, the spontaneous decay of the atom will most probably occur in the fundamental line going $|e,n-1\rangle \rightarrow |g,n-1\rangle$. By repeating the process, it can be shown that the ground state of the oscillator can be achieved (see e.g. Rempe *et al* [93]). This technique, with some modification, can also be applied to macroscopic objects [94].

5.4.5 Laser cooling

A very successful way of cooling atomic gases is the laser cooling technique, the development of which resulted in a Nobel Prize to S Chu, C Cohen-Tannoudji, and W D Phillips [95]. Laser cooling of atoms exploits the fact that if atoms are illuminated resonantly with laser light, the atoms absorb photons, which carry momentum, and therefore receive kicks for every photon absorbed. By slightly detuning the exciting laser towards the red (lower frequency) the absorption of

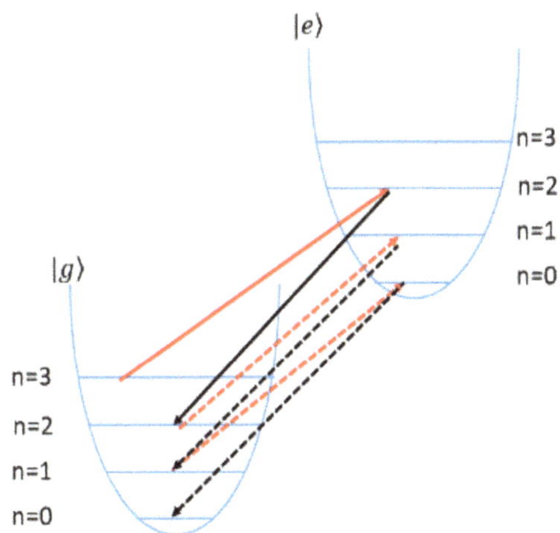

Figure 5.35. Schematic of the transitions addressed in the sideband cooling technique. Red arrows are excitations in the internal atomic levels that remove phonons from the external degree of freedom, and black arrows are spontaneous decays in the internal degree of freedom that typically occur without excitation of the external degree of freedom.

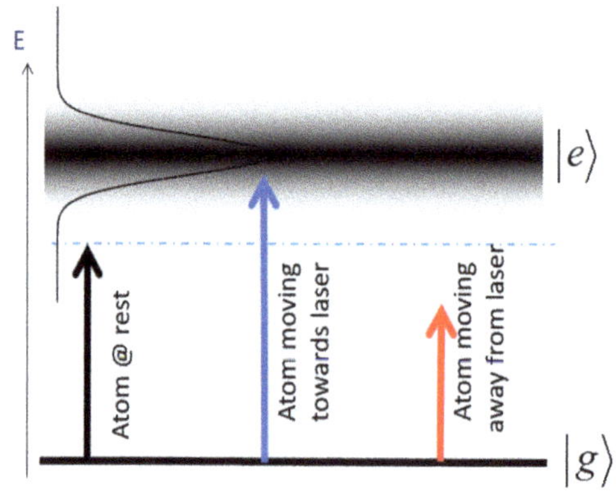

Figure 5.36. Schematic view of the atomic energy levels of a two-level atom. The Gaussian profile graded shadow represents the natural linewidth of the atom and the coloured arrows represent the energy of a laser beam as observed by an atom at rest (black), an atom moving towards the laser source (blue) or an atom moving away from the laser source (red).

photons is made more probable for atoms that move towards the laser source (due to the Doppler effect, the light appears closer to resonance) than for atoms moving in any other direction (see figure 5.36). The later spontaneous reemission of photons happens in a random direction, so on average having a negligible effect. This would then reduce the velocity of the atoms in the direction of the laser source. With a suitable arrangement of six laser sources in the three directions of space, it can be shown that the atoms will see their three-dimensional velocity reduced and therefore the spread of velocities across the atoms in the sample would also be reduced. Thus, the temperature of the cloud would decrease until it reaches the limit where the heating rate due to the random spontaneous emission equalises with the cooling rate due to the laser excitations. This cooling limit is the so-called Doppler limit and is mainly given by the lifetime of the excited state and is of the order of few tens of microKelvins in most cases.

5.4.6 Sympathetic cooling

When the atomic species of interest cannot be effectively cooled by other means, it is still possible to reach extremely low temperatures via confinement in thermal contact, usually mediated by collisions, with a different species whose properties allow much more efficient cooling.

5.4.7 Neutral atom trapping

The trapping of charged particles has been described above, here we give a short introduction to the most common tools for trapping and cooling of neutral particles.

Magneto-optical trap

The workhorse of the cold atom physics community is the magneto-optical trap (MOT) where a linear magnetic field gradient is superimposed over the overlap

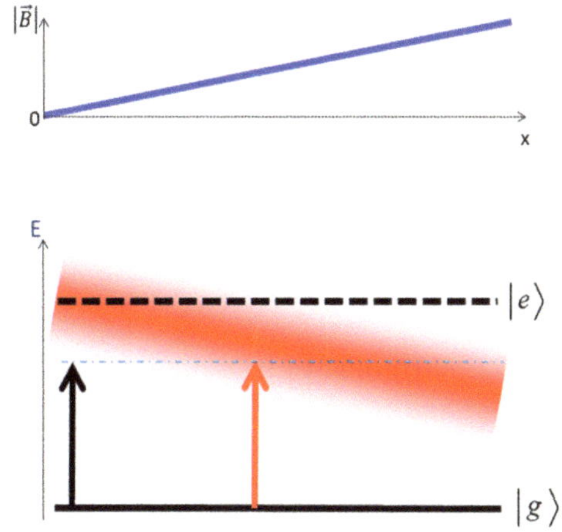

Figure 5.37. Top, variation of magnitude of the applied magnetic field in a MOT with the distance to the centre of the trapping region. Bottom, schematic representation of the energy of the ground $|g\rangle$ and excited $|e\rangle$ state of a two-level atom.

region of the laser beams implementing a three-dimensional laser cooling (see section 5.4.2 on laser cooling). The magnetic field gradient induces a Zeeman shift of the atomic levels thus making the absorption of photons less probable as the magnetic field decreases. In figure 5.36 the energy level of a two-level atom in such a trap is shown. The red graded band represents the dependency of the energy of the excited state $\Delta E_{|e\rangle} = -\vec{\mu}\vec{B}$ (where $\vec{\mu}$ is the magnetic moment of the particular energy level) with the position of the atom due to the Zeeman Effect. The arrows represent the energy of the laser beam as seen by atoms belonging to a particular velocity class. This illustrates that atoms in areas with a higher magnetic field are more likely to absorb photons, and therefore to get momentum kicks towards the magnetic null. This creates a region where the atoms are not only cooled but also trapped (figure 5.37).

Conservative traps

Once the atoms are cooled to the Doppler limit, it is desirable to trap the atoms in a conservative trap where the atoms do not exchange energy with the environment. Two possible implementations of conservative traps are the so-called magnetic traps and optical dipole traps (ODTs). ODTs are also known as optical tweezers when applied to macroscopic objects [96]. In a magnetic trap, the atoms are trapped by means of a strong magnetic field gradient. This magnetic field gradient creates a force on the atoms proportional to their internal magnetic moment, so if B is linear in the position $F \propto -\mu B = -kx$ (where k is a proportionality constant depending on the shape of the magnetic field and internal properties of the atoms) then the atoms will see a harmonic potential. In an ODT the mechanism is very similar but instead of using the magnetic properties of the atom, the electrical dipole moment is used. The ODT uses the fact that a tightly focused laser beam produces an electric field intensity gradient in space, with a maximum at the focused point, and the fact that neutral atoms become polarised if immersed in an electric field. As they become polarised they develop an electric dipole moment that gets attracted to the regions with a high electric field thus producing a linear force on the atoms and therefore subjecting the atoms to a harmonic potential with spring constant k depending on the internal properties of the atom and how tightly focused the laser beam is. In both cases, the magnetic traps and the optical dipole traps, the spring constant k can be tuned experimentally by changing the field parameters.

Evaporative cooling

An atom cloud trapped in a harmonic potential as described earlier can be further cooled using the so-called evaporative cooling technique. This technique relies on selectively removing the atoms with higher kinetic energy from the ensemble and allowing the cloud to re-thermalise, thus leading to an ensemble with lower mean kinetic energy and therefore lower associated temperature. This can be performed by either lowering the potential depth or, in the case of a magnetic trap, applying a so-called radio-frequency knife.

In the radio frequency knife technique, we make use of two facts:

1. The potential is created by a linear magnetic field gradient. This leads to a shift in the energy levels of the atoms that is also linearly dependent on the position, r, of the atoms themselves:

$$\Delta E(r) = -\vec{\mu}\vec{B}(r),$$

   where μ is the magnetic moment of the atom.
2. With the atoms being trapped in a harmonic potential, the amplitude of their motion is linearly dependent on their kinetic energy.

Using these two facts, if the atomic cloud is irradiated with an electromagnetic wave of frequency ν_k, then the atoms have a kinetic energy allowing them to explore the region where their relevant energy level is split by the photon energy of the applied radio frequency $|\Delta E| = \mu B = h\nu_k$ (where h is the Planck constant). The valence electron would then undergo a spin flip bringing the atom from an attractive potential to a repulsive potential and therefore resulting in 'hot' atoms being expelled from the trap.

Using this technique, temperatures below the nano Kelvin range can be achieved [97].

References

[1] Giauque W F and MacDougall D P 1933 *Phys. Rev.* **43** 768
[2] Sydoriak S G, Grilly E R and Hammel E F 1949 *Phys. Rev.* **75** 303
[3] London H 1951 *Proceedings of the International Conference on Low Temperature Physics* (Oxford: Clarendon Laboratory) p 157
[4] Walters G K and Fairbank W M 1956 *Phys. Rev.* **103** 262
[5] Edwards D O *et al* 1965 *Phys. Rev. Lett.* **15** 773
[6] Das P, de Bruyn Ouboter R and Taconis K W 1965 *Low Temp. Phys.* **LT9** 1253
[7] Hall H E, Ford P J and Thomson K 1966 *Cryogenics* **6** 80–8

[8] Bradley D I *et al* 1982 *Cryogenics* **22** 6
[9] V.A.Mikheev *et al* 1984 *Cryogenics* **24** 4
[10] Kiricheck *et al* 2003 *Physica* B 329–33 Part 2
[11] Uhlig K *et al* 1997 *Cryogenics* **37** 5
[12] Uhlig K 2002 *Cryogenics* **42** 2
[13] Frossati G 1978 *J. Phys.* **39** 1578
[14] Vermeulen G A and Frossati G 1987 *Cryogenics* **27** 139
[15] Dm Shvarts *et al* 2004 *Meas. Sci. Technol.* **15** 131–6
[16] Stone N J and Postma H (ed) 1986 *Low-temperature Nuclear Orientation* (Amsterdam: North-Holland)
[17] Aalto M I *et al* 1972 *Cryogenics* **123** 184–91
[18] Roukes M L *et al* 1985 *Phys. Rev. Lett.* **55** 4
[19] Junyun Li *et al* 2000 *Physica* B **280** 1–4
[20] Netsch A *et al* 2006 *AIP Conf. Proc.* **850** 1593–4
[21] Richardson R C and Smith E N 1988 *Experimental Techniques in Condensed Matter Physics at Low Temperatures* (Boca Raton, FL: CRC Press)
[22] White G K and Meeson P J 2002 *Experimental Techniques in Low-Temperature Physics* 4th edn (Monographs on the Physics and Chemistry of Materials vol 59) (Oxford: Oxford University Press)
[23] Ventura G and Risegari L 2008 *The Art of Cryogenics* (Amsterdam: Elsevier)
[24] Cousins D J *et al 1999 J. Low Temp. Phys.* **114** 547–70
[25] Vermeulen G A and Frossati G 1987 *Cryogenics* **27** 139–47
[26] Balz *et al* 2016 *Nat. Phys.* **12** 942–9
[27] Thompson *et al* 2017 *Appl. Phys. Lett.* **110** 162602
[28] den Haan A M J *et al* 2014 *Rev. Sci. Instrum.* **85** 035112
[29] ter Haar E *et al* 1995 *J. Low Temp. Phys.* **99** 151–66
[30] Chawner J M A *et al* 2019 *Sci. Rep.* **9** 19642
[31] Arute F *et al* 2019 *Nature* **574** 505–10
[32] Johnson M *et al* 2011 *Nature* **473** 194–8
[33] Devoret M H and Schoelkopf R J 2013 *Science* **339** 1169–74
[34] Reilly D J 2019 Challenges in scaling-up the control interface of a quantum computer *2019 IEEE Int. Electron Devices Meeting (IEDM) (San Francisco, CA, USA)* 31.7.1–6
[35] Bradley D *et al* 2016 *Nat. Commun.* **7** 10455
[36] Samkharadze N *et al* 2011 *Rev. Sci. Instrum.* **82** 053902
[37] Sarsby M *et al* 2020 *Nat. Commun.* **11** 1492
[38] Tsubota M and Halperin W P (ed) 2009 *Progress in Low Temperature Physics: Quantum Turbulence* vol XVI (Amsterdam: Elsevier)
[39] Skrbek L 2011 *J. Phys.: Conf. Ser.* 318 012004
[40] Bradley D I *et al* 2009 *J. Low Temp. Phys.* **156** 116–31
[41] Hallock R B 1995 The properties of multilayer ^{3}HE–^{4}HE mixture films *Progress in Low Temperature Physics* ed W P Halperin vol 14 (Amsterdam: Elsevier) pp 321–443
[42] Balibar S *et al* 2016 *C. R. Phys.* **17** 264–75
[43] de Waard A *et al* 2006 *Class. Quantum Grav.* **23** S79
[44] Usenko O 2012 *Doctoral Thesis* (Leiden University)
[45] The Cuore Collaboration 2019 *Universe* **5** 10
[46] Teleberg G *et al* 2008 *J. Low Temp. Phys.* **151** 669–74

[47] Azzoni S *et al* 2020 *J. Low Temp. Phys.* **199** 771–9

[48] The QUBIC Collaboration 2011 *Astropart. Phys.* **34** 9

[49] Triqueneaux S *et al* 2006 *Cryogenics* **46** 4

[50] de Klerk D and Steenland M J 1955 *Progress in Low Temperature Physics* (Amsterdam: Elsevier) pp 273–321

[51] Lounasmaa O V 1974 *Experimental Principles and Methods Below 1 K* (New York: Academic)

[52] Kurti N, Robinson F and Simon F *et al* 1956 Nuclear cooling *Nature* **178** 450–3

[53] Enss C and Hunklinger S 2005 *Low-Temperature Physics* (Berlin: Springer)

[54] Pobell F 2007 *Matter and Method at Low Temperatures* (Berlin: Springer)

[55] Duval J-M, Prouvé T, Shirron P, Shinozaki K, Sekimoto Y, Hasebe T and Vermeulen G *et al* 2020 LiteBIRD cryogenic chain: 100 mK cooling with mechanical coolers and adrs. *J. Low Temp. Phys.* **199** 730

[56] Dub L, Duval J M and Luchier N 2014 SAFARI engineering model 50 mK cooler *Cryogenics* **64** 213

[57] Bradley D I, Bradshaw T W, Guenault A M, Keith V, Locke-Scobie B G, Miller I E, Pickett G R and Pratt W P 1982 *Cryogenics* **22** 296

[58] http://emplatform.eu

[59] Jones A T, Scheller C P, Prance J R, Kalyoncu Y B, Zumbühl D M and Haley R P 2020 Progress in cooling nanoelectronic devices to ultra low temperatures *J. Low Temp. Phys.* **201** 772–802

[60] Yao W, Knuuttila T A, Nummila K K, Martikainen J E, Oja A S and Lounasmaa O V 2000 A versatile nuclear demagnetization cryostat for ultralow temperature research *J. Low Temp. Phys.* **120** 121

[61] Bradley D I, Guénault A M, Keith V, Kennedy C J, Miller I E, Mussett S G, Pickett G R and Pratt W P 1984 New methods for nuclear cooling into the microkelvin regime *J. Low Temp. Phys.* **57** 359

[62] Pickett G R 1988 Microkelvin physics *Rep. Prog. Phys.* **51** 1295

[63] Guenault A 2007 *Statistical Physics* (Berlin: Springer)

[64] Glazer M and Wark J 2001 *Statistical Mechanics: A Survival Guide* (Oxford: Oxford University Press)

[65] Zemansky M W and Dittman R H 1996 *Heat and Thermodynamics* (New York: McGraw-Hill)

[66] Shirron P J and McCammon D 2014 Salt pill design and fabrication for adiabatic demagnetization refrigerators *Cryogenics* **62** 163

[67] Shirron P, Canavan E, DiPirro M, Francis J, Jackson M, Tuttle J, King T and Grabowski M 2004 Development of a cryogen-free continuous ADR for the constellation-X mission *Cryogenics* **44** 581

[68] Catarino I and Paine C 2011 ^{3}He gas gap heat switch *Cryogenics* **51** 45

[69] Kimball M O, Shirron P J, Canavan E R, Tuttle J G, Jahromi A E, Dipirro M J, James B L, Sampson M A and Letmate R V 2017 Passive gas-gap heat switches for use in low-temperature cryogenic systems *IOP Conf. Ser.: Meter. Sci. Eng.* **278** 012010

[70] Yao W, Knuuttila T A and Nummila K K *et al* 2000 A versatile nuclear demagnetization cryostat for ultralow temperature research *J. Low Temp. Phys.* **120** 121–50

[71] Knuuttila T A 2000 Nuclear magnetism and superconductivity in rhodium *PhD Thesis* Aalto University https://aaltodoc.aalto.fi/handle/12345678989/2302.

[72] Volovik G 2009 *The Universe in a Helium Droplet* (Oxford: Oxford University Press)
[73] Bradley D I, Fisher S N, Guénault A M, Haley R P, Martin H, Kopu J, Pickett G R, Roberts J E and Tsepelin V 2008 Relic topological defects from brane annihilation simulated in superfluid ^{3}He *Nat. Phys. (N.Y.)* **4** 46
[74] Bradley D I, Fisher S N, Guénault A M, Haley R P, Pickett G R, Potts D and Tsepelin V 2011 Direct measurement of the energy dissipated by quantum turbulence *Nat. Phys. (N.Y.)* **7** 473
[75] Batey G, Casey A, Cuthbert M N, Matthews A J, Saunders J and Shibahara A 2013 A microkelvin cryogen-free experimental platform with integrated noise thermometry *New J. Phys.* **15** 113034
[76] Todoshchenko I, Kaikkonen J-P, Blaauwgeers R, Hakonen P J and Savin A 2014 Dry demagnetization cryostat for sub-millikelvin helium experiments: refrigeration and thermometry *Rev. Sci. Instrum.* **85** 085106
[77] Toda R, Murakawa S and Hiroshi Fukuyama. 2018 Design and expected performance of a compact and continuous nuclear demagnetization refrigerator for sub-mK applications *J. Phys.: Conf. Ser.* **969** 012093
[78] Schmoranzer D, Gazizulin R, Triqueneaux S, Collin E and Fefferman A 2019 Development of a sub-mK continuous nuclear demagnetization refrigerator *J. Low Temp. Phys.* **196** 261
[79] Bradley D I, George R E, Gunnarsson D, Haley R P, Heikkinen H, Pashkin Y A and Penttilä J *et al* 2016 Nanoelectronic primary thermometry below 4 mK *Nat. Commun.* **7** 10455
[80] Palma M, Scheller C P, Maradan D, Feshchenko A V, Meschke M and Zumbühl D M 2017 On-and-off chip cooling of a coulomb blockade thermometer down to 2.8 mK *Appl. Phys. Lett.* **2017** 253105
[81] Sarsby M, Yurttagül N and Geresdi A 2020 500 Microkelvin nanoelectronics *Nat. Commun.* **11** 1492
[82] Dragovan M 1984 Private communicaton
[83] Duband L 2001 Double stage helium sorption coolers *Cryocoolers* vol 11 (New York: Plenum)
[84] Torre J P *et al* 1985 Miniature liquid ^{3}He refrigerator *Rev. Sci. Instrum.* **56** 318
[85] Roubeau P 1969 Adsorption de l'Hélium '4 par le charbon actifColloque international vide et froid, SFITV Grenoble
[86] Duband L *et al* 2018 Sub-kelvin cooling for the BICEP array project *Cryocoolers 20* ed M Ross *Int. Cryocooler Conf.* 347–56
[87] Duband L *et al* 2008 HERSCHEL flight model sorption coolers *Cryogenics* **48** 95–105
[88] Duband L *et al* 2014 SAFARI engineering model 50 mK cooler *Cryogenics* **64** 213–9
[89] Paul W 1990 Electromagnetic traps for charged and neutral particles *Rev. Mod. Phys.* **62** 531
[90] The Nobel Prize in Physics 1989. NobelPrize.org. nobel media AB 2020. Sun. 26 Jul 2020 https://nobelprize.org/prizes/physics/1989/summary/
[91] Paasche *et al* 2003 *Eur. Phys. J.* D **22** 183–8
[92] Peil S and Gabrielse G 1999 Observing the quantum limit of an electron cyclotron: QND measurements of quantum jumps between Fock states *Phys. Rev. Lett.* **83** 1287
[93] Rempe A *et al* 2013 Ground-state cooling of a single atom at the center of an optical cavity *Phys. Rev. Lett.* **110** 022003
[94] Teufel J, Donner T and Li D *et al* 2011 Sideband cooling of micromechanical motion to the quantum ground state *Nature* **475** 359–63

[95] The Nobel Prize in Physics 1997 NobelPrize.org. Nobel Media AB 2020. Sun. 26 Jul 2020 https://nobelprize.org/prizes/physics/1997/summary/
[96] Ashkin A, Dziedzic J M, Bjorkholm J E and Steven Chu 1986 Observation of a single-beam gradient force optical trap for dielectric particles *Opt. Lett.* **11** 288–90
Grimm R, Weidemüller M and Ovchinnikov Y B 2000 Optical Dipole Traps for Neutral Atoms *Advances in Atomic, Molecular and Optical Physics* **42** 95–170
[97] Leanhardt A E, Pasquini T A, Saba M, Schirotzek A, Shin Y, Kielpinski D, Pritchard D E and Ketterle W Cooling Bose–Einstein condensates below 500 picokelvin *Science* **301** 1513–5

Chapter 6

Cryogenics in particle accelerators and fusion reactors

John Weisend, Jaroslaw Fydrych, Andy May and Shrikant Patalwaar

6.1 Overview of requirements

John Weisend

6.1.1 Introduction

Cryogenics plays an important role in particle accelerators and fusion reactors. All current and planned large-scale accelerators and fusion reactors incorporate cryogenics in one or more applications. Table 6.1 describes the use of cryogenics in a selection of accelerator and fusion facilities.

At the same time, the requirements of particle accelerators and fusion reactors have driven research and development in cryogenics, particularly in the areas of large-scale helium refrigeration, cooling with He II (superfluid He) and low-cost, low-heat leak cryostat design.

Applications of cryogenics in accelerators and fusion reactors include:

- Cooling of superconducting (SC) magnets that:
    - Contain, steer and focus particle beams;
    - Confine plasmas in fusion reactors;
    - Provide background magnetic fields in particle detectors;
- Cooling of superconducting radio frequency (SRF) cavities that accelerate particle beams;
- Provision of dense, pure fluids at cryogenic temperatures for:
    - Scattering targets;
    - Moderators;
    - Bubble chambers (now obsolete);
    - Particle detection and measurement;
- Provision of large, clean vacuum spaces via cryo-pumping;
- Cryogenic sample environment for material studies.

These applications, including their history and examples, are discussed below.

doi:10.1088/978-0-7503-2757-2ch6

Table 6.1. Cryogenic applications in a selection of current and planned accelerator and fusion facilities.

Facility Name	Type of Facility	Application of Cryogenics	Operating Temperatures (K)	Status	Reference
LHC	Proton Collider	SC Magnets, SRF, Particle Detection and Identification	1.8, 4.2, 50, 80	Operation	[1]
BEPC II	Electron/Positron Collider	SC Magnets, SRF	4.5, 77	Operation	[2]
XFEL	Free Electron Laser	SRF, SC Magnets	2, 5, 50	Operation	[3]
ESS	Neutron Source	SRF, Moderator, SC Magnets, Sample Environment	2, 4.2, 20, 50	Construction	[4]
SNS	Neutron Source	SRF, Moderator, SC Magnets, Sample Environment	2, 4.2, 20, 50	Operation	[5]
CEBAF	Electron Accelerator	SRF, SC Magnets, Scattering Targets	2, 4.2, 20, 50	Operation	[6]
FRIB	Heavy Ion Accelerator	SRF, SC Magnets	2, 4.2, 50	Commissioning	[7]
LCLS II	Free Electron Laser	SRF, SC Magnets	2, 4.2, 5, 50	Construction	[8]
ISIS	Neutron Source	Moderator, Sample Environment	20, 4.2, 2	Operation	[9]
Spiral2	Ion Accelerator	SRF	4.4	Operation	[10]
JPARC	Neutron source	Moderator, Sample Environment	20, 4.2, 2	Operation	[11]
PIP II	Proton Accelerator	SRF, SC Magnets, Particle Detection and Identification	2, 4.2, 50	Design	[12]
RAON	Heavy Ion Accelerator	SRF, SC Magnets	2, 4.2, 55	Construction	[13]
SHINE	Free Electron Laser	SRF, SC Magnets	2, 4.2, 5, 50	Construction	[14]
FAIR	Heavy Ion Accelerator	SRF, SC Magnets	2, 4.2, 50	Construction	[15]
FCC	Proton Collider	SC Magnets, SRF, Particle Detection and Identification	1.8, 4.2, 50, 80	Design	[16]
ILC	Electron/Positron Collider	SC Magnets, SRF, Particle Detection and Identification	1.8, 4.2, 50	Design	[17]
CEPC	Electron/Positron Collider	SRF	2, 5, 80	Design	[18]
HEPS	Light Source	SC Magnets, SRF	4.2, 80	Design	[19]
ITER	Fusion Reactor	SC Magnets, Cryo-Pumping	4.5, 80	Construction	[20]
EAST	Fusion Reactor	SC Magnets, Cryo-Pumping	4.5, 80	Operation	[21]
SPARC	Fusion Reactor	SC Magnets	20	Design	[22]
CSNS	Neutron Source	Moderator	20	Operation	[23]
W7X	Fusion Reactor	SC Magnets, Cryo-Pumping	3.8, 4.5, 80	Operation	[24]
Super KEKB	Electron/Positron Collider	SRF Cavities, SC Magnets	4.5, 80	Operations	[25]
SST1	Fusion Reactor	SC Magnets	4.5 K, 80 K	Operation	[26]
KSTAR	Fusion Reactor	SC Magnets	4.5 K, 70 K	Operation	[27]

6.1.2 Cooling of superconducting magnets

Rationale

Superconducting magnets are used in particle accelerators and fusion devices because they can provide high magnetic fields (1 T or greater) with a minimum use of electrical power. While water-cooled, resistive copper magnets can in theory be used in many of these machines, the amount of power needed to drive the magnet current and to provide adequate water cooling is far more expensive than using superconducting magnets even with the costs of the associated cryogenic plants. In addition, superconducting magnets in these applications are more reliable and frequently smaller than corresponding resistive magnets. Superconducting magnets provide bending, focusing and steering of charged particles in the accelerator, identification and measurement of particles in particle detectors and plasma confinement in fusion devices.

Typical features of cryogenics for these superconducting magnet systems include: large installations with multiple cryo-plants and complicated distribution systems; operation at 4.2 K and below; high reliability and automated control requirements and advanced cryostat design (see chapter 3).

History and examples

The first significant use of superconducting magnets in accelerators was the K500 cyclotron magnet at the National Superconducting Cyclotron Lab in 1982 [28].

Two years later, a much larger application of superconducting magnets in an accelerator was commissioned at Fermilab in the Energy Saver/Doubler proton synchrotron. This machine [29], later renamed the Tevatron, incorporated many of the features seen in subsequent accelerator superconducting magnet systems. These features include: cooling of the magnets by a subcooled or supercritical flow to avoid gas voids in the magnet winding; heat transfer from the subcooled flow to a two-phase flow via a heat exchanger in the magnets; use of multiple helium refrigeration plants; magnet cryostats that include an actively cooled thermal radiation shield; significant amounts of ancillary piping and connections to allow for cool-down, warm-up and safe venting of helium.

In the early 1990s, the proton synchrotron of the HERA machine at DESY Lab was commissioned. The 5 T, 4.4 K superconducting magnet system for this machine [30] was very similar to that of the Tevatron, with a couple of important differences. The warm iron return yoke of the magnets was moved into the cryostat and operated at the magnet temperature. The thermal radiation shield is now cooled by a liquid helium flow and operates at 40–50 K rather than the 80 K shield cooled by a liquid nitrogen flow used in the Tevatron. Both these approaches have since become part of the standard design for accelerator magnets.

Both the Tevatron and HERA used niobium titanium (NbTi) superconducting magnets operating at about 4.5 K. NbTi can reach significantly higher magnetic fields when cooled to lower temperatures. The Tore Supra Tokamak fusion device, built in Cadarache, France in the 1980s, took advantage of this property to design and operate 9 T NbTi superconducting magnets for plasma confinement. To reach this field, the magnets had to be cooled to 1.8 K. Helium below 2.2 K enters into a second liquid phase know as He II or sometimes more informally 'superfluid helium'. Tore Supra [31] was the first large-scale application of He II. To reach 1.8 K, the pressure of liquid helium is reduced, thus also reducing the saturation temperature of the helium. The best way to achieve this lower pressure in large-scale cryogenic plants is to install cold compressors (essentially cryogenic vacuum pumps) into the cryogenic plants. The requirements of Tora Supra helped motivate the cryogenic industry to develop reliable, efficient cold compressors. Tore Supra also drove experimental studies in engineering applications of He II.

The construction of the Large Hadron Collider (LHC) at CERN during the 1990s also encouraged development of cold compressors and studies of He II behaviour, in particular two-phase flow behaviour [32]. LHC was built in a preexisting tunnel at CERN and in order to maximise the beam energy, 9 T bending magnets were required. As in the case with Tore Supra, NbTi magnets operating at 1.9 K were designed. Characteristics of the LHC cryogenic system are shown in figure 6.1.

The International Tokamak Experimental Reactor (ITER) currently (2021) under construction in Cadarache has a cryogenic system that rivals that of LHC in size. Here though, the design of magnets for fusion systems has evolved to use a Nb_3Sn superconductor which provide higher fields even at 4.5 K so ITER is not a He II system. Additionally, the heat load requirements of fusion magnets lead to designs in which the magnets use cable-in conduit conductors cooled by force flow supercritical helium. Figure 6.2 shows the ITER magnet system. Note the person in the figure that helps convey the immense size of the ITER magnets. The cryogenic system is based on three parallel cryo-plants, each producing up to 25 kW of cooling at 4.5 K.

The SPARC machine [22], currently in design, is another fusion project. This machine uses high temperature superconductors operating at about 20 K and producing a magnetic field of approximately 12 T. This will result in smaller magnets and a smaller machine overall.

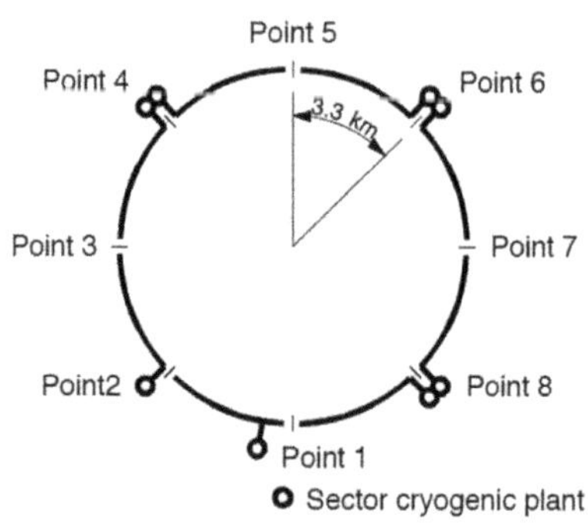

Figure 6.1(a). Distribution of cryo-plants around LHC ring. Reprinted from [80] with permission from Cern.

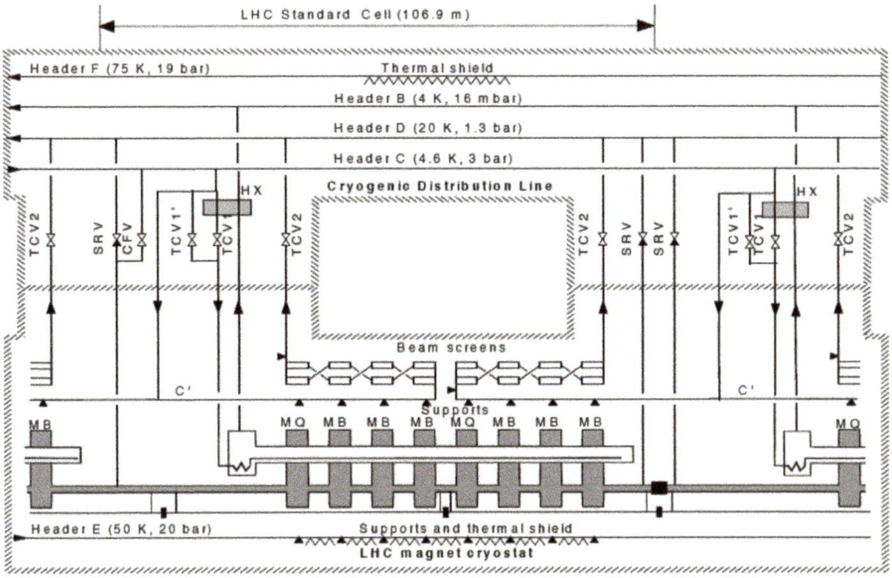

Figure 6.1(b). Cryogenic flow scheme of an LHC lattice cell. Reprinted from [80] with permission from Cern.

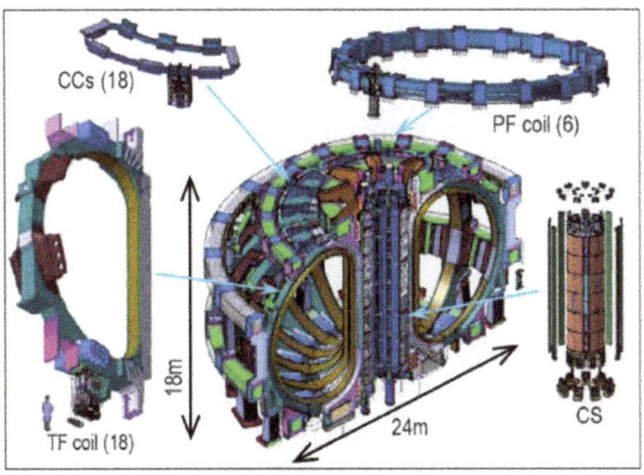

Figure 6.2. The ITER magnet system. Courtesy of ITER.

6.1.3 Cooling of superconducting radio frequency cavities

Rationale

Superconducting radio frequency (SRF) cavities are used to accelerate a charged particle beam. As the technology of these cavities has developed over the last 30 years, cooling of SRF cavities has become an increasingly important application of cryogenics in accelerators. In many linear accelerators, cooling of SRF cavities is the principal (or in some cases the only) application of cryogenics.

The advantage of superconducting over copper RF cavities is that in superconducting cavities far less RF power is dissipated in heating the cavity walls and is thus available for accelerating the beam. Since this is an application of AC superconductivity the wall losses, even in the superconducting cavities, are not zero. In addition to being a function of the material, the wall losses are a function of temperature and RF frequency. The RF frequency used in accelerating cavities is a function of the machine design and the particles being accelerated. Generally speaking, above 500 MHz in RF frequency it is more energy efficient to operate at 2 K or below, even considering the additional energy required to cool below 4.2 K. Below about 300 MHz in RF frequency it is generally more energy efficient to operate at 4.2 K. However, there are additional considerations. Unlike magnets, SRF cavities are typically bath-cooled in saturated helium at a given temperature. Operating below 2.2 K puts us in the He II phase of liquid helium. One of the features of He II is that it doesn't support bulk boiling. This eliminates the issue of bubbles from boiling affecting the tuning of the resonant cavities. For this reason, even some of the systems operating below 100 MHz will also operate at 2 K or below, The FRIB machine [7] is an example of this choice.

History and examples

The first accelerators based on SRF cavities were built at Stanford University in the 1970s operating at 1.8 K. The first large-scale linac using SRF cavities was the CEBAF accelerator [6] built during the early 1990s operating at 2.1 K. At the same time, the LEP machine at CERN was upgraded with SRF cavities operating at 4.5 K. During the mid to late 1990s the TESLA project at DESY developed high performance 1.3 GHz cavities that operated at 2 K or below for use in electron and positron accelerators. This type of cavity has since been used in the XFEL machine at DESY [3], the LCLS II machine at SLAC [8] and the SHINE machine in Shanghai [14] as well as in a number of smaller university-based linacs.

The SRF cavities in almost all the applications mentioned above are made from pure niobium, bath-cooled in saturated helium at 2 K–4.6 K. There is ongoing research on constructing cavities from Nb_3Sn and even on using small cryo-coolers rather than bulk liquid helium to cool the cavities [33].

A typical SRF cavity cryogenic system is shown in figure 6.3. This figure shows a cryomodule from the ESS project [4] containing four SRF cavities and connected to a parallel transfer line. A cryomodule is simply a cryostat containing SRF cavities. The term was coined during the CEBAF construction. In figure 6.3 the cavity is contained in a saturated bath of helium. These baths are connected to a common two-phase line that is pumped on by cold compressors and one warm compressor in the helium plant via a large low-pressure line in the parallel transfer line. The cryomodule also contains an actively cooled thermal radiation shield operating between 40 and 50 K and a number of other lines for heat interception and cool-down/warm-up. The transfer line runs along the length of the linac and the cryomodules are thus cooled in parallel. Between the cryomodules are warm sections containing beam instrumentation, room temperature magnets and vacuum systems. The beam tube runs through the centre of the cryomodules and warm sections.

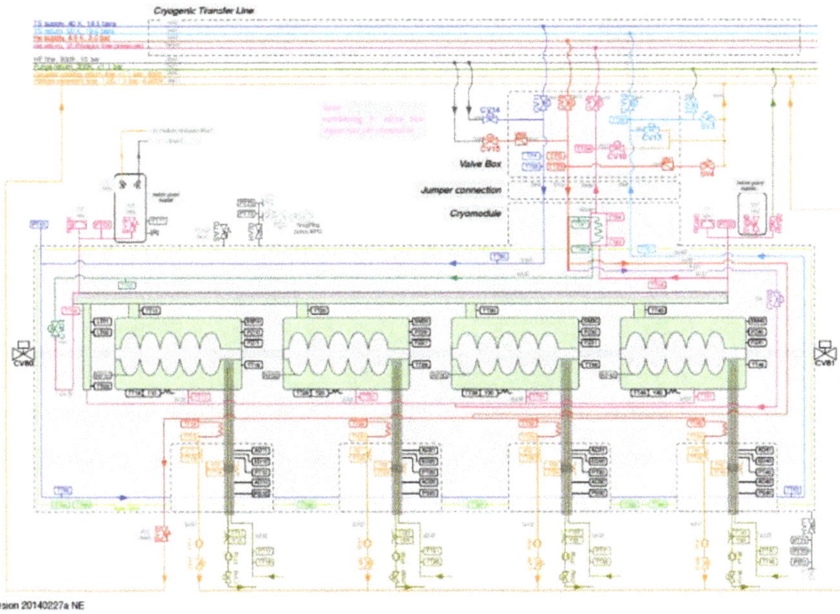

Figure 6.3. Flow schematic for the ESS cryomodule and distribution system. Courtesy of ESS.

This geometry of cryomodules, warm sections and transfer line is very common in SRF linear accelerators and is also seen in the SNS, FRIB and CEBAF accelerators. However, once the size of the machine exceeds about 100 cryomodules, the large number of warm/cold transitions in this design starts to become too costly and inefficient. The TESLA project was meant to develop technology for the future International Linear collider which would have more than 2000 cryomodules. To reduce costs, the parallel transfer line was moved into the cryomodule and a long string of cryomodules was tied together without any warm sections between them. This approach significantly reduces cost and the heat leak into the cryomodules but means that any magnets will now need to be superconducting and that problems with one cryomodule will require warming up multiple cryomodules. This cryomodule design [34] has proved very popular and has been adapted for the XFEL, SHINE and LCLS II machines. Figure 6.4 shows the cross-section of the LCLS II cryomodule. Note that the large diameter, low-pressure pumping line—the Helium Gas Return Pipe (HGRP) has been moved into the cryomodule itself.

6.1.4 Provision of dense pure fluids

Rationale

Liquefying elements such as hydrogen, argon and helium greatly increase the density and thus reduce the volume, of the resulting liquid. This effect is used to make transportation of large quantities of these elements easier but, in physics applications, this effect is also used to increase the mass of an element available within a given volume. In fixed target experiments, where the science is driven by the

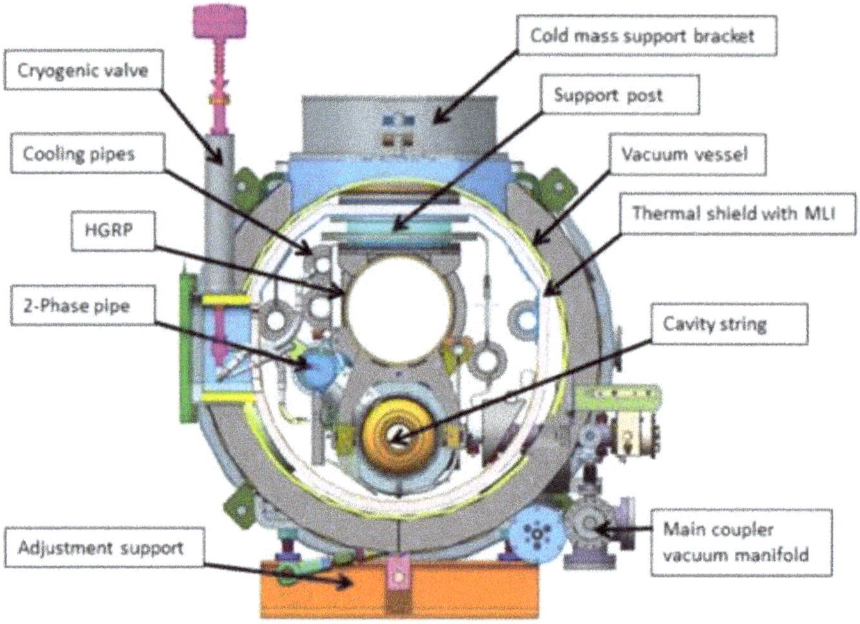

Figure 6.4. Cross-section of LCLS II cryomodule. (Courtesy T Peterson—SLAC.)

interaction of a particle beam with either electrons or protons within a target the use of liquid hydrogen greatly increases the number of interactions within the target resulting in the production of more scientific data.

In moderators, the goal is not to produce interactions to be studied but rather to reduce the energy of a particle (typically a neutron) by scattering it off colder nuclei (typically hydrogen). The lower energy neutrons can then be used for material studies.

A third application of dense pure fluids is in particle detectors where the goal is to have particles under study interact with a dense fluid (such as hydrogen) to detect and measure properties of the particles. The increased density of the fluid increases the interaction rate and thus the scientific output.

All three of these applications require cryogenic systems to liquefy the fluid of choice, and remove the heat deposited in it by particle interactions and the outside environment.

History and examples

Progress in physics has made fixed target experiments increasingly rare. Both a history [34] of fixed target designs used at the SLAC laboratory and a more contemporary example [35] can be found in the reference section.

Hydrogen moderators associated with neutron sources are a much more common application. Almost all neutron sources, both spallation-based and reactor-based have hydrogen moderators associated with them. Figure 6.5 shows a flow schematic of a moderator system at the ISIS spallation source along with some photos of the system [9]. In moderator systems, not only is it necessary to keep the hydrogen a

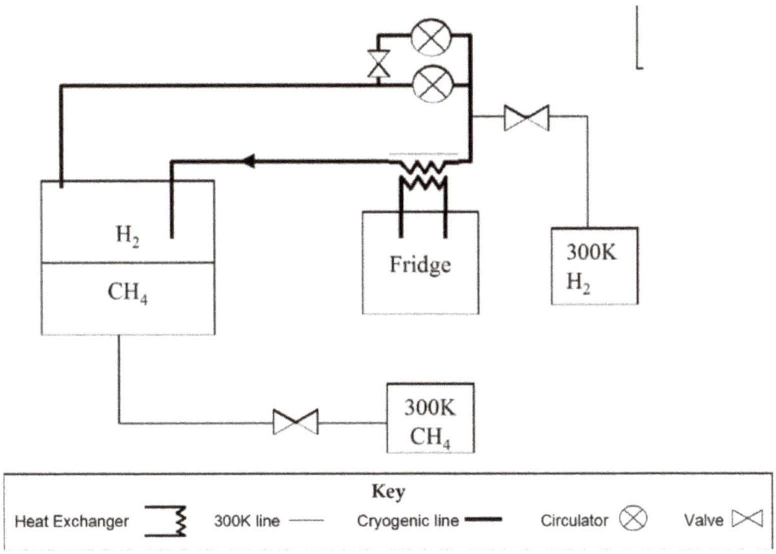

Figure 6.5(a). Schematic of the ISIS TS2 hydrogen moderator system (Courtesy S Gallimore—ISIS.)

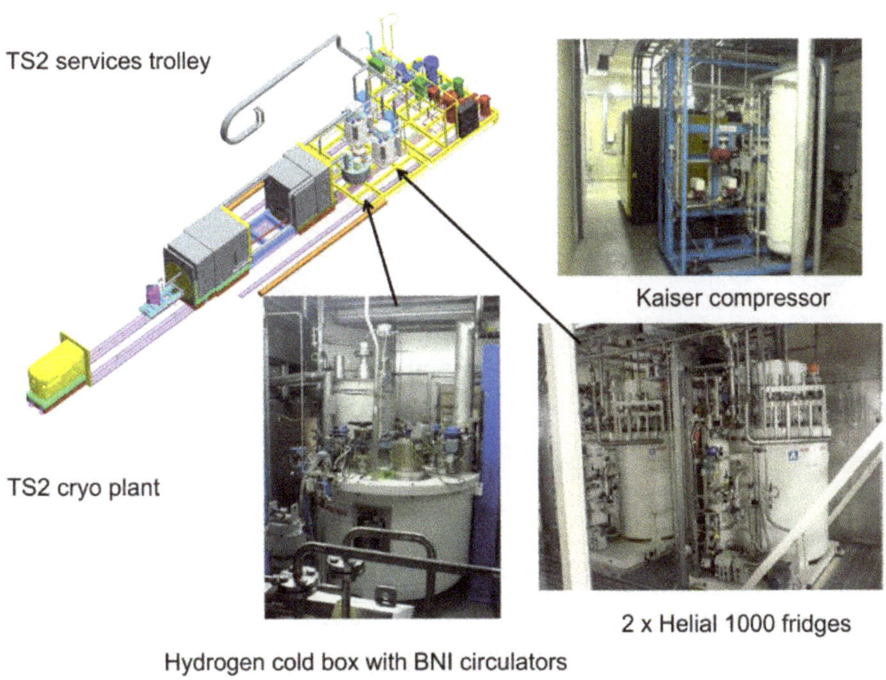

Figure 6.5(b). Equipment layout for the ISIS TS2 hydrogen moderator system (Courtesy S Gallimore—ISIS.)

liquid but, generally, it is required to keep the density variations of the hydrogen to a minimum. Moderator systems typically consist of a closed-cycle loop in which the hydrogen is circulated via a pump. The heat deposited in the hydrogen by the neutron scattering is removed via a heat exchanger cooled by a separate helium refrigerator operating at around 16–17 K. Additional challenges in these moderators include the safety issues associated with hydrogen, the varying heat load with beam operation and the need to keep the hydrogen in its parahydrogen state [36]. Examples of hydrogen moderator systems can also be found at the ESS [4], SNS [5] and JPARC [11] spallation neutron sources. These moderator systems can become quite large; at ESS the neutrons deposit up to 30.2 kW into the 20 K hydrogen moderator requiring a 35 kW helium plant operating at 16 K for cooling.

An example of a cryogenic system supporting the use of a dense fluid for particle detection is given in figure 6.6. This is the Long Baseline Neutrino Facility [38] which will provide cryogenic support for a liquid argon-based neutrino detector deep underground. Challenges for this system include the need to keep the argon extremely pure and the safety risks associated with handling large amounts of cryogenic fluids in underground spaces.

6.1.5 Provision of large clean vacuum spaces

Surfaces at cryogenic temperatures will condense almost all atmospheric gases producing a vacuum in an enclosed space. This technique, known as cryo-pumping, and has a number of advantages including high pumping speeds, the ability to pump

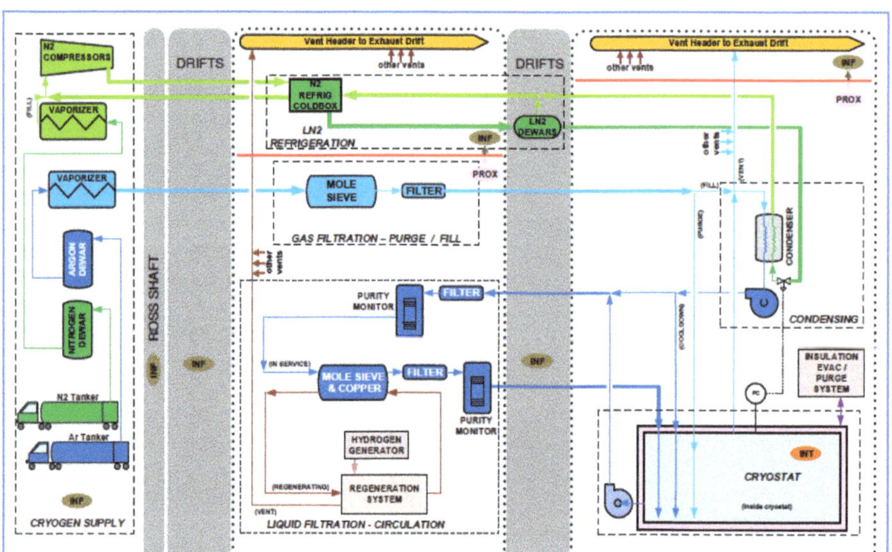

Figure 6.6. Process flow diagram for the Long Baseline Neutrino Facility cryogenic system. Reproduced from [38].

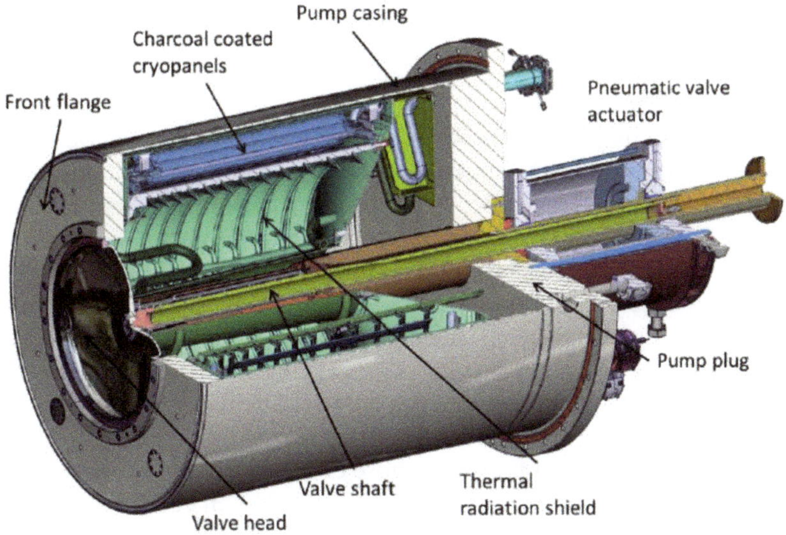

Figure 6.7. View of an ITER Torus cryo-pump. Reprinted from [39], copyright (2013), with permission from Elsevier.

irregularly shaped enclosures and production of vacuum without contamination by hydrocarbons from mechanical pumps. Cryo-pumping is used extensively in the semiconductor manufacturing industry. In scientific facilities, a very significant role of cryo-pumping is the evacuation of plasma chambers in fusion reactors. Figure 6.7 is a view of one of the cryo-pumps found on the ITER reactor [39]. As much of the gas to be pumped out during ITER operation is hydrogen isotopes, this pump is designed to operate below 5 K and uses charcoal adsorbers cooled to this temperature. The pump shown is roughly 1.6 m in diameter and 2 m in length. It has a 11.2 m^2 surface area of activated charcoal. Cryo-pumping at ITER is a major load on the cryogenic system with up to 25 KW at 4.5 K available for this application.

6.1.6 Sample environments

In many scientific facilities, such as light sources and neutron sources, cryogenics plays a role in providing the right environment for the sample being studied. Applications may include superconducting magnets cooled by liquid helium to produce high magnetic fields at the sample and the need for the sample itself to be studied at cryogenic temperatures. Desired sample temperature ranges can vary widely; from 2 to 80 K that can be achieved by small cryo-coolers or liquid helium down to sub milliKelvin levels requiring dilution refrigerators or adiabatic demagnetisation refrigerators. Early estimates at the ESS project showed a requirement for up to 7500 litres of liquid helium a month for use in sample environments.

6.2 Distribution techniques

Jaroslaw Fydrych

Cryogenic facilities listed in table 6.1 are large in size and consist of a number of cryogenic applications such as superconducting magnets, cryomodules and cryogenic vacuum pumps. Therefore, they need to comprise complex cryogenic distribution systems which are intended to transfer cooling power and distribute it among the applications. Usually, the cooling power is transferred by means of cold helium at different thermodynamic states. Some facilities also use liquid nitrogen for cooling the thermal shields. However, due to different prevention measures for oxygen deficiency hazards that may be caused by leakages of helium and nitrogen, it is now recommended that only cold helium is used in the cryogenic distribution system.

Cooling power is produced in one or more cryogenic plants which are composed of two main parts, namely the warm compressor station and cold box. The cold box is a vacuum insulated box housing helium process pipes, heat exchangers, expansion turbines, Joule–Thomson valves (JTVs) and other additional components such as process control valves, filters, absorbers, subcoolers and instrumentation. Here, the compressed helium at ambient temperature coming from warm compressors is subjected to various processes of isentropic and isenthalpic expansion as well as heat and mass transfers which result in cooling and lowering its temperature. Usually, the cold box is designed to produce cold helium at two thermodynamic states, namely subcooled liquid helium at a temperature of 4.5 K and at a pressure between 3 bar and 4 bar absolute and cold helium gas at a temperature between 30 K and 70 K and at a pressure between 10 bar and 20 bar absolute. The subcooled liquid helium in cryogenic jargon is also called supercritical helium.

The cryogenic distribution system is composed of multi-channel transfer lines and special boxes. Multi-channel transfer lines are described in detail in references [40–42]. Their function is to connect two cryogenic apparatuses. They are usually composed of a number of different modules, whose shapes and lengths depend on the cryo-line routing as well as on manufacturing, transportation and installation procedures. There are straight, elbow, angular, zed and tee modules. Their lengths are usually equal to a single or double random length of pipe spool. A typical cryo-line module is composed of an external envelope (vacuum jacket) made of a large-diameter pipe which houses several process lines, thermal shields, internal supports and thermal expansion compensation components such as thin layer metal bellows and flexible metal hoses. The process pipes and thermal shields are usually covered with multi-layer insulation (MLI) layers as in typical cryogenic cryostats. Some modules can also include external supports, vacuum barriers, vacuum pumping ports and pressure relief devices for protecting the external envelope against excessive pressure.

The special boxes of the distribution system are cryogenic devices designed for carrying out specific processes such as splitting or returning cryogen flows, controlling flow rates and subcooling. They are composed of a vacuum jacket which, apart from process pipes and thermal shields, can contain cryogenic valves, heat exchangers, electrical heaters, cryogenic pumps, subcoolers, instrumentation (temperature and pressure sensors, flow meters) and electrical feedthroughs. In general, these boxes are called valve boxes, since they usually have a number of cryogenic valves intended for controlling flow rates of cryogenic fluids to adjacent devices or subsystems. Some specific valve boxes, which are intended for carrying out specific processes, are called distribution boxes, auxiliary boxes, feed boxes, splitting boxes, end boxes, subcooler boxes, etc. These names may differ for different institutions and facilities, since there is no defined common nomenclature.

The layout, architecture and design of a cryogenic distribution system depends on the number and locations of cryogenic users that need to be linked with the cryo-plant cold box, and on the geometrical features of the site infrastructure. They also depend on the required cold temperature and applied cooling method for the specific cryogenic applications. There are two main cold temperature levels, about 4.2–4.5 K and 1.8–2.1 K, and three main cooling methods, i.e. cooling with supercritical helium, saturated superfluid helium (sHe II) and pressurised superfluid helium (pHe II).

The method of applying supercritical helium is used if the temperature required by operation of superconducting components is at the level of liquid helium temperature (4.2 K). Use of saturated liquid helium is not recommended in order to avoid problems in transferring the liquid helium long distances and in collecting heat from the superconductor due to two-phase flow and volumetric evaporation phenomena. Because the critical pressure of helium is very low (2.26 bar absolute), it is technically very easy to produce subcooled liquid helium. Figure 6.8 shows schematically a system using cooling with supercritical helium. In this case, the cryo-plant cold box is equipped with a subcooler, in which the cold process helium at a pressure above the critical value is cooled down to a temperature around 4.5 K. Such supercritical helium can be transferred directly to the cryogenic user if the distance is relatively short. For a longer

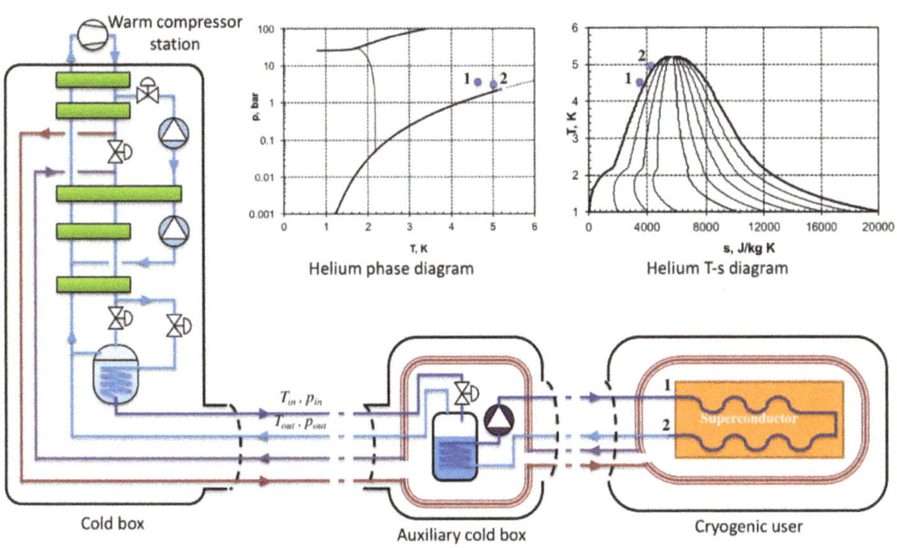

Figure 6.8. Schematic flow diagram of a cryogenic system using cooling with supercritical helium.

distance, the temperature of the supplied helium may be increased above the required value due to unavoidable heat loads along the transfer line. Therefore, there is a need for an auxiliary cold box equipped with another subcooler.

The process helium delivered from the cold box is throttled in a JTV to a pressure close to atmospheric (1.1 bar to 1.3 bar absolute) to produce a liquid helium bath at a temperature of 4.3 K–4.5 K. This bath is used for conditioning another process helium stream. This helium circulates between the auxiliary cold box and the cryogenic application and its flow is driven by a cryogenic pump. The cooling power produced by this system is a product of supercritical helium mass flow rate and the difference in enthalpy between the inlet and outlet to the cryogenic superconducting device; $\dot{Q} = \dot{m}(h_2 - h_1)$.

Apart from the two cold helium circuits, there is a circuit of helium at a higher temperature level. This helium stream is taken from the cold box high pressure circuit, where its temperature is between 30 K to 80 K, and it is used for cooling down the thermal shields of the cryogenic application, auxiliary box and transfer line.

An example facility applying cooling with supercritical helium is the ITER cryogenic system [43, 44]. A simplified diagram is presented in figure 6.9. The system includes three liquid helium plants supplying supercritical helium to five auxiliary cold boxes (ACB) via a cryo-plant termination cold box (CTCB). In each of the ACBs the supercritical helium is throttled at the inlet to a subcooler's phase separator to a pressure of 1 bar absolute. The resultant liquid phase has a temperature of 4.2 K and is used for subcooling process helium used for cooling

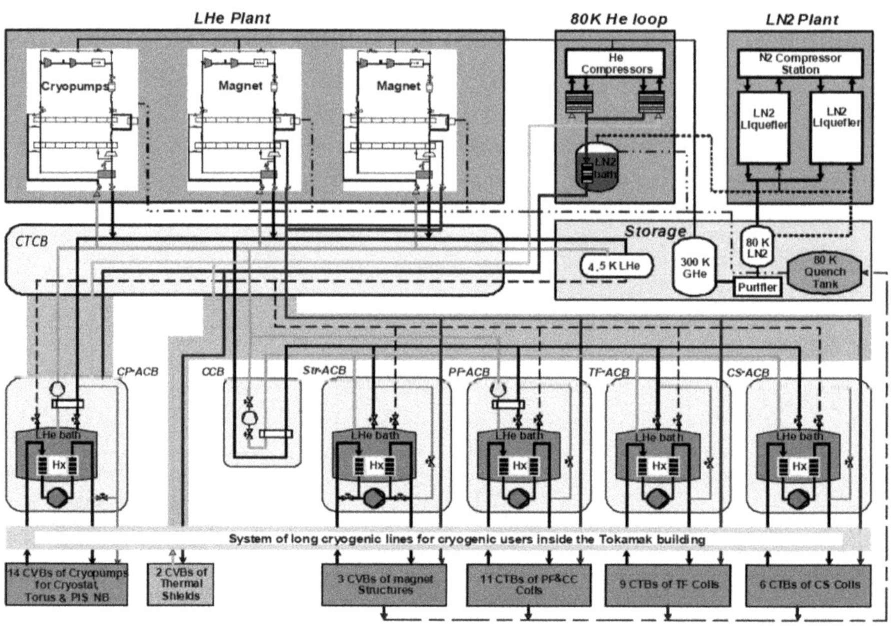

Figure 6.9. Architecture of the ITER cryogenic system. Reproduced from [44], with the permission of AIP Publishing.

the ITER tokamak poloidal field (PF) coils, toroidal field (TF) coils, central solenoid (CS), mechanical structure (Str) and cryogenic vacuum pumps (CP). The helium vapour is taken from the phase separators by a cold compressor located in the cold compressor box (CCB) and directed back to the LHe plants. The subcooled process helium is pumped by the cryogenic pumps of the PF-ACB, TF-ACB and CS-ACB to 26 cold termination boxes (CTB) at the interfaces to the magnet coils and further into their cable channels in conduit conductors. In the cases of CP-ACB and Str-ACB, the pumps supply the process helium to 17 cold valve boxes (CVBs) of the cryo-pumps and magnet structures. In summary, the ITER distribution system is composed of 51 cryo-distribution boxes and 45 cryo-lines (24 single- and 21 multi-channel lines) with a total developed length of 3.7 km [45].

A schematic of a cryogenic system which uses cooling with saturated superfluid helium is shown in figure 6.10. Here, the cryogenic plant supplies supercritical helium at a temperature and pressure of 4.5 K and 3 bar absolute, respectively. Then, the supercritical helium is transferred to the user, where it is precooled to 2.2 K in a counterflow heat exchanger (HX) and later throttled in a JTV down to a pressure of 31.3 mbar absolute to produce superfluid helium at a temperature of 2 K. In this case the superconducting device is immersed in a bath of saturated superfluid helium and the process of cooling comes from the latent heat of liquid helium evaporation. So, the generated cooling power of this system is a function of the helium mass flow rate, enthalpy of the precooled helium and enthalpy of the helium vapour in the superconductor vessel: $\dot{Q} = \dot{m}(h_{3''} - h_3)$.

The helium vapour needs to be recovered from the cryogenic device and transferred back to the cryogenic plant via a large-size return process line designed

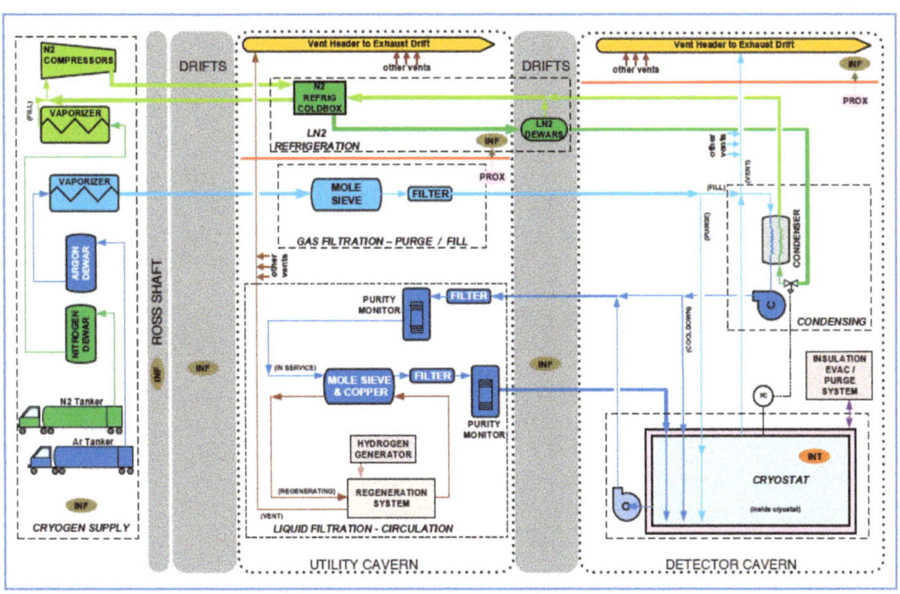

Figure 6.10. Schematic flow diagram of a cryogenic system using cooling with saturated superfluid helium.

for transferring a high volumetric flow of low-density vapour. The flow of vapour is driven by cold compressors (CC) at the cold box, which generate the sub-atmospheric pressure required for reaching 2 K in the helium vessel inside the cryogenic device. In the case of smaller systems, the sub-atmospheric pressure is produced by a set of warm vacuum pumps. The precooling of the 4.5 K helium to 2.2 K can be performed in either a number of small heat exchangers located just upstream of the JTVs or in one central heat exchanger placed close to the cryo-plant [46]. The design with distributed heat exchanges can be found in the cryogenic distribution systems of SNS [5], ESS [47] and PIP-II [48], whilst the layout with a central heat exchanger is used in the XFEL and LCLS-II cryogenic systems [49, 50].

The ESS superconducting linac is a good example of a facility cooling with saturated superfluid helium. Figure 6.11 shows a schematic flow scheme of its cryogenic distribution system (CDS). The system is intended for distributing cooling

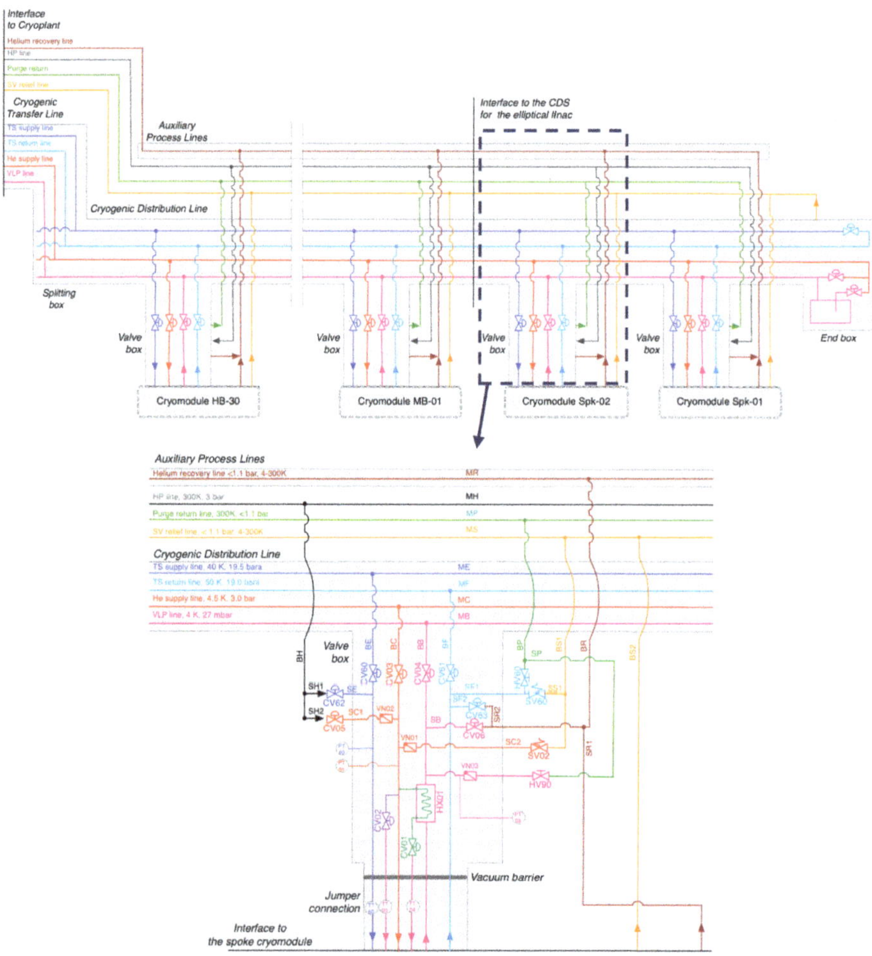

Figure 6.11. Simplified flow scheme of the ESS linac cryogenic distribution system.

power among 43 cryomodules (13 spoke, 9 medium-β and 21 high-β elliptical cavity cryomodules). The nominal operating temperature for both the spoke and elliptical cavities is 2 K and the total heat loads to the cavity cooling loops and their thermal shields are estimated as 3.0 kW and 10.8 kW, respectively. The required cooling powers are provided by the accelerator cryo-plant and delivered in two temperature levels: 4.5 K and 40 K [51]. The 2 K temperature level is produced at each cryomodule by isenthalpic expansion of the 4.5 K supercritical helium, previously subcooled to 2.2 K in a counterflow heat exchanger.

The CDS consists of a cryogenic transfer line with one splitting box, cryogenic distribution line including 43 valve boxes and one end box, and also a set of four auxiliary process lines. The splitting box has a terminal dedicated for an easy and relatively quick connection of another cryogenic distribution line, in case of a future extension of the linac lattice [52]. The main function of the valve boxes is to allow for warming up and cooling down of one or more cryomodules while keeping the others cold at 4.5 K stand-by mode. The end box is for returning the remaining flows of cold helium in the main circuits back to the cryo-plant cold box. These flows are to ensure a proper temperature of the cold helium in the inlets to the last cryomodules.

The valves that allow the closing of the cold circuits in the cryomodules (CV03, CV04, CV60 and CV61) as well as several other control valves are grouped in valve boxes. Each valve box for a spoke cavity cryomodule is equipped with eight cryogenic control valves, one heat exchanger and some other equipment, such as check valves, warm control valves, safety valves, pressure transmitters and temperature sensors. The cryomodule is connected to the valve box via a branch cryo-line, so-called jumper connection. Because the cryomodule insulation vacuum has to be separated from that of the CDS, the jumper connection includes a vacuum barrier. In the elliptical portion of the ESS linac the HX, CV01 (JTV) and CV02 (filling valve) are located in the cryomodule. The ESS cryo-distribution is complex but not very long. Its total developed length is equal to 380 m.

The cold helium vapour is taken back to the cryo-plant by a set of cold compressors. Since the compressors can generate the sub-atmospheric pressure of 27 mbar absolute, the available pressure drop for the VLP line is equal to 4.3 mbar only. Thus, for the specified helium vapour flow rate of 111 g s^{-1} the VLP line needs to be made of large diameter pipes (DN200 and DN250 in the distribution and transfer lines, respectively). For comparison, the He supply line, designed for the supercritical helium flow of 120 g s^{-1}, is made of a DN65 pipe, whilst the TS supply and return lines are DN50 in size. The sizes of these three process lines are too large even for nominal operation conditions, because they are determined by high volumetric flow capacities required at the cool-down and warm-up phases.

The principle of a cryogenic system using cooling with pressurised superfluid helium is shown in figure 6.12. Here, the cryogenic plant also produces supercritical helium. Then, the helium is precooled to 2.2 K in a counterflow heat exchanger and later throttled in a JTV down to a pressure of 16 mbar absolute to produce saturated superfluid helium (sHe II) at a temperature of 1.8 K. The sHe II is not directly used for cooling the superconducting components. It is used for collecting heat from the so-called pressurised superfluid helium (pHe II). The superconductor is immersed in

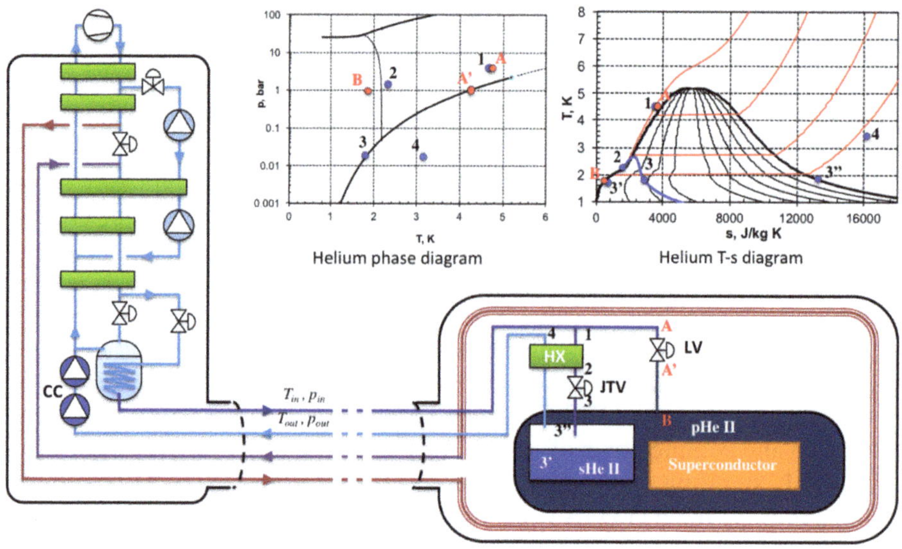

Figure 6.12. Schematic flow diagram of a cryogenic system using cooling with pressurised superfluid helium.

the pHe II bath at a temperature of 1.9 K, due to thermal boundary resistance (Kapitza resistance) resulting in an observed temperature difference of 0.1 K.

The use of pHe II makes the cryogenic system much more complex but it brings several advantages [53]. A significantly reduced quantity of helium under sub-atmospheric pressure limits the risk of air inleaks, which may result in contamination of the process helium and blocking of process pipes and vessels. It eliminates a risk of electrical breakdowns at low voltages that could appear in low-pressure helium vapour due to its low dielectric strength. And finally, the pHe II bath has much higher capacity for thermal stabilisation of superconducting devices, since it can absorb around 10 times more heat than even saturated He II.

An exemplary facility cooling with pressurised superfluid helium is the LHC cryogenic system [54]. It is the largest cryogenic facility worldwide. The system includes eight cryogenic plants producing cooling power for eight sectors of the LHC machine, 3.3 km long each. The cooling power is transferred to the cryogenic users, mainly cryostats of superconducting dipoles and quadrupoles, via a very complex and sophisticated cryogenic distribution system [55]. Figure 6.13 shows the layout of the LHC cryogenic system and the architecture of a so-called feeding point. The feeding point is composed of two cryogenic plants, two 1.8 K refrigeration units, a cryogenic interconnection box and two cryogenic distribution lines. At CERN, the distribution lines are called QRL. Each 1.8 K unit is composed of a cold compressor box and a warm compressor station and has an installed pumping capacity of 125 g s^{-1} at 15 mbar absolute. Each QRL is around 3.3 km long, comprises 26 different service modules (valve boxes), and is terminated with a return module (end box).

Figure 6.14 shows a schematic flow diagram of a typical LHC lattice cell. The cell consists of two strings of eight superconducting magnets each (six dipoles [MB] and

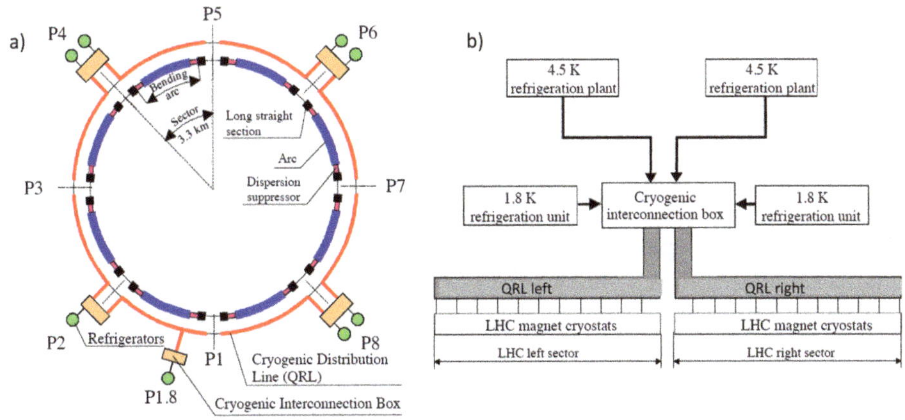

Figure 6.13. Layout of the LHC cryogenic system (a) and architecture of a typical feeding point (b) [55], copyright 2000, Springer Science Business Media New York, with permission of Springer.

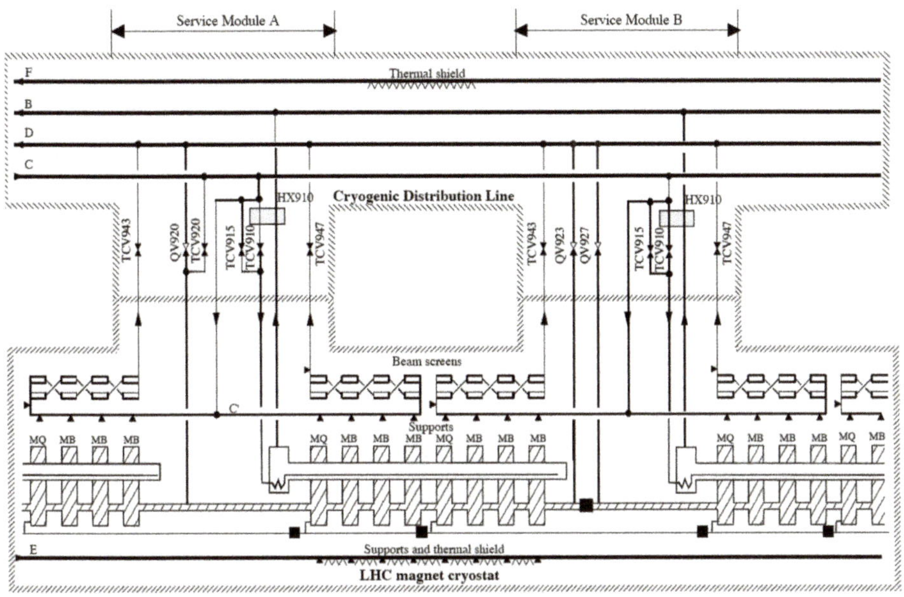

Figure 6.14. Schematic flow diagram of a typical LHC lattice cell [55], copyright 2000, Springer Science Business Media New York, with permission of Springer.

two quadrupoles [MQ]) and a 107 m long section of QRL including two service modules. The magnet coils and yokes are immersed in pHe II bath at 1.9 K and at a pressure slightly above 1 bar absolute. The QRL houses four main process lines (so-called headers). Header C supplies supercritical helium at 4.6 K and 3.6 bar absolute. The 4.6 K helium is precooled in a heat exchanger (HX910), throttled in a JTV (TCV910) and supplied into a biphasic line running along the magnet string.

The sHe II produced collects the heat from the pHe II bath and its vapour is evacuated by cold compressors via a heat exchanger (HX910) and header B. The pHe II bath is only supplied with helium via a filling valve (TCV920) during the final cool-down phase of the machine. In the event of a quench (resistive transition) of the magnets, an excessive pressure in the pHe II bath is reduced by discharge of helium into header D via so-called quench valves (QV820, QV923 and QV927). Service modules also supply the 4.5 K helium to the beam screen cooling loops. This helium flows back to the cryo-plant via header D and the temperature of the beam screen is regulated by temperature control valves (TCV943 and TCV947). Headers E and F are designed for cooling the thermal shields of both the magnet cryostats and QRL at a temperature varying from 50 K to 75 K.

6.3 Cryogenics for superconducting RF cavities

Andy May and Shrikant Patalwaar

6.3.1 Introduction

As introduced in section 6.1.3, superconducting radio frequency (SRF) cavities are widely used in modern high energy accelerators to impart energy to the charged particle beam. The push for increasingly higher beam energies and intensities has been the key driver for the considerable progress in SRF technology and associated cryogenic systems that has been made since the late 1970s [56, 57].

6.3.2 SRF cavity theory and operation

Accelerating structures are resonating electromagnetic cavities, typically operating between 50 MHz and 3 GHz. Cavities may broadly be considered to fall into one of two categories: the first for accelerating charged particles moving at velocities (v) close to the speed of light ($v/c = \beta \approx 1$, e.g., electrons in a high-energy linear accelerator or storage ring), and the second for those moving at low velocities (β <0.3, e.g., heavy ions from a dc high-voltage Van de Graaff accelerator) [58]. We will focus here on cavities of the first type, an example of which is shown in figure 6.15.

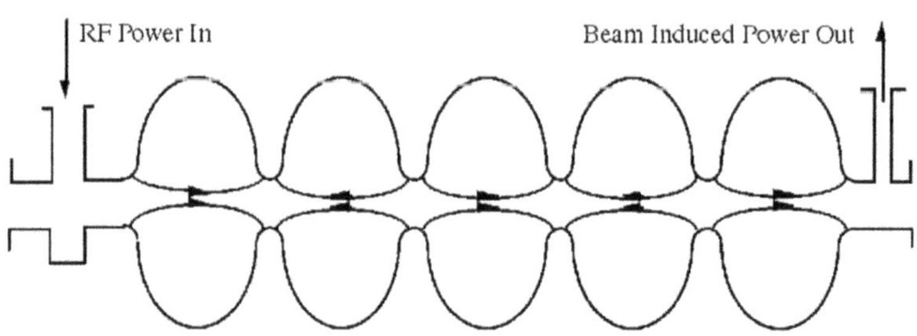

Figure 6.15. A 5-cell high-β accelerating structure. The ports shown are for providing input power to the beam and extracting extract power deposited by the beam in higher-order modes. Arrows indicate the electric field phase on the axis of each cell for the accelerating mode. Reproduced from [59]. Copyright IOP Publishing Ltd. All rights reserved.

It shows how a series of cavities resonating in the TM_{010} fundamental mode may be used to accelerate the beam; as a charged particle traverses the half-wavelength ($\lambda/2$) accelerating gap of a cavity in half an RF period, it continuously experiences an electric field pointing positively in the longitudinal direction. The maximum gradient (E_{acc}) that can be achieved is a key figure of merit for a given accelerating structure and is proportional to the maximum peak electric (E_{pk}) and magnetic (H_{pk}) fields that can be supported on the cavity surface.

Normal-conducting accelerating structures are typically fabricated from copper, operating at around room temperature. Water cooling may be used to extract the heat generated by Ohmic power dissipation in the walls of the cavities, which increases with the square of the accelerating field. For gradients above a few MV/m, especially where the accelerator application calls for continuous wave (CW) or long-pulse fields, normal-conducting structures become acutely uneconomical; the microwave surface resistance of a superconductor is five to six orders of magnitude lower than for copper cavities. Even allowing for the Carnot efficiency of extracting the heat at 2 K [$2/(300 - 2) = 0.0067$] and a typical modern cryo-plant efficiency of 30%, a net gain on the order of 200 [$= 10^5 * 0.0067 * 0.3$] may be achieved using SRF technology.

The current state-of-the-art for large high energy accelerators is bulk niobium SRF cavities, cooled by liquid helium and operating nominally at 4.5 K or 2 K. Examples of current and forthcoming projects employing this type of technology include XFEL, ESS, HL-LHC, LCLS-II, and Diamond Light Source. Niobium is a type-II superconductor with the highest superconducting transition temperature ($T_c = 9.2$ K) and superheating critical field. In the frequency range of interest, the RF fields change suitably quickly that the Meissner state can persist above the lower critical field (H_{c1}) (H_{sh}) of any pure material. Significant efforts are also being devoted to the development of SRF thin film technologies; by supporting higher fields, coated cavities may be able to improve still further on the performance of bulk niobium.

The geometry for a given cavity, as well as the coupler design, is determined by the required beam parameters. The input coupler (also commonly referred to as the fundamental power coupler, FPC) transfers power from the generator to the cavity and thereafter the beam. High order mode (HOM) couplers may be used to damp unwanted beam-induced modes, which can both degrade the quality of the beam and increase power dissipation in the cavity walls. Detailed electromagnetic design of the cavity assembly makes extensive use of a variety of two- and three-dimensional finite element and finite differencing codes.

The manufacturing process for bulk cavities begins with sheet niobium, which is deep-drawn or spun to form half-cells. These are then trimmed and electron-beam welded to assemble complete cavity structures. As surface defects and microscopic contaminants can severely degrade the performance of the cavity, a series of complex and highly optimised chemical etching and ultrapure water high-pressure rinsing (HPR) processes are used to ensure the RF cavity surface is as close to ideal as possible. An example of a completed cavity is shown in figure 6.16.

For accelerator operations, cavities are installed and operated in large cryostats, generally referred to in this context as cryomodules. However, given the sensitivity

Figure 6.16. A 9-cell 1.3 GHz SRF cavity developed for the ILC [59] CERN-PHOTO-8004204. Reprinted with permission from CERN.

of cavity performance to manufacturing defects and contamination, along with the complexity of cryomodule integration, extensive cavity testing is required prior to their installation in an accelerator.

6.3.3 Vertical cavity testing

Following manufacture, the cavity is qualified with low RF power (<200 W) [60] down to 2 K in a specialist vertical test facility (VTF). Examples of VTFs currently in operation include those at CERN [61], DESY [62], Fermilab [63], INFN [64], Jefferson Lab [65], and STFC Daresbury [66].

For large accelerator projects, VTFs may be required to support high cavity testing throughputs, and hence accommodate multiple cavities during a single cool-down run. Typically, the facility is designed around a cavity test insert which supports the cavities in the cryostat; an example of this is shown in figure 6.17. The use of multiple inserts and a common cryostat allows a set of cavities to be tested whilst the next set is prepared on a second insert simultaneously.

Figure 6.17 shows a typical VTF cryostat architecture, sharing many common features with those listed above. Actively cooled thermal shielding at 40–80 K reduces the heat load from the bottom and sides of the vacuum vessel, with a set of passively cooled baffles used to reduce the radiative load from the insert top plate. Cavities may either be tested bare by submerging in a bulk liquid helium bath, or after their helium jackets have been assembled by filling helium into the cavity jackets individually. The cryo-plant supplying liquid helium at 4 K and, in most cases, gaseous helium to cool the shielding is sized to supply the required cryogen flow rates given the size of the VTF. Ultra-high vacuum (UHV) systems are required to reach pressures on the order of 10^{-7} mbar inside the cavities.

Testing at 2 K is achieved by reducing the vapour pressure of the liquid helium to ~30 mbar using a cold compressor or vacuum pump, possibly in combination with a 2 K heat exchanger and JTV to allow continuous operation. Pumps are sized to provide the required cooling power at 2 K. PID control may be used to provide pressure (and hence temperature) stability at the level of ±0.1 mbar (±1 mK) under static loading. Depending on the cooling scheme, a 5 K intercept may be used to

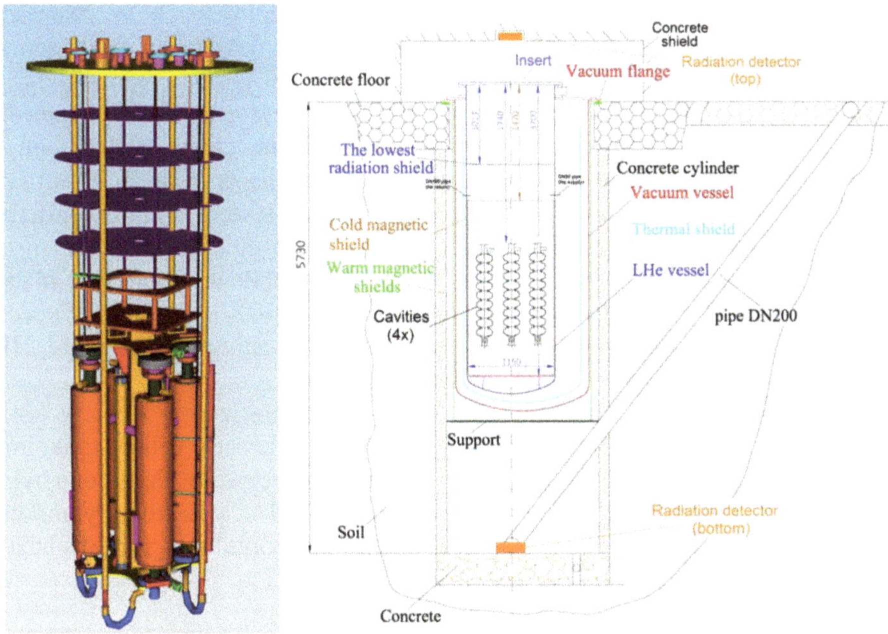

Figure 6.17. 3D model of DESY vertical insert with four jacketed cavities (left) and section view of the insert installed in the test cryostat (right). Reprinted figure with permission from [62], Copyright (2017) by the American Physical Society.

reduce the heat load to the 2 K stage. There are several motivating factors for operating cavities at 2 K rather than 4.5 K. Firstly, the superfluid state of helium below the lambda-point (T_λ = 2.2 K) does not support bulk boiling, thereby minimising microphonics. Secondly, as the RF cavity surface resistance is inversely proportional to its temperature, additional gains in performance may be achieved by operating below 4.5 K. Commonly, bottom fill lines are used for the initial fill (taking advantage of the cold boil-off to more efficiently cool the insert) and top fill lines for subsequent top-ups. Heaters may be employed at the bottom of the helium vessel to expedite boil-off following completion of cavity tests, with warm gas circulation used to speed up the cryostat warming.

As with any cryogenic facility, safety is paramount, and significant efforts are devoted to the understanding of potential failure modes. Commonly, the worst-case scenario is sudden and catastrophic loss of insulating or beam vacuum. In either case, room temperature air impinging on the 2 K stage would be expected to give rise to a heat flux of ~3.8 W cm^{-2} to bare surfaces and ~0.6 W cm^{-2} to those insulated with MLI [67]; safety devices including pressure relief valves and burst discs must be sized accordingly. Oxygen depletion monitors in the VTF areas are also essential safety devices.

To avoid trapping magnetic flux in the superconducting material of the cavities, and thereby degrading their performance, the cavity must be shielded from Earth's and other ambient magnetic fields during cool-down. This may be achieved at the level of <1 μT using some combination of active (i.e., a solenoid), warm passive (typically a high magnetic permeability nickel–iron alloy such as Mu-metal, concentric with the vacuum chamber), and cold passive shielding (typically concentric with cavity jackets).

Given the potential for the production of ionising radiation (specifically, bremsstrahlung x-rays induced by field emission at the cavity surface) during high-power testing, concrete and/or lead shielding of the cryostat is essential to prevent staff exposure, along with a comprehensive system of area radiation monitors and RF power interlocks. Cabling must be routed to allow for instrumentation and cryogenic service access to the cryostat while shielding is in place.

Some proportion of cavities, typically on the order of 30% [62], may be expected to fail to meet the gradient specification during vertical testing. At this stage, HPR processing may be repeated and the cavities retested.

In some research and development programmes, a 'horizontal' activity testing phase is employed. Here, cavities including all ancillary devices (helium tank, input coupler, tuner, and magnetic shielding) may be tested in a cryostat that is of a hybrid design between the VTFs described above and the full accelerator cryomodules described in the following section. Examples of such horizontal test cryostats include those at BESSY [68] and Uppsala University [69].

6.3.4 Cryomodule integration and testing

Once cavity performances have been validated under vertical (and in some cases horizontal) testing, they may then be assembled into cavity strings under ISO4 cleanroom conditions. The strings are then integrated with the complete cryomodules [70], which undergo extensive testing before being installed into the accelerator tunnel. Cryomodule reliability is considered at all stages of design and test to maximise the accelerator availability to experiments. An example of a complete cryomodule is shown in figure 6.18.

The cutaway in figure 6.18 shows the string of four cavities assembled along the beam axis. Each cavity is surrounded by its individual liquid helium tank and cold

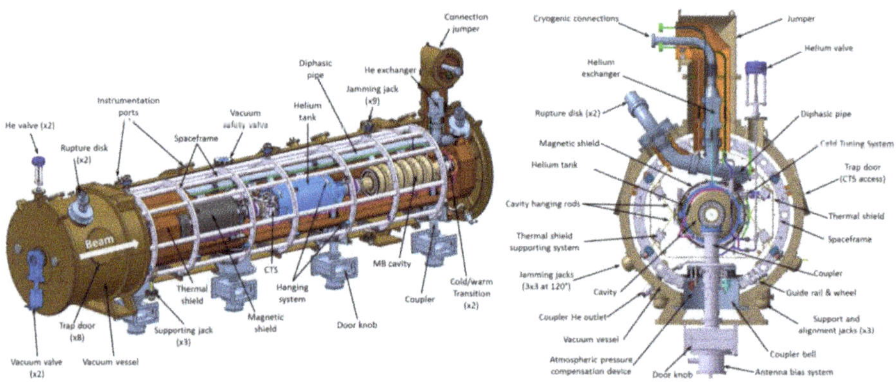

Figure 6.18. Cutaway diagram of the ESS medium-beta cryomodule. Reprinted with permission from [71], CC-BY 3.0..

magnetic shield, and dressed in MLI. Top and bottom fill lines are used to supply liquid helium to the cavity jackets. Heat deposited in this stage evaporates vapour at the surface in the two-phase line above the cavities. Tuners installed on each cavity are used to ensure operation at the correct frequency. Cavities may be detuned by pressure changes and mechanical stresses, requiring ‘slow’ correction by a mechanical tuner, as well as by Lorentz forces which require ‘fast’ correction by a piezoelectric cartridge.

Couplers are installed to provide RF power and HOM pick-up from the cavity and are generally either coaxial or waveguide with ceramic windows to provide environmental isolation for the cavity. A 5 K circuit may also be used for cooling of the beam screens and coupler thermal intercepts. Bellows are used in a variety of locations to allow for movement in tuning, adjustment of couplers, small differences in cryogenic pipe locations and thermal expansion and contraction during cool-down and warm-up. This ensemble of devices is commonly referred to as the cold mass.

The cold mass is surrounded by a thermal radiation shield, usually fabricated from aluminium or copper and cooled to 40–80 K via finger-welded cooling pipes (to avoid excessive stresses). The cold mass and thermal shields are surrounded by a warm magnetic shield concentric with the vacuum chamber. The entire assembly is supported from the vacuum chamber which, as well as serving as the major structural element, also contains the insulating vacuum (typically on the order of 10^{-6} mbar) and provides pressure containment in the event of an internal cryogen line failure. Vacuum chambers, along with shield cooling lines and LHe circuits, are designed and tested in accordance with PED (Europe), ASME codes (US), or equivalent [72].

Structural elements such as posts, tension members, and space frames, are required to support the cold mass under mechanical loading from shipping and handling, cool-down, fluid flow, and ground motion. Commonly used approaches include using a spaceframe [71] or strongback [73], where the cold mass alignment is done outside and translated horizontally into the vacuum vessel; and top loading [74], where the cavity string is directly supported by the power couplers and is lowered vertically into the vacuum vessel via a full-length demountable top lid. Support elements must contribute as low a heat load to the cold stages as possible; accordingly, these are typically fabricated from high-strength, low-conductivity materials including fibre-epoxy composites such as G-10 and G-11, stainless steel, and titanium.

Extensive finite element analysis and experimental efforts are employed to understand the complete transient and steady-state thermal and mechanical behaviour of the cryomodule. Vibrations can significantly affect cavity tuning; the resonant frequency of the cryomodule structure must be designed to be far away from the repetition rate of the accelerator.

Conductive heat loading through each structural support can be as low as 40 mW from 40 K to 2 K. Typical radiative heat fluxes are 1 W m^{-2} for the radiation shield at 40 K dressed with 30 layers of MLI and 100 mW m^{-2} for the cavities at 2 K with 10 layers of MLI. Other contributions to the conductive heat load include through the input coupler, beam line cold-to-warm transition, view ports and relief lines.

These contributions constitute the static load. Dynamic loading is the additional heat dissipated in the cryomodule when the cavities are powered, chiefly in the cavity and input coupler; these may be calculated using coupled electromagnetic and thermal analysis. Typically, the cavity dynamic load dominates the 2 K circuit, and the thermal shield static load dominates at 70 K. If focusing magnets are used in the cryomodule, it is also vital to account for both the static and dynamic heat load from the current leads.

Following the completion of the cryomodule assembly, extensive high power RF testing including pressure validation is carried out in a radiation shielded bunker at 2 K. Examples of test bunker facilities include those at CEA-Saclay [75], CERN [61], DESY [76], ESS [77], Fermilab [78] and Jefferson Lab [79].

Cryogens are supplied and recovered both in the test bunker and accelerator tunnel by a complex distribution system as described in section 6.2. The flows to a given cryomodule are controlled by a cold valve box, which interfaces with the module via a jumper connection. To avoid sub-atmospheric leakage into the helium circuits, it is standard practice to supply supercritical (4.5 K, 3 bar) helium to the module, where it is then precooled (to 2.2 K) in a counterflow heat exchanger and throttled through a JTV (down to ~30 mbar) in order to immerse the cavities in saturated superfluid. Pumping of the 2.2 K vapour is provided by an external cold compressor/vacuum pump. A schematic of such a distribution scheme is shown in figure 6.10. Cold helium gas is also supplied for the shield cooling; liquid nitrogen is avoided where possible due to the potential for the production of solid nitrogen in the case of a leak to a surface at liquid helium temperature. It is vital in determining the required cryo-plant capacity for cold testing and cryomodule operation to include appropriate factors of safety with respect to the total heat loading. Additional interfaces to be made with the cryomodule include input RF power from the amplifiers, RF signals for control, HOM power extraction, electrical feeds for motors and actuators, and instrumentation signals including temperature, vacuum, pressure, and liquid level.

While cooling, fast cool-downs (in the order of >20 K h^{-1}) are required to inhibit the formation of hydrides in the so-called 'Q-disease' regime between 90 and 175 K, which can significantly degrade cavity performance. Once the cryomodule is cold, the heat loading to the thermal shield circuit can be determined by measuring the inlet and outlet temperatures and pressures, calculating the associated increase in enthalpy, and multiplying by the measured mass flow rate. The heat loading to the 2 K circuit can be determined by measuring the mass flow rate at the cold compressor/vacuum pumps, subtracting the vapour generated at J–T expansion, and multiplying by the latent heat of evaporation of helium at 2 K. The cavity position when cold may be reviewed using a wire position monitor or similar. Alignment mechanisms are required, with tolerance requirements on the order of <0.5 mm RMS and <1 mrad.

Extensive low- and high-power RF testing of the couplers and cavities is then required. Following validation, the cryomodule can then be integrated into the accelerator complex. During accelerator operations, test bunkers also are used for repair and maintenance of cryomodules, as well as various research and development activities.

References

[1] Lebrun P 2000 Cryogenics for the Large Hadron Collider *IEEE Trans. Appl. Supercond.* **10** 1

[2] Jia X L *et al* 2004 Cryogenic facility design in BEPC II superconducting upgrade *Proc. 20th Intl. Cryo. Engr. Conf.*

[3] Bozhko Y, Escherich K, Jensch K, Petersen B, Schnautz T and Sellmann D 2017 Commissioning and first cooldown of XFEL linac *IOP Conf. Series: Materials Science and Engineering* **278** 012100

[4] Weisend J G II, Arnold P, Fydrych J, Hees W, Jurns J M and Wang X L 2015 Cryogenics at the European Spallation Source *Phys. Proc.* **67** 27–34

[5] Arenius D, Chronis W, Creel J, Dixon K, Ganni V and Knudsen P 2004 Cryogenic system for the spallation neutron source *Advances in Cryogenic Engineering, Transactions of the Cryogenic Engineering Conf. - CEC (Anchorage, Alaska, 22-26 September 2003) AIP Conf. Proc.* vol 710 (Melville NY: American Institute of Physics)

[6] Arenius D, Creel J, Gannil V, Dixonl K and Knudsenl P 2010 Overview and status of the 12 GeV cryogenic system upgrade at JLab *Transactions of the Cryogenic Engineering Conf. - CEC Advances in Cryogenic Engineering, AIP Conf. Proc.* **1218** 1087–91

[7] Casagrande F *et al* 2020 FRIB cryogenic system status *IOP Conf. Ser.: Mater. Sci. Eng.* **755** 012089

[8] Ravindranath V 2017 Process simulations for the LCLS II cryogenic systems *Ser.: Mater. Sci. Eng.* **278** 012106

[9] Thomason J W G 2019 The ISIS spallation neutron and muon source—the first thirty-three years *Nuclear Inst. Methods Phys. Res.* A **917** 61–7

[10] Vassal A *et al* 2019 Dynamic modeling and control of the SPIRAL2 cryomodules *IOP Conf. Ser.: Mater. Sci. Eng.* **502** 012111

[11] Tatsumoto H *et al* 2010 *Adv. Cryog. Eng.* **55A** 297–304

[12] Chakravarty A *et al* 2017 PIP-II cryogenic system and the evolution of superfluid helium cryogenic plant specifications *IOP Conf. Ser.: Mater. Sci. Eng.* **278** 012092

[13] Yoon S *et al* 2017 Specification of the 2nd cryogenic plant for RAON *IOP Conf. Ser.: Mater. Sci. Eng.* **278** 012103

[14] SHINE Cryogenic System Conceptual Design Report 2019

[15] Kauschke M *et al* 2019 Options for a reliable cryogenic supply for the FAIR facility *IOP Conf. Ser.: Mater. Sci. Eng.* **502** 012128

[16] Chorowski M *et al* 2017 Towards the conceptual design of the cryogenic system of the Future Circular Collider (FCC) *IOP Conf. Ser.: Mater. Sci. Eng.* **278** 012097

[17] Nakai H *et al* 2017 Cryogenic system configuration for the International Linear Collider (ILC) at mountainous site *IOP Conf. Ser.: Mater. Sci. Eng.* **171** 012036

[18] Han R *et al* 2019 Design of 650 MHz 2-cell cavity cryomodule for the CEPC accelerator *IOP Conf. Ser.: Mater. Sci. Eng.* **502** 012085

[19] Ge R *et al* 2019 Concept design of cryogenic system for the high energy photon source at IHEP *Proc. of the 15th IIR Int. Conf.: Prague, Czech Republic (April 8-11, 2019)*

[20] Fauve E *et al* 2017 ITER cryoplant infrastructures *IOP Conf. Ser.: Mater. Sci. Eng.* **171** 012008

[21] Zhou Z W *et al* 2019 Fault modes and criticality analysis of the EAST cryogenic system based on operating experience over 10 years *IOP Conf. Ser.: Mater. Sci. Eng.* **502** 012121

[22] Whyte D G, Minervini J, LaBombard B, Marmar E, Bromberg L and Greenwald M 2016 Smaller and sooner: exploiting high magnetic fields from new superconductors for a more attractive fusion energy development path *J. Fusion Energy* **35** 41-53

[23] He C *et al* 2019 Development of commissioning and operation of CSNS cryogenic system *IOP Conf. Ser.: Mater. Sci. Eng.* **502** 012017
[24] Nagel M *et al* 2019 Concept for the cryo distribution for the Wendelstein 7-X cryo vacuum pumps *IOP Conf. Ser.: Mater. Sci. Eng.* **502** 012109
[25] Zong Z *et al* 2019 Commissioning of the superKEKB final focusing superconducting magnet cryogenic systems *IOP Conf. Series: Materials Science and Engineering* **502** 012119
[26] Panchal P *et al* 2019 Cryogenic process optimization for simultaneous cool down of the TF and PF superconducting coils of SST-1 Tokamak *IOP Conf. Ser.: Mater. Sci. Eng.* **502** 012106
[27] Kim Y S *et al* 2005 Design of cryogenic system for the KSTAR device *Proceedings of the Twentieth International Cryogenic Engineering Conference (ICEC20)* ed L Zhang, L Lin and G Chen (Amsterdam: Elsevier) pp 99–102
[28] Blosser H G 1979 The Michigan State University superconducting cyclotron program *Proc. 8th Intl. Conf. on Cyclotrons and Their Applications* (Piscataway, NJ: IEEE)
[29] Edwards H T 1985 The tevatron energy doubler: a superconducting accelerator *Ann. Rev. Nucl. Part. Sci.* **35** 605–60
[30] Lierl H and Wolff S 1996 Superconducting magnet and cryogenic system of HERA *J. Cryogen. Supercond. Soc. Jpn* **31** 360–72
[31] Claudet G *et al* 1986 Design of the cryogenic system for the Tore Supra tokamak *Cryogenics* **26** 443–9
[32] Rousset B *et al* 2000 He II two phase flow in an inclinable 22 m long line *Adv. Cryo. Engr* **45B** 1009–16
[33] Dhuley R *et al* 2020 Demonstration of cw accelerating gradients on a cryogen-free, cryocooler conduction-cooled SRF cavity *IOP Conf. Series: Mater. Sci. Eng.* **755** 012136
[34] Weisend J G II (ed) 2016 *Cryostat Design* (Berlin: Springer)
[35] Weisend J G II *et al* 2006 Liquid hydrogen target experience at SLAC *Adv. Cryo. Engr* **51B** 1043–52
[36] Covrig S D *et al* 2005 The cryogenic target for the G^0 experiment at Jefferson lab *Nucl. Instrum. Methods Phys. Res. Sect.* A **551** 218–35
[37] Jurns J *et al* 2015 Spallation target cryogenic cooling design challenges at the European Spallation Source *IOP Conf. Ser.: Mater. Sci. Eng.* **101** 012082
[38] Montanari D *et al* 2017 Status of the LBNF cryogenic system *IOP Conf. Ser.: Mater. Sci. Eng.* **278** 012117
[39] Dremel M *et al* 2017 The new build to print design of the ITER Torus cryopump *Fusion Eng. Des.* **88** 760–3
[40] Fydrych J 2016 *Cryogenic Transfer Line, in Cryostat Design, Case Studies, Principles and Engineering International Cryogenics Monograph Series* (Berlin: Springer) pp 241–74
[41] Rusinski E *et al* 2014 Selected aspects related to the calculations and design of a cryogenic transfer line *Arch. Civil Mech. Eng.* **14** 231–41
[42] Fydrych J *et al* 2010 Design methodology of long complex helium cryogenic transfer lines *Adv. Cryogen. Eng. Adv. Cryogen. Eng. Trans. CEC* **55** 1103–10
[43] Kalinin V *et al* 2006 ITER cryogenic system *Fusion Eng. Design* **81** 2589–95
[44] Serio L *et al* 2010 Challenges for cryogenics at ITER *Adv. Cryogenic Eng. Trans. CEC* **55** 651–62
[45] Badgujar S *et al* 2014 Progress and present status of ITER cryoline system *Adv. Cryogenic Eng. Trans. CEC, AIP Conf. Proc.* **1573** 848–55

[46] Fydrych J and Pietrowicz S 2019 A comparative thermodynamic analysis of helium distribution systems with central and distributed precooling heat exchangers *IOP Conf. Ser.: Mater. Sci. Eng.* **502** 848–55
[47] Fydrych J *et al* 2015 Cryogenic distribution system for the ESS superconducting proton linac *Phys. Proc.* **67** 828–33
[48] Lebedev V 2017 The PIP-II conceptual design report 2017 section 3.4.3 The cryogenic distribution system
[49] Lierl H *et al* 2004 Conceptual layout of the European X-FEL linear accelerator cryogenic supply *Proc. of LINAC 2004, Lübeck, Germany* 225–7
[50] Dalesandro A *et al* 2017 Thermodynamic analyses of the LCLS-II cryogenic distribution system *IEEE Trans. Appl. Supercond.* **27** 0500804
[51] Wang X *et al* 2015 Specification of the ESS accelerator cryoplant *Phys. Proc.* **67** 89–94
[52] Lindroos M *et al* 2013 The European spallation source *Proc. of PAC2013* 1448–52
[53] Lebrun P and Tavian L 2014 Cooling with superfluid helium, proceedings of the CAS-CERN accelerator school: superconductivity for accelerators ed R Bailey *CERN, Geneva* 453–76
[54] Lebrun P 2000 Cryogenics for Large Hadron Collider *IEEE Trans. Appl. Supercond.* **10** 1500–6
[55] Erdt W *et al* 2000 The cryogenic distribution line for the LHC: functional specification and conceptual design *Adv. Cryogenic Eng. Trans. CEC* **45** 1387–94
[56] Padamsee H, Knobloch J and Hays T 2008 *RF Superconductivity for Accelerators* vol 2011 (New York: Wiley-VCH))
[57] Padamsee H 2017 50 years of success for SRF accelerators—a review *Supercond. Sci. Technol.* **30** 053003
[58] Padamsee H 2001 The science and technology of superconducting cavities for accelerators *Supercond. Sci. Technol.* **14** R28
[59] Yamamoto A 2009 Superconducting RF cavity development for the International Linear Collider *IEEE Trans. Appl. Supercond.* **19** 1387–93
[60] Powers T 2006 Theory and practice of cavity RF test systems. No. JLAB-ACE-05-475; DOE/ER/40150-4011 *Thomas Jefferson National Accelerator Facility, Newport News, VA* https://osti.gov/biblio/890534
[61] Chambrillon J, Therasse M, Brunner O, Maesen P, Pirotte O, Vullierme B and Weingarten W 2011 CERN SRF assembling and test facilities *Proc. SRF11, Chicago* https://accelconf.web.cern.ch/accelconf/SRF2011/papers/tupo058.pdf
[62] Reschke D, Gubarev V, Schaffran J, Steder L, Walker N, Wenskat M and Monaco L 2017 Performance in the vertical test of the 832 nine-cell 1.3 GHz cavities for the European x-ray free electron laser *Phys. Rev. Accel. Beams* **20** 042004
[63] Ozelis J P *et al* 2007 Design and commissioning of Fermilab's vertical test stand for ILC SRF cavities *2007 IEEE Particle Accelerator Conference (PAC)* (Piscataway, NJ: IEEE) pp 2283–85 https://osti.gov/servlets/purl/915547
[64] Michelato P *et al* 2015 *INFN Milano-LASA Activities for ESS* http://inspirehep.net/record/1482174/files/thpb010.pdf
[65] Geng R L 2011 Overview of ILC high gradient cavity R&D at Jefferson Lab *Proc. of SRF11, (Chicago, IL)* 74 https://accelconf.web.cern.ch/SRF2011/papers/mopo012.pdf
[66] Pattalwar S *et al* 2019 Cryogenic performance of the vertical cryostat for qualifying ess srf high-beta cavities *19th Int. Conf. on RF Superconductivity (SRF'19) (Dresden, Germany, 30*

June–5 July 2019) (Geneva: JACOW Publishing) http://www.jacow.org/srf2019/papers/thp027.pdf

[67] Bizel-Bizellot L *et al* 2019 Cryogenic safety considerations for vertical test facility for qualifying high-β SRF cavities for the European Spallation Source *IOP Conf. Ser.: Mater. Sci. Eng.* **502** 012074

[68] Knobloch J, Anders W, Borninkhof J, Jung S, Martin M, Neumann A, Pflückhahn D and Schuster M 2004 Status of the HoBiCaT superconducting cavity test facility at BESSY *EPAC04, Lucerne, Switzerland* 970 http://accelconf.web.cern.ch/AccelConf/e04/PAPERS/TUPKF008.PDF

[69] Kern R S, Ekelöf T, Gajewski K, Hermansson L, Ruber R J M Y, Ziemann V, Bujard P, Chevalier N R, Junquera T and Thermeau J P 2014 The HNOSS horizontal cryostat and helium liquefaction plant at FREIA *Proc. of the 5th Int. Particle Accelerator Conf. (Dresden)* https://doi.org/10.18429/JACoW-IPAC2014-WEPRI110

[70] Pagani C and Pierini P 2004 Cryomodule design, assembly and alignment *SUP'04* http://accelconf.web.cern.ch/accelconf/SRF2005/papers/sup04.pdf

[71] Darve C *et al* 2017 ESS superconducting RF collaboration *No. IPAC-2017-MOPVA090* http://wwwsrf.mi.infn.it/publications/papers/2017/files/mopva090.pdf

[72] Peterson T J *et al* 2012 A survey of pressure vessel code compliance methods for superconducting radio frequency cryomodules *AIP Conf. Proc.* **1434** 1575–82

[73] Roger V and Saravan Kumar Chandrasekaran D P 2019 Design strategy of the PIP-II cryomodules *19th Int. Conf. on RF Superconductivity (SRF'19) (Dresden, Germany, 30 June-5 July 2019)* (Geneva: JACOW Publishing) : https://accelconf.web.cern.ch/srf2019/papers/mop094.pdf

[74] Parma V *et al* 2014 Status of the superconducting proton linac (SPL) cryo-module *No. CERN-ACC-2014-0341* http://ipnwww.in2p3.fr/srf2013/papers/mop085.pdf

[75] Madec C *et al* 2013 The challenges to assemble 100 cryomodules for E-XFEL *SRF2013, Paris* **810** http://ipnwww.in2p3.fr/srf2013/papers/thioa02.pdf

[76] Anashin V *et al* 2015 Experience with cryogenic operation of accelerator module test facility during testing of one third of XFEL cryomodules *IOP Conf. Ser. Mater. Sci. Eng.* **101** 012139

[77] Hees W *et al* 2015 The ESS cryomodule test stand *Phys. Proc.* **67** 791–95

[78] Harms E *et al* 2017 Commissioning and first results from the Fermilab Cryomodule Test Stand. *28th Linear Accelerator Conf. (LINAC'16) (East Lansing, MI, 25–30 September 2016)* (Geneva: JACOW Publishing) https://arxiv.org/pdf/1805.02725

[79] Powers T *et al* 2003 Upgrade to cryomodule test facility at Jefferson Lab. No. JLAB-ACE-03–188; DOE/ER/40150-2552. Thomas Jefferson National Accelerator Facility, Newport News, VA https://osti.gov/servlets/purl/816434

[80] Brüning O S, Collier P, Lebrun P, Myers S, Ranko O, Poole J and Proudlock P 2004 *LHC Design Report V1: The LHC Main Ring* (Geneva: CERN)

IOP Publishing

Cryogenics
Fundamentals, foundations and applications
Tom Bradshaw, Beth Evans and John Vandore

Chapter 7

Propulsion, energy storage and renewables

John Vandore, Peter Malkin, Tom Bradshaw, James Fesmire, Toby Peters, Gareth Brett and Stuart Bradley

7.1 Introduction

John Vandore

The role of Cryogenics in the world of energy plays its part directly in the liquefaction of fuels such as Liquefied Natural Gas (LNG) to facilitate their storage and transport in systems which have been well established for decades. Space propulsion has also been supported with liquid cryogenics like hydrogen and oxygen. More recently, technologies have been developed to exploit the properties of liquid air (or liquid nitrogen) in propulsion and energy storage with examples from the Dearman engine and Highview Power Storage. A further, new and exciting strand of association with energy derives from superconducting electric power applications; in electric aircraft and in wind turbines.

This chapter brings together an exceptional group of experts and expert insights, ranging across different technology readiness levels (TRLs). We start and finish with superconducting electric power, in electric aircraft and in superconducting wind turbines by Peter Malkin and Stuart Bradley. Tom Bradshaw delivers a section on LNG. James Fesmire from NASA delivers a very upbeat piece on hydrogen and the prospect for cryogenics. Toby Peters tells the story of the Dearman engine, which leads to liquid air energy storage, addressed by Gareth Brett. At the time of compiling this edition, during the COVID-19 pandemic, Toby has been particularly active concerning the cold chain for vaccine distribution, to ensure it is delivered in the most efficient and sustainable way, providing a lasting COVID legacy which could pivot to other cold chain needs such as in food, so we have included this section too.

doi:10.1088/978-0-7503-2757-2ch7

7.2 Electric aircraft and electric propulsion

Peter Malkin

7.2.1 Introduction

It may seem perhaps curious to identify electric aircraft as a possible application of superconductivity and associated cryogenics [1, 2]. This may seem even more unlikely given the lack of use of this type of technology for ground-based power systems and given the additional difficulties of having cryogenics on aircraft. However, if we look at the limited applications of superconductivity to date it is clear that the only major application area is medical imaging. The key driver for this has been that only superconductors can enable this technology and it has only recently become obvious that this is also true of electrical power for aircraft. This section will try and explain why this is the case and what technical options are being put in place to achieve this. The difficulty here is that this work is highly integrated and requires understanding of several very different technical areas. This section will try and cover the basic principles of these very different areas which cover technologies as varied as power systems to aerodynamics etc. Hence time and space do not allow a detailed analysis so we will just attempt to cover the relevant basics in these areas. If the reader wishes more detail please refer to the bibliography.

The aviation industry has relied on gradual improvements to aircraft design to improve efficiencies over the last 50 years of passenger flight. One of the key improvements has been a growth in the 'by-pass ratio' of the gas turbine engines. This means that the engine core gets smaller and the by-pass fan becomes bigger. Other improvements have been, for example, to move to carbon composite airframes to produce lighter aircraft. However, these improvements are now approaching their limits where the latest generation of engines are not able to offset the increased drag and weight of the larger and larger engine units. This may have been acceptable but the other major change in the last twenty years or so has been the growth in demand for air travel. Prior to the COVID-19 situation this growth had continued at over 5% per year and this is forecast to continue. See figure 7.1.

Aircraft powered by jet fuel of course emit significant quantities of CO_2, although the industry argues that this is less than 5% of total man-made global emissions, this takes no account of the effect of this going directly into the upper atmosphere nor indeed the other emissions consisting of NOx and, of course, noise. Most importantly many people are now choosing not to fly where possible in order to help reduce global emissions and this effect is growing even without taking into account the COVID effects. It has been apparent for some time that pressure would grow on aircraft emissions and this is why the civil aerospace industry is looking towards electric flight as a possible solution.

7.2.2 Electric propulsion issues

The industry has investigated both battery—electric and hybrid electric propulsion systems.

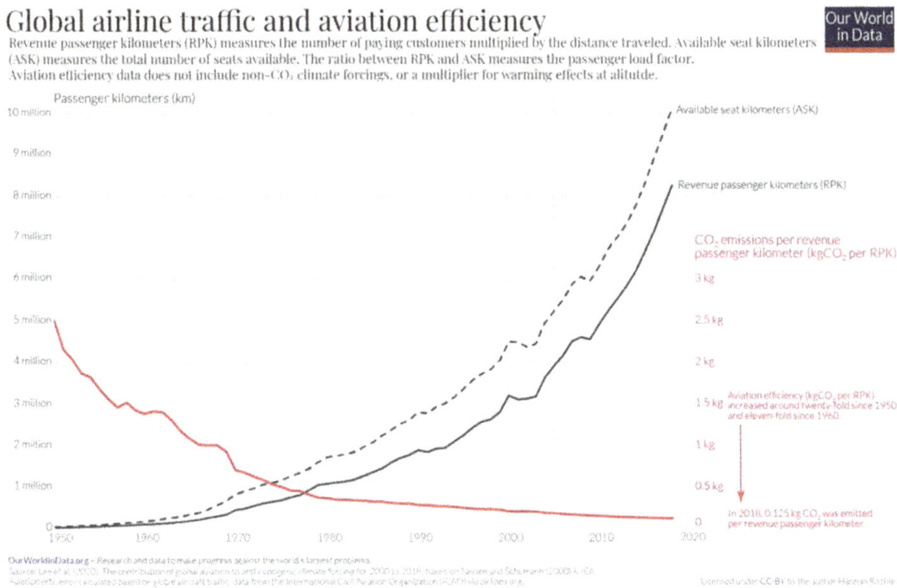

Figure 7.1. CO_2 emissions—impact of growth in air travel. Reprinted with permission from [17], CC BY 4.0.).

Pure electric propulsion

It may seem obvious that a possible solution is to use an electric battery supply propelling an electric motor to power an aircraft. In fact, that is what has already occurred in the field of model aircraft giving quiet efficient propulsion systems that offer zero emissions. Indeed, one of my ex-colleagues started me thinking about this many years ago with a question that if one can make model electric planes why not larger passenger carrying ones? There are two key explanations, the first and obvious one is the battery energy density but the second is key to this whole area and that is that conventional electric power systems do not scale linearly with power density.

Clearly the key difficulty in the use of electric systems at these multi megawatt power levels required for aircraft propulsion are those of power density and efficiency. Whilst these have been mainly associated with electric motors, the issues apply to the whole electric system.

Despite this there is now considerable interest in small vertical take-off electric aircraft with over 200 companies worldwide developing these currently. The key reason is of course the substantial improvement brought by the use of lithium ion batteries. This has been stimulated by the market driver of solving say traffic congestion around airports rather than improving emissions. However, to date these companies have not yet solved the second problem and that is the electrical systems (and machines) power densities (figure 7.2).

Even with these relatively small aircraft, problems have arisen with both of the issues outlined above. Small aircraft with power levels around 1 MW or less, and hence limits to both battery problems and electrical system power density, may however be workable.

Figure 7.2. Some examples of recent designs of electric VTOL aircraft (credit: Chesky/Shutterstock).

It is clear that scaling up to larger aircraft would be problematic with this approach.

Hybrid electric

An obvious solution for propulsion could be the use of a hybrid electric power system (figure 7.2 for example) where we retain a 'prime mover' (almost certainly a gas turbine) but use this fully or partially drive an alternator and then power an electrical system [3]. This is clearly possible but as we retain a gas turbine where is the advantage?

As has been shown in marine applications this approach can give much better flexibility over the operating cycle and crucially for aircraft can improve 'aerodynamic' effects. This is because of the fact that electrical machines scale with torque and not power. This implies that it is possible to run a larger number of small machines at higher rotational speeds. This in turn, creates the possibility of a distributed propulsion system where we could place these electrical propulsors say along the trailing edge of the wings. This firstly removes the weight and drag effects of the increasingly large fans of the current high-by-pass ratio gas turbines. (see figure 7.3). In addition, the distributed propulsors can now be located where they have a positive aerodynamic effect by accelerating the boundary layer flow over the wing. This is known as boundary layer ingestion (BLI), which is known to give benefits in cruise by effectively reducing drag. Other arrangements are also possible, and this gives much more 'flexibility' in the aircraft design [4].

Hydrogen fuels

Recently, there has been renewed interest in the possibility of replacing jet fuel (kerosene) with hydrogen. This would immediately create zero-carbon flight and much work is currently ongoing to investigate the production of hydrogen from renewable or carbon-free sources.

However, it might seem that it is easy to replace the existing fuel for hydrogen without the need to change the aircraft design or propulsion system.

The first issue though is that hydrogen is light but occupies a large volume. So much so that compressed gas tanks would still be far too large to go on a aircraft.

Figure 7.3. A typical by-pass gas turbine engine (credit: Dysanovic, CC BY 2.0, https://www.flickr.com/photos/dysanovic/2678713012/).

Therefore, to come close to achieving the volume densities required it is evident that the hydrogen would have to be stored on board as liquid hydrogen (LH2) which even so would occupy around three times the volume of kerosene tanks (albeit less than one third of the weight) which requires significantly modified aircraft designs. Work is currently underway to design hydrogen fuelled aircraft using either modified gas turbines or fuel cells.

Even so, burning hydrogen in gas turbines would create an additional emission problem, that of contrails. These are the water vapour trails that aircraft leave at high altitudes which are now thought to be a problem for global warming. Hydrogen is likely to increase these very significantly.

Fuel cells

Another option for propulsion using hydrogen is to utilise fuel cells to generate electricity. This could then power the aircraft via an electrical propulsion system. These fuel cells can use either solid oxide or proton exchange membrane (PEM) techniques, however, the latter are currently undergoing rapid development to increase the power density. Recent reports suggest significant improvements in power density of PEMs although it is difficult to obtain figures as this has become extremely competitive recently.

For those who think this approach is too complex to be practical I would remind the reader that it is possible to purchase cars using this technology today. For example, the Hyundai Nexo is available with a fuel cell powering a 120 kW motor. Note that it may also be possible to use both techniques together in a hydrogen powered aircraft. However, in all cases an electrical power system will be required (figure 7.4).

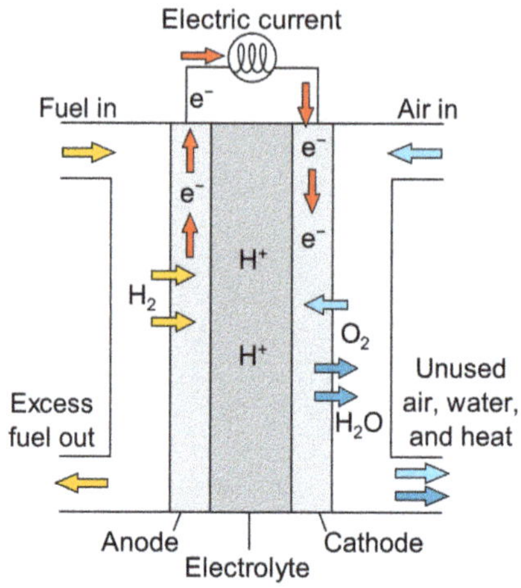

Figure 7.4. A proton exchange membrane fuel cell. This Proton Exchange Fuel Cell Diagram image has been obtained by the author from the Wikimedia website, where it is stated to have been released into the public domain. It is included within this article on that basis.

Simple gas turbine propulsion using hydrogen

It is of course possible to adapt a gas turbine engine to burn hydrogen rather than jet fuel. We have argued that some modifications to the aircraft would still be necessary in the previous section, but nevertheless this is by far the simplest approach. However, in the view of the author, whilst this approach may happen for an intervening period this will not last very long.

There are several reasons for this

- Aircraft drag. The volume of the tanks will cause increased drag on a tube and wing design. Hydrogen is more costly than jet fuel resulting in a poor economic return.
- Increased contrail effects and NOx emissions. Burning H_2 produces no CO_2 but can produce increased NOx and more water vapour.
- Customer impact. The passengers may not be confident in an aircraft that looks very similar to existing aircraft but is genuinely running on LH2 with reduced emissions.
- Fuel heating. Considerable energy would have to be expended in pre-heating the fuel to be fed into the engines. This is a wasteful process as the energy of compression cannot be recovered.

7.2.3 Electrical power systems

Power density

I stated earlier that one of the essential problems for electrical aircraft was the fact that scaling up the power system was not straightforward. This is because the system must become more complex at higher powers. This may seem odd to the reader given the wide use of these systems at high (multi-megawatt) power levels. It is of course possible to design high power electrical systems for propulsion functions, as shown in figure 7.5 with an 80 MW electrical system (or part of it).

Figure 7.5. Image of part of the 80 MW electrical propulsion system on QM2 consisting of 4 × 21.5 MW Rolls-Royce Mermaid Engines (credit: Image courtesy of Kongsberg Maritime).

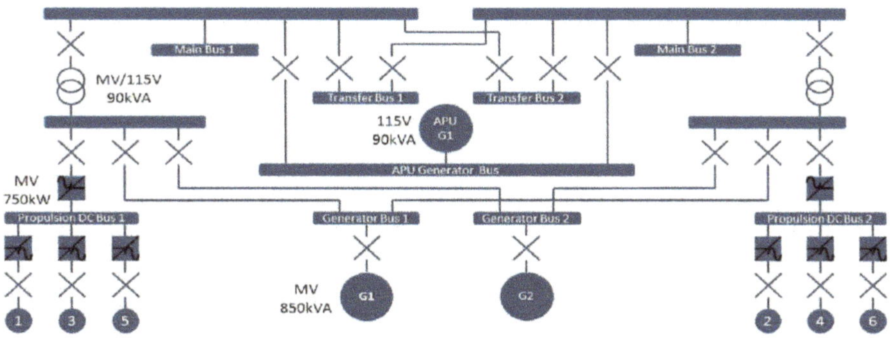

Figure 7.6. A power circuit layout for a small/medium aircraft system—the crosses represent circuit breakers. Reprinted courtesy of Pete Malkin.

It is very clear from this image that power density is a problem. Note however that marine applications are far less concerned about weight and volume than those for aerospace and the motor size is defined by the torque (and not power) requirement. In addition as these motors are direct drive, the rotational speeds are very much constrained by cavitation effects. These speed limitations create the need for very heavy and high torque machines.

However, the problem goes far beyond motors and drive densities as at these power levels complex switching and protection systems are required to prevent damaging fault arcs. This creates a much more complex system. A typical system diagram for aerospace is shown in figure 7.6 where the complexity is evident [5].

This leads not only to power density problems but also each device adds to losses which pose both cooling and efficiency issues.

Voltage withstand

There is also a further issue here which is associated with scaling these systems. As electrical systems are scaled up to handle more power this implies of course increasing the current levels. However, cable-sizes increase with the square of

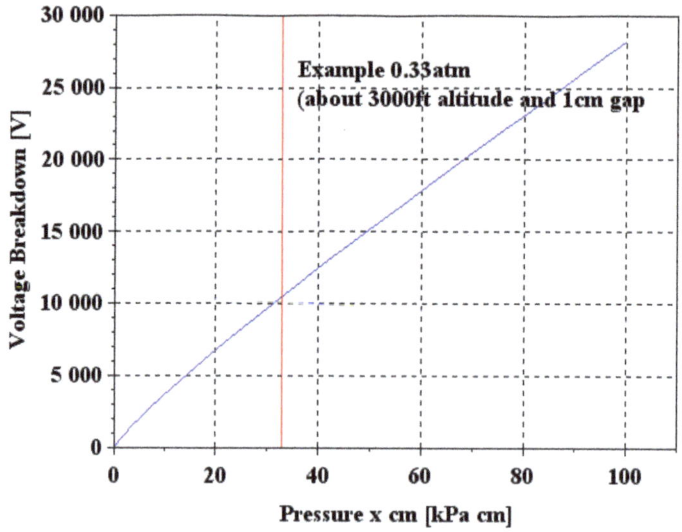

Figure 7.7. Paschen's law of voltage breakdown in air.

current and hence grow to large and heavy cables very quickly. The traditional solution is to increase the operating voltage. This of course allows for power increases without the large and expensive cables. It is indeed because of this approach that is used by most electricity transmission grids which operate at very high voltages. For example the UK grid transfers electricity at 400 kV and the Marine applications would operate at between 12 kV and 24 kV typically.

Presumably therefore we could use this approach for aircraft applications. However there is a fundamental problem here. This is that the insulating properties of the surrounding air that allow us to increase the voltage in power networks depend on pressure. At typical cruise altitudes the air pressure has dropped by almost seventy percent compared with sea level. Voltage breakdown depends on local electric field but in a reasonably uniform field it depends on the applied voltage.

The dependence is defined by Paschen's Law, see figure 7.7.

This means unless we take very extensive precautions with electric fields we are limited in the voltage we can use without seeing breakdown or pre-breakdown discharges such as corona.

In practice, for large systems we would like to use voltages between 12 kV and 36 kV but the best possible option would give around only one tenth of this voltage to a few KV, which means large power systems cables would be very heavy and not practical.

Fault management

There is also a further problem related to safety at these high powers which is known as 'fault management'.

If faults arise in a high-power electrical network, they can give rise to large short-circuit currents. These fault currents can be very large and can degenerate into fault arcs. These are very dangerous and destructive phenomena and must be averted at

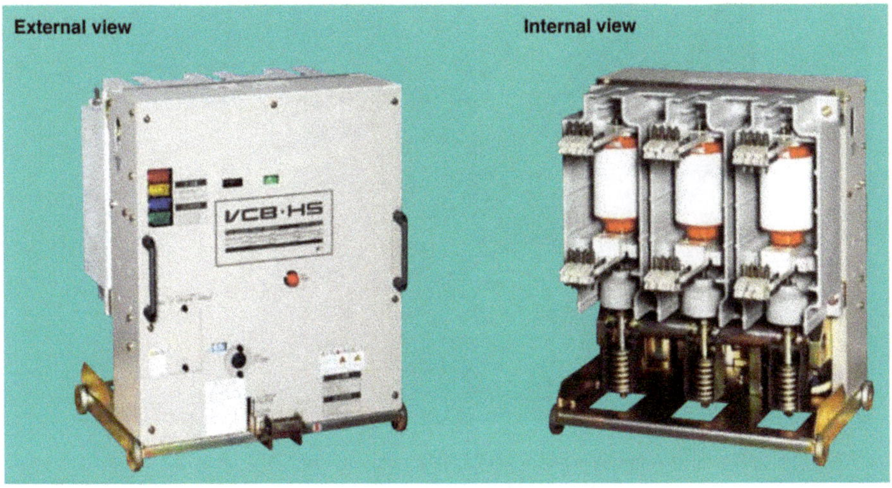

Figure 7.8. A typical circuit-breaker module using vacuum interrupters (credit: kind permission of Fuji Electric).

all costs for a transport application. These are normally controlled by a combination of circuit breakers (figure 7.8) with a control system specifically designed to detect, isolate and remove faults of this type, known as protection systems [6]. This is a fairly specialised area but suffice to say that these circuit breakers are limited in performance (maximum is around 12 KV/40 KA AC) and are generally large and heavy devices. As shown in figure 7.6, a large number of these are required, in order to protect every circuit zone and ensure that fault arcs do not occur. These add to cost complexity and weight of power networks [7].

7.2.4 Cryogenic superconducting electric power systems

Clearly, if LH2 is adopted this would be an ideal match to cryogenic and superconducting electrical power systems. As explained earlier, in order to have maximum volumetric and gravimetric density this would have to be in the form of LH2.

Given that LH2 has a temperature of approximately 25 K this is an ideal temperature to create a superconducting power network.

This section is not designed to give the reader a background in high temperature superconducting materials as this could take up considerable space and time.

Cables

It is well known that these materials can result in a significant increase in current density with modern materials showing between one and two orders of magnitude improvement in current densities compared with say copper at room temperature. This is illustrated in figure 7.9.

It is evident from this diagram that HTS cables and wires in our HTS power system will have very significant improvements in power density and indeed, given that the DC resistance of these materials is zero, a significant reduction in losses, thereby increasing our system efficiency.

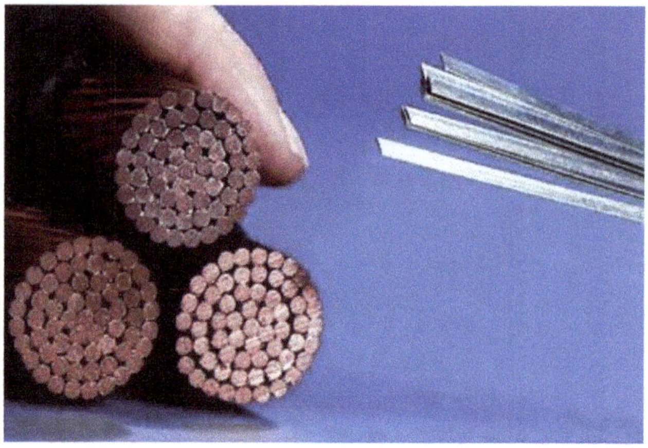

Figure 7.9. High temperature superconductor (HTS) wires can conduct 150 times the electricity of conventional copper conductors (credit: photos provided courtesy of AMSC).

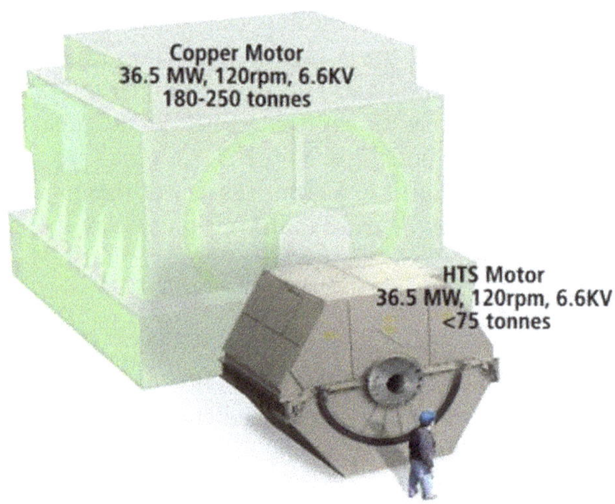

Figure 7.10. A motor for marine requirements at 120 rpm in conventional and HTS format (credit: photos provided courtesy of AMSC).

Electrical machines

It is also well known that this increased current density can improve the performance of electrical machines. Fundamentally, torque generation from electrical machines is due to the vector product of $\mathbf{J} \times \mathbf{B}$ where J represents the current density in the coil and B the magnetic field in the air gap. The superior current density of HTS materials will clearly increase both these parameters which should create very large improvements in both power density and efficiency of electrical machines.

Figure 7.10 indicates a factor of around 60% weight reduction, however, this is for an early form of HTS motor and it is widely believed that this is a very conservative figure. Again, with machines, quoting power densities can be very misleading as these scale with torque and not power. However, assuming a much higher rotational speed for aerospace and using more recent machine topologies, some authors claim improvements in power densities of between ten and twenty times compared with modern conventional machines.

There are several practical issues that are being addressed currently in the design of these machines. The first problem is that whilst HTS materials have zero DC losses, AC losses can still exist. As machines are essentially AC devices these can prove difficult. Also by definition we are dealing with HTS materials in very high magnetic fields which can push these materials towards a quench situation and/or induce high AC losses. To date most HTS machines have a superconducting stator and are synchronous type machines (as in figure 7.10) which limits their performance. However, many workers report on progress towards either superconducting stator or indeed fully superconducting machines which, in theory should give further improvements in performance.

A second issue is that the performance of many materials at these low temperatures is not as well understood as one might expect. One example is that of iron, which is used to focus and concentrate the magnetic field in 'conventional' machines. However, at LH2 temperatures iron loses its magnetic properties to an extent that it can no longer be used for this purpose and other design solutions must be adopted.

Hence it is obvious that HTS power networks can result in both major gains in power densities and efficiencies in both cables and electrical machines but in fact very significant gains are possible with this approach

System voltages

In section 7.2.3 the problems of scaling electrical power networks were discussed. With conventional networks we were obliged to increase voltages for larger powers to avoid very large and expensive cabling and bus-bars. Given the changes in current densities outlined with HTS wires this is clearly no longer an issue. However, this does mean that we have to deal with very high current HTS cables which at multi-megawatt levels would be carrying tens of kA of current. This will create significant magnetic fields and therefore electromagnetic forces on adjacent conductors which means these would have to be mechanically supported in some way [8].

Against this we can now operate a high power system at low voltages which makes life very much simpler in many ways, not the least of which is of course we are no longer concerned by voltage breakdown effects.

Fault management—HTS fault current limiters

However the high normal currents outlined in section 7.2.3 do pose another problem and that is fault or short-circuit currents. Most circuit breakers can only interrupt around 40 kA and short-circuit currents in these networks could easily reach much higher values.

Fortunately, HTS also provides a solution here by using HTS fault current limiters or FCLs. These devices consist simply of a coil of superconducting wire or tape designed so that if a certain value of current is exceeded it will undergo a ‘quench’ process. This quench is the natural transition back for a superconductor to a normal conductor if one of the critical values (current in this case) is exceeded (see, for example [9]). In many superconducting applications quenches can be a major issue (in steady-state magnets for example) to be avoided at all costs, but here we wish to utilise this natural process.

The quench can happen in a very short time (typically in microseconds) and causes a very steep increase in resistance. If this is designed to coincide with a fault current the current growth will be damped down by the increase in resistance.

This may seem a simple process, but it has very significant implications in that:

1. As this depends only on the property of the material there are no control circuits necessary.
2. It is essentially fail-safe.
3. Even more significantly, the process is very fast, and this means that positioned correctly an FCL can ensure that the network will never see a full fault current and unlike a normal circuit-breaker network does not have to be designed to cope with these currents. This can result in wide-scale weight reduction and simplification of the power network!

These devices have been built and tested at full scale in ground-based distribution systems and have been shown to operate successfully. See reference [9].

However, for these to be used in aircraft designs they need to be adapted to aerospace requirements of weight and volume and this work is currently in progress. An outline ground-based design is shown in figure 7.11. Figure 7.12 also shows fault current limiter operation.

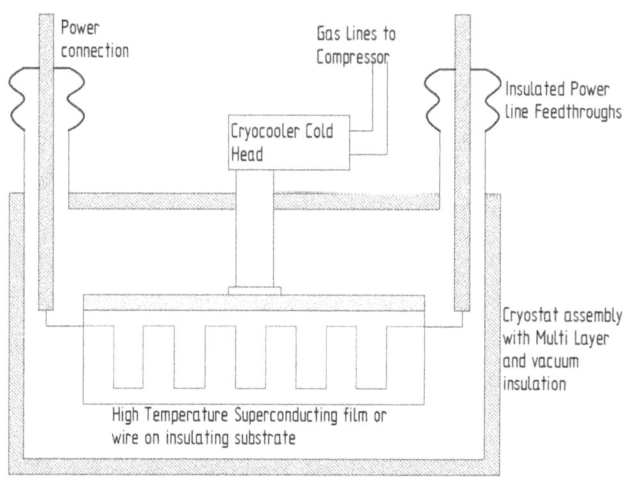

Figure 7.11. Basic layout of a superconducting fault limiter.

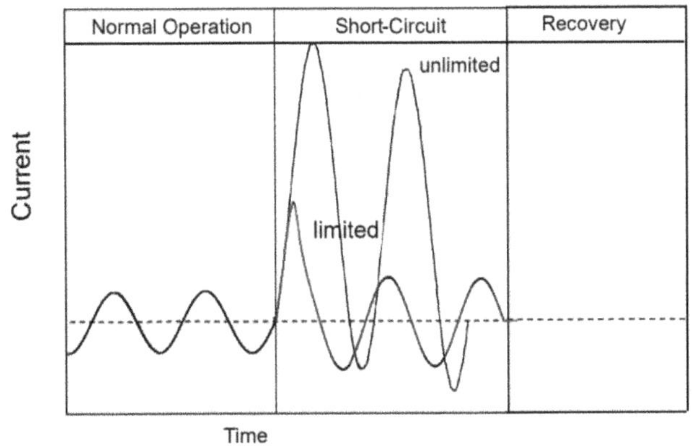

Figure 7.12. Illustration of fault current limiter operation (credit: by kind permission of Mathius Noe, KIT).

Solid state circuit breakers (SSBs)

I am often asked why we cannot use semiconductor devices for switching and interruption in power networks. After all, they can switch off large currents and are widely used for this in other types of applications.

This is because of the extreme requirements placed on a circuit-breaker device. Essentially this must create the absolute minimum loss in normal operation and may not see a fault in many years and then operate, clear, or limit the fault and withstand the fault energy. This means that the device must be more reliable than the probability of fault occurrence, have zero losses in normal operation and then survive a fault current and withstand any recovery voltage after the fault. As these SSBs require many elements to operate in high current and voltage networks solid state switching devices have many components and in these arduous conditions are not sufficiently reliable enough yet to carry out this function. Even more seriously, as they rely on semiconductors, these are neither perfect conductors nor insulators and hence will create significant losses in normal conditions and may not withstand fault conditions. Various approaches have been proposed to overcome these shortcomings but in all cases these still have significant issues, not least in weight, complexity and volume constraints.

7.2.5 Superconducting propulsion systems

It is clear from the previous arguments that using a superconducting power system not only resolves the problem of power density but also resolves the other network issues of voltage withstand and fault management.

In terms of efficiency in fact the situation is even better. If we are now using LH2 or similar fuel we can actually use the losses from the electrical system to actually heat and evaporate the fuel to pass into either the fuel cells or gas turbine. Hence our losses actually become useful, as illustrated in figure 7.13.

This means that we can also scale the system relatively easily creating the possibility of large, long-range aircraft using this principle.

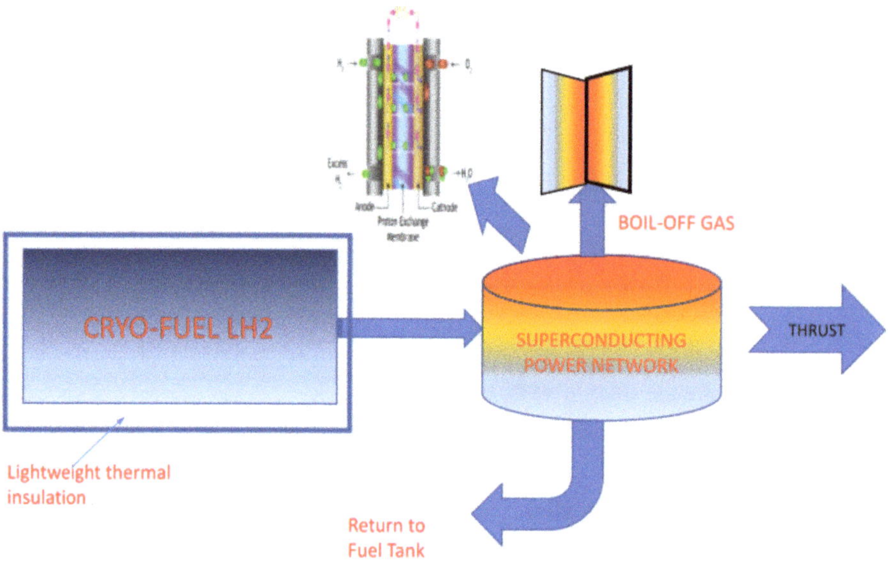

Figure 7.13. A superconducting aircraft propulsion system assuming LH2 fuel.

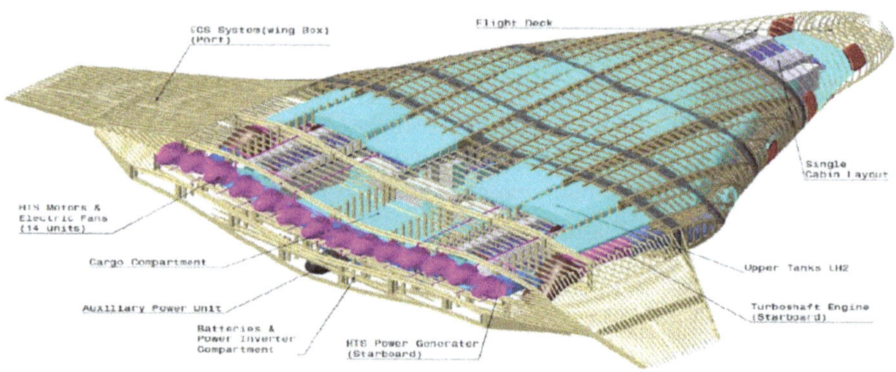

Figure 7.14. An example of a blended wing aircraft design with superconducting distributed propulsion system (courtesy of Cranfield University).

The use of such an electrical system will allow the aircraft designer to use a distributed system with relatively low drag to give maximum design flexibility; see figure 7.14 for example.

It may also be possible to use this power system in new ways to reduce drag—research is currently ongoing to investigate these new possibilities.

Challenges of this approach

- We have focussed on the benefits of this technology, but we must also point out some of the technical issues associated as well.
- The first is that of complexity of design and system integration.

- Secondly, to date no one has yet built and tested a superconducting power system of this type of size.
- Also, whilst we have removed the issue of voltage, the use of very high currents poses problems of high magnetic fields and forces.
- Safety issues of the cryogenic system, such as H_2 leaks, need to be addressed.
- The costs of such a complex system will be high both for design and operation.

Currently none of these issues seem to be a major barrier to this technology but it is clear that a significant amount of R&D work will be required.

7.2.6 Cryogenic systems and fuel tanks

The first response of many cryogenic engineers to this type of propulsion system requirement is to say that as we are dealing with a cryogenic fluid, we will inevitably need double vacuum insulated metal tanks which will be complex and heavy for this type of application. This not the case when we reflect on the nature of this application. As shown in figure 7.13, we are using the LH_2 as fuel and essentially need to evaporate this into gaseous form to feed into either the gas turbines or fuel cells. Indeed, the cooling function is a by-product of this approach and as mentioned in section 7.2.5 we are effectively using the electrical losses to aid this evaporation process. In the vast majority of applications we wish to keep the cryogenic fluid in this state for as long as possible, whereas here, beyond a few hours, the opposite is true—we actually need the LH2 to evaporate, albeit in a controlled way. Hence the tanks' functions are to contain the LH2 for a few hours and prevent effects such as icing being a problem.

Safety and LH2 handling processes

Many people still believe that hydrogen is too dangerous to use on commercial aircraft. Indeed at most conferences this is often the starting question and indeed the Hindenburg Airship disaster is often cited. For younger readers this was a hydrogen-filled airship that caught fire in 1937 and all on board were killed.

Hydrogen is of course flammable, as is jet fuel currently in use. Such are the dangers in jet fuel that every large commercial aircraft flying today has complex nitrogen enrichment systems built into its fuel tanks to ensure the explosive mixture in half empty tanks does not explode. Hence the truth is that hydrogen carries around the same overall risks as jet fuels. On the positive side, hydrogen evaporates and disperses far more readily than jet fuel and is therefore less dangerous in the open air. However, extreme precautions must be taken indoors (say in a service hanger) as hydrogen could leak and pool under the roof, forming a potentially explosive layer. Another downside is the propensity for leaks, and care must be taken again to avoid internal build-up of the gas.

Overall, with the correct precautions hydrogen is a safe fuel when compared with other options.

Handling in airports

This technology will of course require major airports to change to distributing and handling LH2 from jet fuel. This will require considerable changes at airports with large cryogenic tanks and either tankers or some underground pumping system. In terms of safety, it is vital that all H_2 is purged before placing the aircraft in a hanger as this could lead to a dangerous gas build-up under the roof, but this should not be too onerous.

Also note, between flights it should be possible to use a very small amount of cryogenic liquid such as LN2 to keep tanks reasonably cool between refilling with LH2.

7.2.7 Summary

The use of HTS power systems in aircraft can give significant benefits beyond the gains in power density and efficiency. If aircraft are designed to run on LH2 fuel as is currently forecast, this approach will become the only approach to enable effective distributed propulsion. Indeed (in the view of the author) this area is likely to become a major application area for cryogenic and superconducting technologies.

Despite the complexities and difficulties, this approach seems highly likely to result in very low emission aircraft travel and it is hoped will make a significant contribution towards limiting the effects of global warming.

Acknowledgements

The author expresses his thanks to Cranfield University for the use and some of the figures and in particular to Professor Howard Smith for his kind permission to use his study summarised in figure 7.14. He also expresses his thanks to the School of Engineering at Newcastle University for their support.

7.3 Cryogenics in LNG and biomethane

Tom Bradshaw

LNG is an essential component of the UK's energy strategy. The demand for natural gas peaked in the 2000–2005 time period and then saw a decline, but still serves to service mostly the domestic and electricity generation sectors (figure 7.15).

7.3.1 The UK perspective

The first trial shipment of LNG from America was in 1959 on the vessel Methane Pioneer [11]. The ship was fitted with five tanks with balsa wood supports and insulated with plywood and urethane. One of the technical team on board was Dennis Rooke, who later became Chairman of British Gas. He was credited with developing Britain's gas distribution system [12]. These first trial shipments were stored in two 2200 m^3 perlite insulated aluminium tanks.

The UK was also the first country to commercially import LNG from the Hassi R'Mel gas fields in Algeria to Canvey Island—East of London [13] on 12th October 1964 on the ships Methane Princess and later the Methane Progress [11]. The two

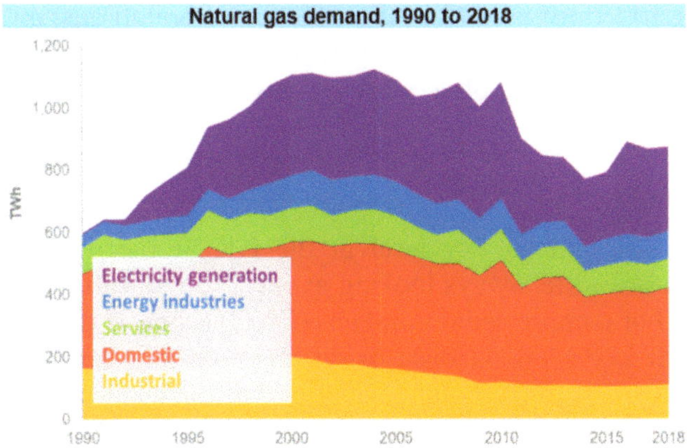

Figure 7.15. Natural Gas demand in the UK [10] (reproduced under Open Government Licence v3.0).

ships had aluminium tanks insulated with balsa wood, plywood and glass fibre. Mineral wool was also used to insulate the top of the tanks. As a safety measure, the regions surrounding the tanks were purged with nitrogen gas supplies from liquid nitrogen stores elsewhere on the vessel.

Soon after followed France, Spain Italy and then the rest of Europe. It is interesting to note that the driver to import LNG was to reduce the dependence on coal following heavy pollution in the 'Great Smog' of 1952 which resulted in 4000 direct fatalities. Legislation followed that encouraged the use of gas for cooking and heating. Now legislation is restricting the use of fossil fuels in their entirety.

The world trade in LNG in 2013 was 237 million tonnes, most of the expansion in this trade is going to the Asian market. Cryogenic storage onboard ships is vital to this trade. On land, storage is handled in large cryogenic vessels. An example is the Dragon storage facility at Milford Haven in Wales. This receives LNG from tankers and stores it in two storage tanks, each 44 m tall and 73 m in diameter. When required, the liquid is gasified and pumped into the National Grid. Also at Milford Haven is the South Hook facility which stores liquid in five tanks that are each the volume of the Albert Hall. Another storage facility is at the Isle of Grain east of London on the Medway which has a capacity of 1 million cubic metres of LNG.

More recently, ethane is being transported in refrigerated vessels as it is being produced in large quantities in the US by the shale oil industry. It is used as a feedstock for production of ethylene [14].

7.3.2 Storage and control

On tankers, LNG is kept liquid and the amount of methane released into the atmosphere is kept to a minimum by the use of refrigerators. This is vital as methane (the primary component of LNG) is a greenhouse gas with a high global warming potential (GWP). The GWP is a measure of how much heat the gas traps in the Earth's atmosphere relative to carbon dioxide. The GWP for methane is 84 and resides in the atmosphere for over 12 years, so keeping fugitive emissions low is vital.

A common refrigerator used (for reliquefaction of boiloff gas) on these tankers is a reverse Brayton cycle [15]. This cycle relies on turbines for the isentropic compression and expansion of a working gas and is capable of long life and relatively quiet operation.

7.3.3 Motive power

Although this is at quite a low level in the UK at the moment, hauliers and distribution industries are always looking at ways to reduce their carbon footprint [16]. Using LNG or compressed natural gas (CNG) is one of those ways. In use, the truck running off LNG has nitrogen oxide emissions 75% less than a comparable diesel engine. A perhaps unexpected benefit is that the engine is much quieter.

The UK now has an expanding network of LNG and CNG refuelling stations. Several leading distributors run fleets of trucks operating on LNG—these have an on-board cryostat to store the fuel. Some configurations have a dual fuel capability with diesel to offer security and ease range anxiety. The use of LNG is restricted to high mileage heavy duty lorries that are in almost constant use. On smaller cars, the evaporative losses when not being used would make the system uneconomical and polluting. LNG cryostats for automotive use are supplied by CHART Industries and mounted on Volvo and Iveco trucks for example.

7.4 Hydrogen—energy carrier of the future

James Fesmire

Hydrogen is no longer the top candidate to be the energy carrier of the future. Hydrogen *is* the energy carrier of the future. To be more specific, liquid hydrogen is the basis for the logistics of moving large amounts of energy around the world, or from any geographical region to another. Batteries of lithium ion and gas bottles of hydrogen will always have their place for small items, but to store up and move around large amounts of energy, liquid hydrogen will be at the heart of the matter. To keep things simple, liquid hydrogen in this context is meant to cover any cryogenic state of hydrogen, whether it be liquefied hydrogen (LH2), cryo-compressed (or supercritical hydrogen), or cryo-adsorbed hydrogen, as the main point of cryogenics is as a means of providing a practical, high-density method of energy storage (figure 7.16).

The use of LH2 both as a rocket fuel and as a supply for electrical power began about the same time around 1959 by NASA. These are two very different things, as one involves combustion and the other electrochemistry, but both applications made possible the space programs to come. Most everyone recognizes the use of big tanks of liquid hydrogen and liquid oxygen (aka, space launch vehicles) for rocket propulsion. But the vital importance of electrical power in space, also made possible by long-duration storage tanks of LH2, is obscured and sometimes overlooked. This dichotomy is finally beginning to change. For example, in 2019 Blue Origin announced plans for a lunar lander system that will have an LH2 tank to service both the propulsion and the electrical power needs on the Moon. The boiloff from

Figure 7.16. A view of the liquid hydrogen storage tank and transfer line at the Launch Complex 39B of NASA's Kennedy Space Center in Florida. The cryogenic systems will service the Space Launch System (SLS) rocket and the Artemis Moon to Mars exploration program. Credit: NASA/Ben Smegelsky.

the LH2 tank will serve to cool the liquid oxygen tank and to provide feedstock for the electrical power system.

There are certainly places and applications for 'all of the above' in the energy world, but hydrogen is king because of its simplicity, versatility, and potential to be completely clean. As the most abundant element on Earth, and in the known Universe, it can be obtained from anywhere and just about anything. Hydrogen can be combusted as a fuel, exhausting water as the by-product, but the beautiful thing is that it is now used directly for electric power. There is no fuel burning and no moving parts, just pure electrical power produced by the hydrogen electrical cell (or 'fuel cell'). These electrochemical cells have moved into the practical realms of cost and functionality for electricity supply of everything. Well, most everything. What in the world today is going away from being run on electricity? Not much. More and more electricity, at scales from micro to mega, means the demand for more and more hydrogen. And that means liquefied hydrogen.

Most do not realize that hydrogen production on an industrial scale also began in the early 1900s, followed by industrial scale liquefaction starting in the 1950s. By the 1980s, hydrogen production was a mature business. For example, more than 50 tanker trailers of liquid hydrogen (LH2) were required for every on-time launch of the Space Shuttle (135 flights over 30 years). In 2018, global hydrogen production was around 70 million tons per year. That much hydrogen, in liquid form, represents about one billion cubic meters (or, in gas form, about 30 trillion cubic feet). That hydrogen is piped and trucked around, and you never hear about it, a good thing from the safety perspective. One cup of water weighs about the same as one gallon of LH2. When spilled or leaked, hydrogen is so lightweight and fast moving that it disperses into the atmosphere very rapidly. Hydrogen can produce powerful engines (which is why it is so useful) but it is hard to achieve this.

From the 1930s it was recognized that hydrogen and oxygen would be the ultimate combination for fuel and oxidizer to power rocket propulsion. Hydrogen is an extremely light and powerful rocket propellant. It has the lowest molecular weight of any known substance and, in combination with oxygen, burns with extreme intensity. Liquid hydrogen yields the highest specific impulse, or relative efficiency, of any known rocket propellant. Research and development of the handling of liquid hydrogen in the 1940s and 1950s paved the way for NASA space programs of the 1960s which enabled the Apollo 11 Moon landing in 1969. liquid hydrogen also enabled living in space through the hydrogen electric cell ('fuel cell') which provided the necessary electrical power and water. Despite early setbacks and daunting engineering challenges, the taming of liquid hydrogen proved to be one of NASA's most significant technical accomplishments.

Today, liquid hydrogen is the signature fuel for space launch and exploration, both government and commercial. In addition to the Saturn V upper stages for the Apollo missions, LH2 has been successfully used since 1963 in the evolution of upper stage Centaur rocket engines. The modern Centaur is used today by United Launch Alliance with its Atlas V and Delta IV launch vehicles. The Space Shuttle, with its three main engines, was successfully powered by LH2 for all 135 flights. The European Space Agency developed the highly successful liquid hydrogen stage for the Ariane rocket in the 1970s which continues today with the Ariane 5. Blue Origin has dedicated its rocket development to use LH2 technology and plans the maiden flight of New Glenn from Florida in 2022. United Launch Alliance, with its plans for the new Vulcan Centaur launch vehicle, includes the next generation of LH2-fueled upper stages.

NASA's new Space Launch System (SLS) heavy lift rocket for the Artemis program has recently completed its maiden flight. The SLS LH2 tank makes up the bulk of the rocket, holding 2033 m^3 (537 000 gallons) of LH_2 in its 8.4 m diameter by 40 m height. Construction of an additional LH2 storage tank at Kennedy Space Center's Launch Complex 39B is also nearing completion to provide a total on-site storage capacity of 8000 m^3 (2 100 000 gallons). Research and development are also underway by NASA for exploring the Moon and establishing long-term residence. As evidenced by the NASA Tipping Point selections in late 2020, dedicating nearly 70% of the total \$370 M funding to the topic of Cryogenic Fluid Management Technology Demonstration, a key technological challenge is the first long-term storage and use of large amounts of liquid hydrogen in space (figure 7.17).

Hydrogen is stored in different formats depending on the need. The cryogenic formats include liquid, cryo-compressed (supercritical), and cryo-adsorbed (molecular). The cryogenic formats provide the ultimate in density but also offer key ancillary benefits such as purity and process cooling. For comparison, the density of liquid is 71 kg m^{-3} at its 20 K normal boiling point while a 700 bar gaseous storage tank is only about 41 kg m^{-3}. Moving hydrogen around is a challenge in any case, but the storage and transfer of LH_2 is now worked out by mature engineering standards and safety codes. Even the storage tank boiloff losses and tanker offloading losses are now a thing of the past with the incorporation of active refrigeration technology.

The big problem with LH_2 has always been the energy-intensive cost of liquefaction. With the advent of renewable energy from the Sun, wind, and other

Figure 7.17. The core stage for the first flight of NASA's Space Launch System rocket is seen in the B-2 Test Stand during a hot fire test January 16, 2021, at NASA's Stennis Space Center near Bay St Louis, Mississippi (credit: NASA Television).

sources we are now beginning to see the dream of clean and abundant energy emerge into reality. The problem then becomes not that of the high cost of liquefaction but the inability to store all that free energy when it is not being consumed. But…wait a minute…we can just liquefy and keep the hydrogen for when it is needed. The LH2 is the battery for storing the excess energy, like electricity in liquid form. And in liquid format, we can move it around whenever and wherever it is needed.

Even the boiloff problem for cars, trucks, boats, and other vehicles is solved. If the vehicle is electric, then it probably also has a battery in addition to its hydrogen electric cell. The boiloff charges the battery. Hydrogen equals electricity (via the electrochemical cell). Electricity equals hydrogen (via the electrolyzer). What more perfect synergy can there be in the energy world?

LH_2 continues to lead the way in the aerospace world of propulsion (figure 7.17). The ultimate chemical rocket is still the one that combines oxygen and hydrogen to make thrust and steam. Space exploration also relies on LH2 for electrical power and drinking water. Why LH2? The same reason that filling stations around the world are now moving away from gaseous storage to liquid storage: for the density, logistical practicality, economic effectiveness, and performance that the cryogenic systems provide. Cryogenics is all about energy, and energy density, and energy is a lot about economics.

Research on energy storage abounds, and yet LH2 makes a strong argument for being the king of all. Although one technology is rarely optimal for all situations, LH2 systems are now being developed to work at scales from small drones to mega-scale carrier ships. In the middle is a lot of other things like semi-trucks, ferry boats, airplanes, and backup power generators. Could it be that all other energy storage methods will fall away? I do not think so, as there are millions of different situations to cover, but just look at it the other way around. Is LH2 storage likely to fall away at either end of the spectrum? Not a chance.

The driver for hydrogen is renewable energy. The vector for putting that energy to use is hydrogen. The future is bright for cryogenic engineering, product development, and manufacturing around hydrogen liquefaction, storage, transfer, servicing, and global transport.

From propulsion to energy storage to renewables, as well as a host of industrial applications, hydrogen is versatile and valuable. To use it at scale and for its full potential, liquid hydrogen must be made, dealt with around the world, and put to work for the end-use applications. After ¾ of a century, the art of cryogenic engineering for LH2 systems may be just beginning.

7.5 Peter Dearman's liquid air engine

Toby Peters

From garage invention to global game-changing innovation

An engine that runs on nothing but air does sounds like a hoax. That's what Lord Sugar thought when Peter Dearman wrote him a one line email at the Mirror to try to get some publicity for his car running on liquid air.

'Dear Sir Alan, I have invented a car engine which uses liquid nitrogen or air as fuel. It uses no fossil fuel, produces no pollution and is cheap to manufacture.'

Lord Sugar replied in his column 'I was not sure whether the email was one of my pals trying to wind me up until I looked at your website .... I'm no expert, but if there are any engineers who wish to have a peep, look at the website and get in touch with Peter. But if you [Peter] are trying to pull the wool over their eyes, forget it, as any credible organisation will just have a good laugh.'

But Peter is a classic British garage inventor who is neither mad nor a hoaxer; just sees the world very differently. Over the years he has had a variety of inventions including a portable resuscitator that is still used in ambulances today. His most impressive achievement, though, was the Vauxhall Nova, which ran at more than 30 mph with an engine he cobbled together from a used coca-cola bottle, a beer keg of liquid nitrogen, some copper pipes and a lot of string.

Fifteen years on, Peter's invention has spawned several zero-emission technologies with $10 millions of investment; a multi-million Euro European-wide research project (CryoHub) to develop and investigate the potential of large-scale cryogenic energy storage at refrigerated warehouses and food factories, and its own research centre at University of Birmingham (Birmingham Centre for Cryogenic Energy Storage), with the potential to solve stubborn global energy problems. This in turn has also led to a whole new way of thinking about cooling and its central role in achieving many of the UN's Sustainable Development Goals.

The engine Peter invented in his garden shed runs not on petrol or diesel but on liquid air—or liquid nitrogen. It operates by the vaporization and expansion of cryogenic fluids. Ambient or low grade waste heat is used as an energy source with the cryogen providing both the working fluid and heat sink. Air liquefies and shrinks in volume 700 times when refrigerated to −196 °C, but boils and expands 700-fold when exposed to ambient temperatures, and this drives the piston. In short it works like a steam engine only more than 300 °C colder.

Cryogenic engines have been around for about a hundred years, but the Dearman engine adds a clever twist to make the process more efficient—a heat exchange fluid. In a car that runs on heat, you need something to remove the heat and keep it cool—a radiator. In a car that runs on cold, you need the opposite—something to provide heat. Prior cryogenic expansion engines have worked on an open Rankine cycle—i.e. similar to a traditional steam engine but with the cryogenic fluid pumped to operating pressure and vaporized through a heat exchanger, before expansion in the engine cylinder.

This approach had a number of drawbacks as the heat exchanger must be large to cope with the heat transfer rates and heavy to withstand the high pressure. Additionally, little heat transfer occurs in the expansion stage in the cylinder (near adiabatic expansion) reducing the work output. The breakthrough for Peter came in 1999 whilst he was watching an episode of Tomorrow's World, in which the presenter visited the University of Washington to report on a converted mail truck running on liquid nitrogen. Peter spotted the missing ingredient and added a mix of anti-freeze and water as a 'heat exchange fluid' allowing for heat to be introduced to the cryogenic fluid inside the engine. Direct contact heat transfer continuing throughout the expansion stroke gives rise to a more efficient near-isothermal expansion, a feat previous attempts with liquid nitrogen engines have struggled to achieve. After each expansion cycle the heat exchange fluid is recovered from the exhaust and reheated to ambient temperature via a heat exchanger similar to a conventional radiator.

It is obvious, but as Peter says, 'It's obvious, but it's only obvious once you've seen it'.

Yet the opportunity was far greater. Liquid nitrogen can also be efficiently (near-isothermally) expanded through a multi-stage turbine to drive an electricity generator. Since nitrogen is liquefied in industrial scale plants powered by electricity, the two processes could be combined to form a grid-scale energy storage device warehousing off-peak nuclear and wind power. The clever inventive step building on Peter's engine was to both ensure near-isothermal expansion—by harnessing the waste heat of the compression stage of liquefaction to provide inter-heating between the turbine stages—but also to capture and recycle the waste cold to drive the liquefaction process to more than double the round-trip efficiency. This is the basis of liquid air energy storage (LAES) developed by Peter with Professors Yulong Ding and Toby Peters and retired businessman Ferd Berger (Dearman, Berger and Peters were the founders of Highview Power).

Highview Power Storage tested and demonstrated a small-scale energy storage plant in Slough for several years before securing a £10M grant to move to a larger 5 MW/15 MWh pilot plant connected to the national grid at Bury, near Manchester. The plant demonstrated how LAES can provide a number of reserve, grid balancing and regulation services and the company is now building a 50 MW, 250 MWh 'cryogenic battery'—the largest non-geological energy storage plant in Europe.

With LAES technology now being proved at scale, it paves the way for the wider adoption of LAES technology globally. True long-duration energy storage is critical: to enable the broader deployment of renewable energy; overcome the intermittency of solar and wind energy; help smooth peaks and troughs in demand; and provide the UK with a stable and secure source of homegrown energy.

Liquid air not only expands when it boils, but also gives off large amounts of cold. In LAES electricity to electricity plants, the waste cold—and heat—given off by the system are recycled internally to raise its efficiency. But in vehicles, 'waste' cold is itself a useful product. Because the Dearman engine produces so much cold as well as shaft power, its first application, tested with Sainbury's in 2015, was a zero-emission transport refrigeration unit (TRU) to displace the highly polluting secondary diesel engines used to power refrigeration on lorries and trailers today. It is not widely known, but a diesel TRU can consumer 20% of the vehicle's total fuel usage, emit up to 30 times as much particulate matter (PM) as the propulsion engine hauling it around, and 165 times more than the official emissions of Euro 6 diesel car. Replacing a diesel TRU with a Dearman liquid air unit would slash the vehicle's total emissions of NOX by over 70% and of PM by over 90%, as well as reducing its CO_2.

The Dearman Engine was from 2014 to 2018 further road-tested as zero-emission transport emission unit running on liquid nitrogen with Unilever, Marks & Spencer and others. As Paul Polman, then CEO of Unilever tweeted following a six month trial in 2018 in Holland, in which a Dearman-cooled trailer successfully delivered premium ice cream and other frozen grocery products 'Great partnership with @Unilever and @Dearmanltd working to our goal of making the transport system as green as possible—win for the environment and for business #carbonfree'.

But there is more. We are seeing an increase in LNG. The Shell LNG Outlook, highlights that global demand for LNG grew by 12.5% to 359 million tonnes in 2019, a significant increase that bolsters LNG's growing role in the transition to a lower-carbon energy system.

However, an extraordinary amount of the energy contained in LNG is simply thrown away, missing out on an important quality. LNG is extremely cold as it is natural gas that has been refrigerated to −162 °C—consuming large amounts of energy—to make it compact enough to transport by tanker. But this 'coolth' or cold energy is normally discarded during re-gasification at the import terminal.

Supermarkets, data centres, air conditioning and industrial processes—as well as logistics—all rely heavily on cooling. Cooling in fact already causes 10% of global CO_2 emissions, three times more than aviation and shipping combined and we are deploying more than 13 new cooling devices every second. The IPCC has projected, for example, that the energy consumed by air conditioning worldwide could rise 33-fold to 10 000 TWh by the end of the century, which is about half the total electricity consumed for all purposes in 2010. We equally require today at least a five-fold increase in transport refrigeration vehicles to meet demand. Less than 15% of food which should be refrigerated globally actually is. If this boom in cooling demand continues to be met by conventional technologies, it will mean massive additional emissions of CO_2, NOx and PM. Given the growth in demand for cooling, there is therefore an urgent need to 'green' the provision of cold, and one major opportunity could be to recycle the waste cold of LNG to provide lower carbon, zero-emission cooling services.

Of the 111 LNG import terminals worldwide, only 23 do any form of cold recovery. Even here the use of the waste cold is usually limited to industrial plants close to the terminal, and only at times when LNG is actually being re-gasified, which in many cases occurs only intermittently. These factors limit the amount of cold that can be recycled, but this could be raised by converting it into novel energy vectors that store and transport it for use on demand such as liquid air (LAIR) or liquid nitrogen (LIN). Recycling waste cold in this way would produce cheap, low carbon, zero-emission cryogenic 'fuel' to provide distributed cold and power for vehicles and buildings. LNG waste cold could be a leapfrog to provide integrated cooling that is both cheaper and zero-emission in many applications across food and pharmaceutical cold chains and the built environment.

As we have seen, technologies are already being developed that could exploit large amounts of LNG waste cold. The Dearman engine, for example, runs on liquid air or liquid nitrogen to deliver distributed clean cold and power to vehicles and buildings through a range of applications. Obviously, it is all about economics. But analysis shows the Dearman approach could generate revenues both from sales of LAIR by producers and savings for the vehicle operator worth more than 10% of the value of LNG at current market prices.

It is vital that the world's rapidly expanding cooling infrastructure should be clean. And it is only logical we stop wasting the 'packaging' of LNG and start to exploit the economic and environmental benefits of this waste cold.

In developing countries, cooling—or its absence—causes a unique set of problems. Here the lack of adequate cold storage and refrigerated transport causes two million vaccine preventable deaths each year, and the loss of 200 million tonnes of food. Food waste depresses farm incomes, raises food prices and worsens hunger. Such colossal food waste has equally large knock-on effects: the FAO estimates that food wastage occupies a land area the size of Mexico; consumes 250 km^3 of water per year, three times the volume of Lake Geneva; and causes 3.3 billion tonnes of CO_2 emissions, making it the third biggest emitter after the US and China.

So there is an urgent need to resolve the cooling dilemma. The good news is that although the environmental, economic and social challenges caused by cooling are horribly entangled, this also means a single solution can tackle many of them at once.

Liquid air does not simply represent a collection of clever technologies, it is a powerful new energy vector capable of storing and moving both cold and power, and of providing many different cooling and energy storage services from a single 'tank of cold'—the basis of a new circular 'cold economy' to radically reduce the environmental damage and cost of cooling.

And all from Peter Dearman's garage in Bishop Stortord.

7.6 Liquid air energy storage (LAES) and Highview Power's 'CRYOBattery'

Gareth Brett

Highview Power is a designer and developer of large-scale energy storage solutions using liquid air or nitrogen as the storage medium. The company has been developing this technology since 2005 and is now at the commercialisation stage, having operated a 300 kW/2.5 MWh pilot plant at SSE's Slough Heat and Power site for over two years, which was partly funded by a UK government grant, and a larger 5 MW/15 MWh system at Viridor's Pilsworth landfill site from 2018 to 2020 (also partially funded by the UK government).

The company is privately funded, has won several awards for its innovations (including the IET sustainability innovation award in 2012, Business Green Technology of the year 2018 and the 2019 Ashden Award for Energy Innovation).

7.6.1 Introduction

In tomorrow's net zero electricity systems, energy storage (and in particular, long duration energy storage) will play an increasingly important role in helping transmission and distribution system operators ensure that the system remains in balance.

Highview's innovative system uses existing air separation and liquefaction technologies (and the associated turbo-machinery) as the basis for a large-scale, low-cost, long-duration, configurable and readily locatable storage system which uses air as its working fluid and energy storage medium.

LAES as a concept has been around for some time and has held an appeal as a potential alternative to pumped hydro or cavern based compressed air energy storage (CAES), both of which have restrictive requirements for specific geography.

7.6.2 Highview LAES system

Highview's LAES system can be described as a thermo-mechanical system where the energy is stored in thermal form, but its extraction and charging is carried out by mechanical means.

In summary, inlet air is compressed, dried and cleaned and the heat of compression stored (in hot water and/or molten salts for later use on the power recovery cycle). The compressed air is expanded in a modified Claude cycle refrigerator (whose performance is enhanced by the cold recycled from the evaporation part of the cycle) and the resulting condensate stored in low pressure insulated tanks.

When the stored energy is required, the liquid air is pumped to high pressure and vapourised (the resultant high-grade cold being stored for later use in the liquefaction cycle). It is then heated with the stored heat from the compression cycle and expanded in a multi-stage process gas expander with inter-stage reheating, which drives the generator (figures 7.18 and 7.19).

The storage of heat and high-grade cold permits this part of the cycle to interface with other processes which may have similar process conditions and thereby present opportunities for efficiency enhancements.

As an example, instead of storing high grade cold produced by the system, external cold, such as that available from a LNG re-gasification terminal, could be used which may offer the prospect of improved efficiency from the liquefier. Similarly, waste heat from a host process (for example exhaust heat from landfill gas engines) could be used to provide further heating of the stored air which would offer additional power per stored kg of air, thereby presenting improved efficiency.

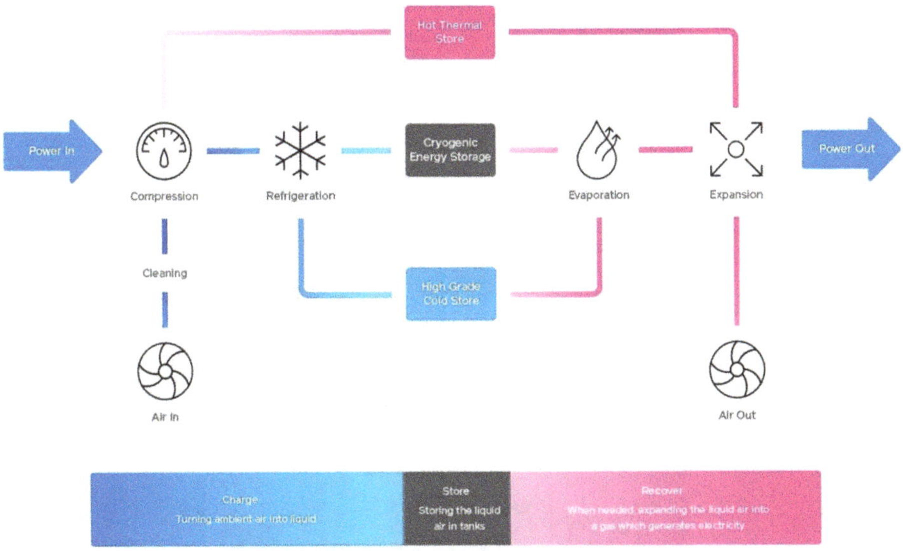

Figure 7.18. Conceptual model of Highview LAES system. Reprinted with permission from Highview Power.

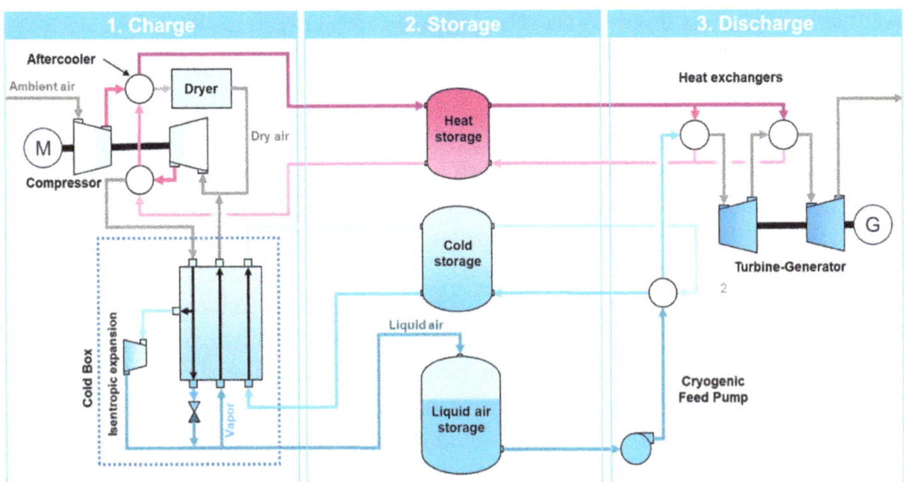

Figure 7.19. Simplified process flow diagram of a Highview standalone LAES system. Reprinted with permission from Highview Power.

A photograph of the original Slough pilot plant and the larger Pilsworth plant are shown in figure 7.20.

Performance

In general, LAES performs better at larger scale. The turbo-machinery is more efficient and the standing losses (thermal leakage) from the storage tanks are lower. In addition, the specific cost of systems is lower at larger scale, improving the system economics.

Figure 7.20. Photographs of Highview's Pilot Plant at Slough Heat and Power and the demonstration plant at Pilworth Landfill Site near Bury UK. Reprinted with permission from Highview Power.

The key indicators for storage system performance usually relate to: Energy density; AC to AC round-trip efficiency; flexibility and response time; and standing losses.

- Energy density—the amount of liquid air (in kg) required to deliver one MWh of electricity back to the grid after losses. 10 000 kg (10 T) per MWh is readily achievable.
- AC to AC round-trip efficiency—discharging energy/charging energy. For a mature standalone unit at the 50 MW/>4 h scale, efficiencies up to 60% are available. Efficiencies >70% can be achieved with the right combination of host site offering waste cold and/or heat.
- Response time—provided that the cryogenic feed pump(s) are kept cold and the turbine oil warm, starting the power turbine to synchronisation is usually achievable in around 5 min with loading rate a function of the turbine size (typically 5 MW per second). On the charging side, provided the liquefier is cold (which usually means it has been at operating temperature within 24 h) starting takes ~10 min—although the compressors will be loaded shortly after starting. If the liquefier has not been operated recently and is at room temperature, a cooling cycle could take around 2 h or more, depending on the liquefier size.
- Standing losses—a liquid air system loses its stored energy through heat leakage into the main cryogenic storage tank which causes boiloff. This is counteracted with insulation and losses of less than 0.2% per day are readily achievable.

Scale

LAES is one of the only storage technologies which can offer large-scale storage without geographical restrictions. Using pre-existing, commercially available components, systems in the 100 MW charge/discharge capacity and GWh scale for storage are readily deliverable with the major equipment available from recognised original equipment manufacturers (OEMs).

7.6.3 Conclusions

Highview's LAES represents a large-scale energy storage system which is deliverable utilising technology which is available today with existing non-combustion based compressor and expander technology at its heart.

The independent nature of the major components allows for a high level of configurability, which in turn permits a level of system design flexibility which will allow its integration with a wide variety of host processes, resulting in efficiency benefits related to the use of waste heat and/or cold.

The nature of the equipment employed in the system is likely to result in a long system lifetime, which is compatible with the expectations of the utility based nature of its applications.

7.7 Looking to the future—Covid-19: pathway to a lasting legacy for clean cooling

Toby Peters

7.7.1 Introduction

The COVID-19 logistics response could be the biggest, single refrigeration challenge the world has ever faced, but rapid expansion must not be based on environmentally harmful technologies. By the time this book is published we hope to be moving out of lock-down to a new normal. Here we set out a series of interventions that could help deliver a cold chain legacy that is both resilient and sustainable. As you read this book in years to come, look back and see what we did right … and wrong. Did we use the crisis to create a positive lasting legacy?

The COVID-19 logistics response could be the biggest, single refrigeration challenge the world has ever faced. Cold chains are energy intensive and rely on refrigerants, often with high global warming potential. Given most of the technologies deployed today will still be in operation in the next decade, this rapid expansion must as far as possible not be based on technologies that are environmentally harmful. But we can go further. Sustainable cold chains that use novel technologies can unlock modal shifts and allow us to think about opportunities in broader temperature opportunities. This opens the door to developing a lasting COVID-19 cold chain legacy that is both resilient and sustainable.

At a minimum, more financing focused on solar-powered refrigerators and freezers can support health facilities in developing countries that suffer from a lack of reliable electricity and do not rely on diesel systems or high GWP refrigerants. However, it is worth noting that WHO still lists equipment using R404A, which has a GWP of 3922, the highest of all the commonly used refrigerants and has been banned in Europe with only recovered or reclaimed R4040A able to be used until 2030.

Intervention 1—Fridges purchased today will likely remain in field for all of this decade. We don't just need energy consumption and refrigerants listed in equipment specifications, but sustainability embedded into pre-qualification criteria for equipment.

Drones are already proving their value for modal shift and rapid delivery of pharmaceutical products in rural locations. Combined with new mobile rechargeable micro-chillers and new business models, they could offer new integrated logistics solutions. At the same time, off-grid and pop-up mobile cold rooms could enable more cost-efficient campaign approaches; critically with the potential to be repurposed once the substantial but temporary need for covid-19 immunisation is addressed. Globally we have less than 15% of the cold chain capacity actually required today, let alone to meet future demand.

Intervention 2—A cooling-as-a-service financing model to enable optimised asset utilisation is highly relevant due to the sudden but temporary need for a substantial and unprecedented deployment of cooling and energy systems for COVID-19 vaccine distribution, and the long lifespan of the equipment versus the potential timeline for vaccine needs. We need to better explore the opportunity for cooling-as-a-service to allow new services including the re-purposing of technology to enable optimised asset utilisation by addressing a multitude of cooling needs.

We have a once in a generation opportunity presented by an urgent demand for new cold chains to go further and leapfrog to new strategies that make accelerated progress toward the Sustainable Development Goals and clean cooling. At the systems level, multi-sectoral synergies could be exploited by creating cold chains that would provide integrated services at different temperature bands, for example, across food, agriculture as well as health. This might increase flexibility and operational efficiency of cold chains, especially in agriculture-dominated developing countries where a substantial amount of food flows from rural to urban communities, and the temperature-sensitive medical products, such as vaccines, flow from urban to rural.

Intervention 3—Explore the opportunity of 'Community Cooling Hubs' supported by appropriate technology and business models to provide resilient, affordable and sustainable cooling dependent services to rural communities by aggregating demand to optimise energy and resource use (figure 7.21).

Intervention 4—Regardless of the strategy, large volumes of new equipment will be deployed. While each country is quantifying the number of health workers required to immunise their population, we must not miss the fact that a lack of qualified in-field refrigeration technicians will lead to long response times to equipment malfunctions and broken cold chains.

Data management and monitoring technologies will play a vital role ensuring the timely delivery of quality COVID-19 vaccines to their intended recipients; the chain of custody and vaccine quality from the point of manufacture through to the point

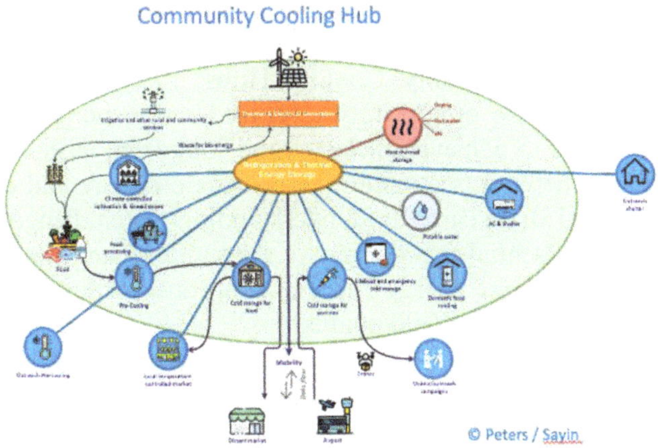

Figure 7.21. Community Cooling Hub. Copyright Toby Peters / Leyla Sayin.

of delivery; people receive the vital second dose, and to avoid double vaccination. It is also key to detect cold-chain equipment malfunction as well as enable predictive maintenance. It can identify and manage risks in real time, such as power outages, weather conditions, transport delays and disruptions. It will help protect against counterfeit vaccines, ensure security and prevent theft as well as provide records for auditing waste management. And with specialist transport containers required for the ultra-cold vaccines, tracking to ensure their fast return may determine the speed at which the supply chain can function.

Intervention 5—Ensure we invest in joined-up and embedded data capture and management alongside deploying hardware. Usage and performance data is critical in developing the capacity to run self-organising models (see Intervention 7) which in turn are vital to developing resilient self-healing systems.

Intervention 6—Let's not forget that alongside vaccines, disposable syringes, personal protective equipment (PPE) and other vaccination supplies will all need to be shipped out to every health centre and rural outreach point. As we plan the capacity to move everything out, we also need to ensure we are ready to deliver the sustainable and safe waste management and disposal recycling.

Where do we start?

In rushing to deploy hardware, an underestimation of the scale of the cooling demand, the system complexities and its impact on energy consumption risks contributing to a lack of ambition in policies, infrastructures, and technology developments as well as capacity building. It could ultimately have far-reaching social, economic, and environmental consequences.

Importantly, a focus on per capita equipment penetration rates pre-supposes a top-down technical solution to the cooling needs. This approach ignores the possibility of demand mitigation through the redesign of systems, the aggregation of demand, modal shifts, the use of waste or currently untapped resources, as well as support for existing cooling practices and behaviours. Understanding the challenges from a whole-system perspective is essential for identifying gaps and seizing the opportunities to align renewable energy and energy-efficient solutions with mass vaccination of COVID-19 and future needs of cold chains.

As described with Community Cooling Hubs, the scale of the COVID-19 response provides a unique opportunity to consider the energy systems and cold chains as a whole and to achieve efficiency and long-term sustainability gains by identifying synergies between sectors and areas where energy and cooling demand can be aggregated and share capacity. Understanding the challenges from a whole-system perspective is essential for identifying gaps and seizing the opportunities to align renewable energy and energy-efficient solutions with mass vaccination of COVID-19 and future needs of cold chains.

After identifying the needs in each sector, the next step in our approach is to explore how to use, make, store, move, manage and finance cooling within the boundaries of the energy and thermal resources available, sustainability goals, targets and commitments, as well as costs. To facilitate analysis and optimization of cooling options, potential interventions are classified into five key 'intervention classes'. Alongside 'reduce', 'shift' and 'improve', we can expand these concepts further by recognising the needs of vulnerable populations at the centre and the efficiency gains through aggregation of cooling needs.

Intervention 7—Use the demand for cooling to transition to a 'from-the-ground-up' participative, collaborative, holistic approach to sustainable cooling design and provision. This will allow communities, governments and other jurisdictions to understand how they use, and will use, cooling and consider ways to reduce demand for cooling as well as explore the use of non-electric energy resources; rather than simply focussing on the energy efficiency of the technology choices (figure 7.22).

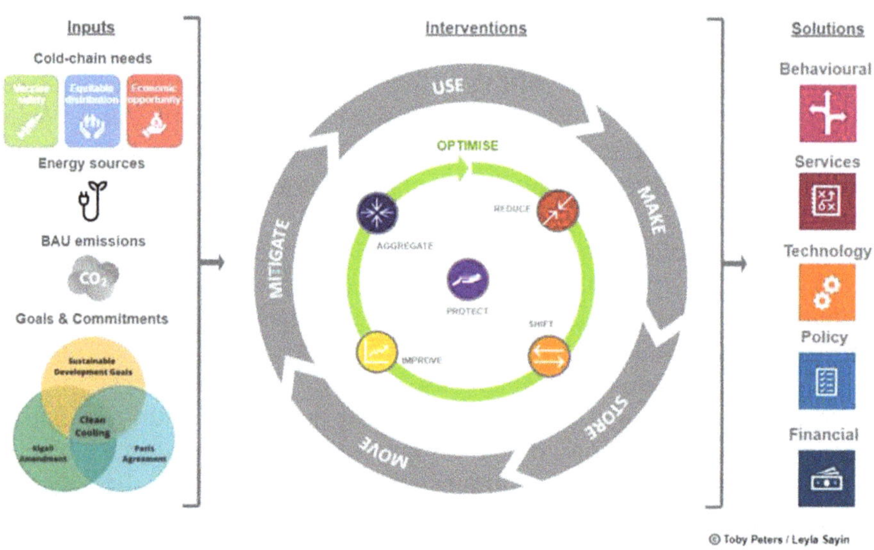

Figure 7.22. Cold chain inputs, interventions and solutions. Copyright Toby Peters / Leyla Sayin.

The biggest hurdle in both designing the current system and what COVID-19 vaccination cold chain configuration secures the greatest legacy is that the systems we need to investigate—cold chains for COVID-19 mass immunisation as opposed to routine immunisation, don't exist yet. We need to understand the system for the impacts of new levels of volume and velocity.

A modelling approach—essentially creating a virtual world which combines both existing capacity, with demand profiles, infrastructure, climate and new technology choices—will therefore be the most efficient and fastest route to a robust solution. Modelling allows us to develop parallel universes in which we can experiment at speeds not achievable in the real world. It can rapidly test thousands of scenarios, self-organising to define the optimum—and most energy efficiency and sustainable—investment strategies.

Looking forward, along with meeting the needs for COVID-19 immunisation today, the models can be exercised to quantify what value can be developed in each country by considerating how a range of COVID-19 vaccine distribution solutions could be redeployed in other cold chain applications.

Intervention 8—where it all starts. Measure thousands of times in a matter of days before deploying thousands of pieces of equipment with multi-year life. Using virtual systems populated with intelligent, decision making capability and driven by clear objectives and high powered computing environments will allow very many model runs with each run being slightly different to the previous and identify robust solutions—in short, those solutions that prove to be better than others in more scenarios. This is important for all of us.

In the short term, it will be the poor who will face the most significant challenges in accessing a vaccine for COVID-19 and cooling will become a serious issue of equity underpinning access to vaccines. But as governments and NGOs develop distribution plans, cold chains must both guarantee the health response and draw on the most sustainable solutions to build long-term resilience. These choices and the costs and environmental impacts will impact all of us for decades to come. We need to get them deployed quickly but we also need to ensure they are right. An intelligent approach to developing smart cold chains for vaccine distribution capability that can be transferred to other supply chains presents the greatest economic opportunity to developing a lasting COVID-19 cold chain legacy that is both resilient and sustainable.

7.7.2 Systems approach to cooling

We take a systems approach to cooling, defining the six elements of a cooling system as use/mitigate, make, store, move, use/manage and finance/regulate. Each of which is summarised below.

Use/mitigate: Mitigating active cooling lies at the core of clean cooling. It is essential to taking a resource-focussed approach and exploring opportunities to harness 'free' cooling, such as predominant breeze/wind airflows, large bodies of surface water, underground aquifers etc. Cooling needs can also be reduced through the application of passive approaches and techniques such as insulation, shading, natural ventilation, coatings and treatments and encouraging behavioural change like relaxing business dress codes in the workplace.

Make: Increasing equipment efficiency and replacing high GWP refrigerants with natural or ultra-low GWP refrigerants.

Store: Storing energy during off-peak time periods, 'wrong time' renewable energy, waste heat/cold can reduce energy costs, increase resilience against energy supply disruptions and reduce reliance on polluting diesel generators. It may be more efficient and sustainable to store energy thermally in ice or phase change materials rather than as chemical energy in batteries.

Move: Cooling systems that incorporate new energy vectors and materials to move thermal energy; not just chilled water for district cooling but also for example, liquid air or nitrogen can be used for cooling and generate mechanical energy.

Use/manage: Integrating smart systems for real-time monitoring of cooling needs and performance (sensors), load adjustments (controls) and integrated system management and storage (communications). Aggregating demands to further reduce the overall cooling requirement.

Finance/regulate: It is important to improve human decision making and modification of behaviour through smarter financing options (such as pay-for-use). It is also important to offer instant visibility of system status and consequences, integration of external costs/benefits (carbon pricing), markets for alternative energy vectors (such as waste heat) and other incentives.

7.7.3 Birmingham centre for cryogenic energy storage

Established in 2013 with a £12M grant under the Eight Great Technologies capital programme, the Birmingham Centre for Cryogenic Energy Storage (now expanded into the Birmingham Centre for Energy Storage) brings together research expertise from across the university to drive innovation from the laboratory to market. At the centre researchers are looking at ways to use cryogenic liquids and novel materials for more efficient and effective alternatives to meet the growing demand for cooling systems. The centre focusses on a portfolio of research activities including: novel storage materials with an aim to further increase energy storage density and lifespan; new thermodynamic cycles and processes; systems integration, control and optimisation; demonstration and testing to carry out detailed study for both component and system level performance improvement; cold economy, investigation of the societal and economic impacts of cooling and the cold chain for the UK and abroad. The University of Birmingham and the research arm of the State Grid Corporation of China (SGCC) have recently unveiled a laboratory on energy storage research in thermal and cryogenic energy storage and their integration and application in energy networks.

7.8 Superconducting electrical machines for wind turbines

Stuart Bradley

7.8.1 Wind turbine drivelines

Wind turbine drivelines convert the slow-speed rotary motion of the blades and hub to electrical energy at the right frequency and voltage for connection to the electrical grid. The minimum viable driveline consists of the prime mover—the blades and rotor, a main shaft and bearings, a generator and power converter. These drivelines have changed rapidly from the modestly sized gear-driven conventional standard-speed generators of the early part of the century to the giant, direct-drive generator devices that we see in service offshore. Nominal power ratings have risen from around 2 MW for onshore devices to 12 MW for offshore machines, with driveline weights per unit power rising from circa 10 kW/tonne to 25 kW/tonne.

7.8.2 Driveline developments

For offshore wind applications, direct drive generators (DDGs) without a speed step-up gearbox dominate the market due to their lower failure impact and improved power density at power ratings above 5 MW. The impact of this technology has been to reduce the levelised cost of energy and to make good use of the employed materials. However, existing DDG designs often use rare-earth element materials such as neodymium to produce strong magnetic fields in the air-gap between rotor and stator. These elements have high and volatile costs, severe environmental impacts from their mining and processing, and need careful design and handling during assembly and operation.

7.8.3 Offshore wind turbines in extreme environments

The transition to a zero-carbon electricity grid by 2050 requires timely and impactful action, with renewable energy resources being made economically viable and ubiquitous. Locating wind turbines in a wider geographical distribution and in more energetic sites would reduce the risk of winter anti-cyclone weather stalling electricity delivery when most needed. Even higher power rating wind turbines, above 20 MW per machine, could provide lower cost of electricity by having a higher power density per unit cost and placing them in the most advantageous sites, or using design optimisation for the site conditions. Since power is proportional to swept area and wind velocity, longer blades are needed for high power ratings. The ultimate length of these blades depends on the material stiffness and the blade tip speed, with the maximum blade tip speed needing to be no more than 100 m s^{-1}. Therefore, the longer blades needed for high-power have to turn slower, and this means that the driveline is subjected to very high torque.

High power rating turbines, able to withstand high torque and extreme weather events, require bigger electrical generators. The existing topology and materials could go much larger in rating by having serial-drivelines or scaling the generator diameter and stator length. As generator diameter increases, the stator mechanical stiffness resisting the magnetic forces will reduce. This means that the stator-to-rotor air-gap can close, which would cause catastrophic damage. To remedy this, stiffness is added using structural materials, increasing the generator mass, and the weight per unit power will drop. There is an optimum design cost point, where the generator dimensions, air-gap and magnetic forces are balanced, and beyond this, capital costs will rise. An advanced generator concept is needed to break these constraints.

7.8.4 Superconducting generator design features

An advanced generator needs to be powerful but lightweight, reliable and cost-effective, robust and adaptable. A solution is to use superconducting materials to replace copper or aluminium as electrical conductors to create the magnetic field. Being efficient and compact devices, these attributes carry into the machine design. If applied to a low-speed direct-drive generator, the optimum arrangement would apply superconducting coil designs in both rotor and stator, since that would maximise the amount of superconductive material. Both stator and rotor would be within a cryostat and shield. The superconducting operating temperature imposes a refrigeration load on the machine, which can be thought of as a loss, and the material choice will be critical to this loss, and therefore the total energy produced. Refrigeration complexity is a technical risk, and the robustness and reliability of this plant needs risk mitigation and technology development for application in offshore wind turbines.

Alternating current causes losses within the armature coils and these losses need careful design and control to prevent loss of superconductivity, but the rotor field windings are proven to be viable. A critical challenge is to transfer torque from room temperature to superconducting temperature, a difference of over 200 K, without transferring an unworkable amount of heat. The removal of heat and control of the rotor magnetic field from a cold environment are associated difficulties that have cost and risk implications. Potential mitigations include the application of superconductivity to the stator only, or adding control and cooling design features on the rotor, both weakening the case for superconductivity since they will impact the machine size and mass. Studies show that a reduction in mass of more than 60% is possible for a fully superconducting machine, and more than 45% for a part superconducting machine. The mass-reduction impacts on the whole wind turbine design give a reduction in levelised cost of energy (LCoE) of between 15% and 32%, with the generator total energy output being similar to existing designs.

The fully superconducting generator output could be very low frequency or direct current, exported to a collector station or the offshore grid. A conventional copper stator design will be able to export electricity akin to existing offshore wind turbines.

7.8.5 Superconducting bulk materials

Replacing rare-earth elements used for rotor excitation within direct-drive generators could help to reduce commercial risk caused by supply chain length and price volatility. These elements could be replaced by adopting superconducting bulk materials. These bulk materials are moulded in place, activated by electrical fields, and until they reach critical temperatures or are subject to demagnetisation, remain magnetic. This simplification is impactful on operational and capital costs, and since the size and mass advantages are maintained, the impact on LCoE is increased. The research priorities for the application of these materials include magnetic field activation and control, material degradation and total wind turbine system cost modelling and reduction.

7.8.6 Summary

The use of superconductivity in wind turbine applications is most applicable to ratings above 15 MW to 20 MW, where the advantages of size and mass reduction outweigh their additional cost and complexity. The most likely generator topology to be implemented is direct-drive, low-speed and high-torque with bulk superconducting materials, and this opportunity gives encouragement to the development of offshore wind turbines for the lowest cost and greatest energy harvest.

References

[1] Malkin P and Pagonis M 2014 Superconducting electric power systems for hybrid electric aircraft *Aircr. Eng. Aerosp. Technol.* **86** 515–18

[2] Malkin P and Pagonis M 2013 The design of fully superconducting power networks for future aircraft propulsion *Joint Propulsion Conf. San Jose 49th AIAA/ASME/ASEE Joint Propulsion Conf*

[3] Smith H 2014 Airframe integration for an LH2 hybrid-electric propulsion system *Aircr. Eng. Aerosp. Technol.* **86** 562–67

[4] Kim H D, Felder J L, Tong M T and Armstrong M 2013 Revolutionary aeropropulsion concept for sustainable aviation: turboelectric propulsion *21st Int. Symp. on Air Breathing Engines, (Busan, Korea)*

[5] Peter M and Jason H 2017 Aircraft electrical design considerations *SEE MEA Conf. (Toulouse)*

[6] Callavik M, Blomberg A, Häfner J ü and Jacobson B 2012 The Hybrid HVDC Breaker – an innovation breakthrough enabling reliable HVDC grids *ABB Grid Systems, Technical Paper*

[7] Malkin P 1989 The vacuum arc and vacuum interruption – a review *J. Phys. D: Appl. Phys.* **22** 1005–19

[8] Cotton I, Nelms A and Husband M 2008 Higher voltage aircraft power systems *IEEE Aerosp. Electron. Syst. Mag.* **23** 25–32

[9] Song W, Pei X, Xi J and Zeng X 2021 A novel helical superconducting fault current limiter for electric propulsion aircraft *IEEE Trans. Transport. Elect.* **7** 276–86

[10] UK Energy in Brief 2019 Department for business *Energy and Industrial Strategy* www.gov.uk/government/statistics/uk-energy-in-brief-2019

[11] 'LNG Shipping at 50' A Commemorative SIGTTO/GIIGNL Publication, October 2014 Society of International Gas Tanker and Terminal Operators Ltd (SIGTTO)

[12] The Independent, Obituary, Monday 8 September 2008

[13] https://canveyisland.org/places-3/places-2-3/the-seafront/canvey-island-methane-terminal

[14] https://maritimeprofessional.com/news/ethane-transport-technology-268133

[15] Chang H M 2012 Modified reverse-Brayton cycles for efficient liquefaction of natural gas *Proc. 17th Cryocooler Conf. (Los Angeles, CA)* p 435

[16] https://ctc-n.org/technologies/liquefied-natural-gas-trucks-and-cars

[17] Richie H 2020 Climate change and flying: what share of global CO_2 emissions come from aviation? *Our World in Data* (accessed November 2021)

IOP Publishing

Cryogenics
Fundamentals, foundations and applications
Tom Bradshaw, Beth Evans and John Vandore

Chapter 8

Life science and healthcare

Martin Fray, Dave Clarke, Marisa Martin-Fernandez, Louise Cuff, Agnieszka Piotrowska, Richard Amos and Russell Thomas

John Vandore

This chapter opens with a masterful overview by Martin Fray. In the process, Martin introduces subjects like cryosurgery and cryotherapy, both of which are given expanded treatment by top people in these fields—Louise Cuff in cryosurgery and Agnieszka Piotrowska in cryotherapy. Other specialisations for which we have outstanding contributors are cryo-electron microscopy, covered by David Clarke and Marisa Martin-Fernandez, and proton radiotherapy by Richard Amos and Russell Thomas. This chapter is a prize in bringing together such a wealth of expertise and bringing together such exceptional insights. It may be the first time that a chapter on cryogenics has embraced topics ranging from genital warts to the Nobel Prize—David and Marisa note that the Nobel-Prize winning technology of Cryo-TEM is now the method of choice for structure determination of viruses and protein complexes, and the potential for more revolutionary developments to come. This is an exciting field.

8.1 Cryogenics and its application to the biological sciences

Martin Fray

Cryogenics can be described as the production and behaviour of materials at very low temperatures. However, at what point on the temperature scale refrigeration ends and cryogenics begins is not well defined.

In the natural world adaptations to extreme cold have arisen in many species. Seeds from many types of plant need a period of sub-zero temperature to break dormancy ahead of germination. Arctic mammals lay on thick layers of blubbers, counter current heat exchange systems have evolved to conserve heat, as in birds' legs, some insects with natural antifreeze chemicals in their blood (haemolymph), and penguins which tolerate the Antarctic winter while incubating an egg with their

doi:10.1088/978-0-7503-2757-2ch8

feet. However, none of these adaptations permit the formation of ice crystals inside living cells. Even the wood frog (*Lithobates sylvaticus*) has evolved mechanisms that prevent ice formation inside cells despite being able to survive several freeze/thaw cycles during the North American winter. Ice crystal formation in living cells causes irreparable damage to the cellular architecture, cell membranes and organelles resulting in cell death—frost bite.

In a biomedical/agricultural setting cryogenics is used in a variety of applications. For example, cryogenic technologies can be used in hospitals for magnetic resonance imaging (MRI) machines which use liquid helium (−269 °C or 4 K) for cooling, cryosurgery/therapy, freezing blood products and biological samples. The discovery that semen can be frozen has led to its widespread use in human fertility clinics and has facilitated the large-scale application of semen cryopreservation in the cattle industry, which is now highly dependent on the use of frozen semen and artificial insemination for cattle breeding.

Cryopreservation is the process that preserves biological materials by cooling them to very low temperatures under controlled conditions. It is not simply a case of putting the samples in the deep freezer. It is a versatile and well proven technique that can safely preserve biological materials for many years, if not in perpetuity. Alternatively, it can be used for short-term holding, for example to allow transfer of bull semen from high merit breeding stock without having to transport live animals. However, there are limitations to cryopreservation. This section will give an overview of cryobiology: the applications of cryogenics to the biological world.

In practice, cryobiology is the study of biological materials or systems at temperatures below normal. The materials or systems studied may include proteins, cells, tissues, organs, or whole organisms. Temperatures may range from moderately hypothermic conditions to temperatures approaching absolute zero (minus 273.15 °C). You could argue there are six major areas of cryobiology:

- Cold adaptation of micro-organisms, plants, animals and vertebrates;
- Supercooling as applied to biological systems;
- Cryosurgery and cryotherapy;
- Freeze-drying (lyophilisation) of pharmaceuticals and their transportation;
- Cryopreservation of cells, tissues, embryos and germplasm used in IVF programmes;
- Organ preservation.

We know LN_2 (−196 °C or 77 K) can be used to maintain the architecture of delicate items but they do become very brittle which can be demonstrated by shattering frozen flowers against a hard surface. In contrast, if the flower is warmed up slowly the petals will droop because the cells have been damaged by the ice crystals. The aim of cryopreservation in the biological context is to freeze and thaw living materials in such a way that no damage is done, and the tissue/organism remains viable.

8.1.1 Preserving biological materials in the laboratory

In this section mouse embryo freezing will be used as an example to describe (in general terms) the principles behind cryopreservation and will focus on (a) the need to allow penetration of cryo-protective agents (CPAs) and some of their properties, (b) the processes required to carefully control the dehydration process, (c) the use of a combination of CPAs and extra-cellular ice crystals to draw out water from cellular systems or (d) the exposure of cells to very high concentrations of CPAs before plunging them directly in LN_2 to induce vitrification.

The potential to preserve biological materials has enormous benefits to human health, animal production, scientific enquiry and conservation programmes. Although some biological materials present the cryobiologist with seemingly intractable problems. He or she is faced with the same issue, i.e., how to freeze living cells without the ice crystals damaging the cells. When it comes to developing successful cryopreservation techniques, size really does matter. Smaller samples allow for better penetration of the CPA and they are more likely to survive a freeze/thaw cycle.

Many plant species have worked this out for themselves by dehydrating seeds to such an extent they can be placed directly in a refrigerated environment for long-term storage. This phenomenon is being exploited by plant conservationists, e.g. Kew Garden's Millennium Seed Bank, which aims to offer a safety net for plant species at risk of extinction; and the Svalbard Global Seed Vault which has been hewn out of a mountain inside the Arctic circle for the sole purpose of preserving the world's crop diversity.

Cryobiologists needs to mimic this process when they attempt to freeze down cells. In fact, controlled dehydration is the key to a successful cryopreservation protocol. Displacement of intra-cellular water can be achieved by placing the cells in CPAs, i.e. cryo-protective solutions. Cryoprotectants are water soluble chemicals that lower the freezing point of water. Outside of cryobiology they are referred to as 'antifreeze'. The first effective CPA was discovered over 70 years ago following the accidental but rather fortuitous discovery that glycerol could be used to successfully freeze fowl and mouse sperm [1]. A wide range of CPAs is now available that can be separated into those that pass through the cell membrane and permeate the cell like glycerol, propylene glycol and DMSO. Or non-permeating CPAs that don't pass through the cell membrane. Typically, sugars like sucrose and raffinose fall into the class of non-permeating CPAs.

Cells may be subject to two main types of injury during the freezing process—(a) being physically squashed/damaged as crystals grow in the extra- and intra-cellular compartments, or (b) through solution effects which arise as the growing ice crystals (formed from pure water) create ever increasing concentrations of solutes in the residual pockets of liquid between the crystals. These high solute concentrations cause chemical and osmotic damage. For cells to survive they must be able to tolerate conditions in these pockets of unfrozen liquid until the freezing process is complete.

Conventional cryopreservation techniques typically use 1.0 to 1.5 Molar solution of CPAs which can be added to cell suspensions/embryos as a single step. This produces the classic shrink/swell response as the osmotically active CPA initially draws water out of the cell and then after a few minutes establishes an equilibrium and the cells regain their shape. The sample can now be loaded into a cryo-vessel and prepared for freezing.

Cryopreservation by slow freezing is typically achieved at freezing rates of 0.1 to 1.0 °C min^{-1} which allows water to leave the cells as the freezing process advances. The samples are supercooled and at a predetermined temperature (−7 °C for embryos) extra-cellular ice crystal formation may be induced by touching the cryo-vessel with a pair of tongs cooled in LN_2, a process called seeding or nucleation. After which the cooling process is continued. The slow freezing rate is important because it allows time for water inside the cell to leave and be replaced by the CPA. This ensures the concentration of CPA inside the cell rises in step with the extra-cellular concentration, a key requirement if intra-cellular ice crystals are to be stopped from developing.

Only salts and other dissolved molecules prevent water from freezing completely. In the absence of a CPA, almost the entire volume of water freezes during the cooling process. However, when a CPA is added, the fluid's overall freezing point is lowered. Even so, the water content will eventually start to freeze. As it does so, the concentration of the CPA steadily increases as the ice crystals grow. Ice crystal growth stops when the CPA becomes concentrated enough for the melting point of the fluid to equal the surrounding temperature. Which means the CPA eventually reaches a temperature that cannot be frozen, i.e. no more ice can grow as the temperature is lowered.

The high initial concentration of the CPA means there is more room for cells to survive between the ice crystals. If the temperature continues to fall, the remaining unfrozen pockets of solution pass through their 'glass transition temperature,' and solidify into a glass-like state, i.e., vitrification occurs. Cells survive freezing by existing inside the glassy solid between ice crystals. As a natural extension to this phenomenon, the larger the starting concentration of the CPA, the larger the unfrozen volume will be at the end of the freezing process. However, there is a caveat here. High concentrations of CPAs can be quite toxic to cells so what might be gained by creating a larger unfrozen void could be lost through greater cytotoxicity.

The overall process described above is often referred to as two-step or controlled rate freezing, because after placing the cells or embryos into the CPA, the temperature is gradually lowered at a controlled rate before the samples are plunged into LN_2 for long-term storage. However, an alternative approach termed vitrification can be adopted.

Vitrification means 'turn to glass' and is a process that occurs when water is cooled too quickly for it to form ice crystals. It avoids the mechanical distortions caused by ice and allows salts and other molecules to remain undisturbed *in situ*. Vitrification avoids the major causes of damage induced by freezing. Nevertheless, this comes at the cost of CPA toxicity because vitrification relies on initial exposure to high concentrations (approx. 5 to 6 Molar) of permeating CPA which displace some intra-cellular water. These CPAs must be added in multiple steps otherwise they will cause osmotic shock which would damage the cells. Equilibration in the CPA is followed by a very rapid cooling, often by plunging directly into LN_2 and the whole sample is effectively turned into the equivalent of a glassy pocket of concentrated CPA. At storage temperature below the glass transition temperature of water (approx −137 °C), all water-borne intra-cellular reactions cease, and no biological degradation should occur. Studies have also shown that biological samples like embryos are not affected by background radiation [2] and should remain viable for thousands of years.

Vitrification works very well for pre-implantation embryos and it is hoped that it will present a solution for preserving whole organs. However, organised tissues sustain more damage than isolated cells during freezing. Unlike cell suspensions, bulky tissues do not have the potential to sequester themselves into unfrozen pockets as the ice crystals grow. They also undergo different heat and mass transfer effects which makes it more difficult to achieve the rapid cooling and warming rates required for successful vitrification. For an organ to resume normal function post thawing, all the diverse cell types it contains must function normally after cooling/warming. For this to be achieved it is more beneficial to perfuse organs with CPA rather than simply submerge them in a CPA solution. Perfusion is best achieved via the circulatory system which takes advantage of the network of capillaries that ramify throughout the tissues.

What can be brought back from the deep?

To some extent freezing or vitrification is the easy part. All biological samples can be frozen. The hard part is bringing the material back from the freezer in a viable condition. Thawing/warming protocols need as much care as the initial freeze step. Here the rise in temperature needs to be compatible with the cooling step to prevent crystals from forming and protein precipitation. In essence, this means very rapid warming rates for vitrified samples and a more measured thawing rate for samples frozen in a controlled rate device. Once the sample has returned to its optimal temperature an osmotic buffer, e.g. sucrose or mannitol, is applied to assist with the removed of the CPA from the intra-cellular compartments. These are referred to as non-permeating cryoprotectants. The role of an osmotic buffer is to establish a concentration gradient that helps draw the CPA out of the cells and thus reduce the swelling as cells imbibe water as they are passed through repeated wash solutions, culminating in transfer to the desired isotonic holding medium. Placing cells loaded with CPA directly in an isotonic solution would cause the cells to swell excessively and burst.

Over the years we have come to understand more about which biological materials will tolerate freezing, i.e., micro-organisms, embryos, sperm, thin tissues, cell suspensions and blood products, as well as seeds. As new technologies emerge, the benefits of cryopreservation are extending. Cells and micro-organisms harvested from a variety of tissues and species are available from commercial enterprises and individual laboratories as well as well-funded global repositories like the American Type Culture Collection which hold over 60 000 authenticated materials. An important medical advance which cuts across many disciplines is the isolation of stem cells harvested from embryos, blood and organs. Stem cells are considered potential goldmines in the field of regenerative medicine. Tissue engineering, gene therapy, regenerative medicine and cell transplantation are largely dependent on our ability to preserve, store and transport these cells without compromising their genetic or cellular potential.

Unfortunately, not all biological material retains its viability once it has been recovered from the freezer, e.g. fruit fly gametes, entire human organs and whole multi-organ species. In this context, it would be beneficial to understand why there are benefits to extending the range of materials that can be frozen.

For example, we know that gametes from numerous vertebrate species can be frozen (man, domestic ruminants, rodents, fish) but this is not universally possible. It is still very difficult to freeze pig embryos and sperm. This may not sound like a big problem, but as a valuable production animal, access to a bio-bank would help preserve and distribute valued genetic traits. Without these frozen reserves, the pig industry has to rely on highly coordinated cold transport chains or live animal movements in order to disseminate genetic traits.

Although good progress is being made, many endangered species have proved intractable to cryopreservation, frustrating the efforts of many conservationists to establish gene banks for species facing extinction [3]. The materials held in gene banks are extremely valuable, not only do they allow for the study of evolutionary biology and genetics, but the embryos and germplasm they hold means that the individuals they represent remain part of the breeding population, long after their death.

Refrigeration allows safe collection and storage of biological products. For example, after the white blood cells have been filtered off, the other cellular components of donated blood (red blood cells and platelets) are separated and quarantined in cold storage. This enables the blood to be tested before use, though the refrigerated products can only be stored for a few weeks and may have to be discarded before use.

The fruit fly (Drosophila melanogaster) may be considered a nuisance organism by some, but is one of the leading biomedical models used for understanding the biology of animal systems. Cryopreserving this organism or even inducing temperature-induced diapause of larval stages has been unsuccessful. This means that live experimental fly colonies have to be turned over every two weeks or so. As a consequence, keeping even a modest number of strains is an expensive and time-consuming activity. More importantly, with each generation, the fly stocks are likely to accumulate hidden mutations—a process called genetic drift, which is the foundation of the evolutionary process.

Renal transplantation is well established as the optimal form of renal replacement therapy, but there are a limited number of organs available for transplantation. Similarly, there is an urgent need for access to other donated organs and tissues, e.g. ovary, liver, testis to help survivors of injury, disease and radio/chemo-therapy. Reliable, sub-zero organ preservation has the potential to transform this area of clinical need. Technical advances are being made with whole organ cryopreservation and there have been some notable successes. For example, the successful cryogenic storage of rat [4] and sheep [5] ovaries, with the birth of live offspring following transplantation. New approaches using mechanical perfusion systems and vitrification protocols are also being explored [6] but we should not underestimate the problems encountered when freezing whole organs.

This brings us to the subject of cryonics; the low-temperature storage of human corpses and severed heads in the hope that one day the person or the content of their brains could be revived. One would not want to discourage genuine research into cryopreservation of organs or even whole organisms. It is possible to speculate that some as yet undeveloped technology may emerge in the future, but we are a long way off successfully preserving multi-organ systems at the present time.

8.1.2 Other applications

Cryogenics is used in a diverse array of biological systems. For example, osmometers which are used for checking the osmolality of culture media employ freezing point depression to measure concentration. This is possible because the freezing point of aqueous solutions is directly proportional to solute concentration.

Another application is freeze-drying (lyophilisation) which involves freezing the material, lowering the pressure, and removing the ice by sublimation. This contrasts with dehydration which removes water by evaporation. Freeze-drying maintains the shape of the original product, which helps to preserve the quality of the original product after rehydration. Freeze dried products can be stored for long periods of time frozen or at ambient temperature, and just need rehydrating to regain their original form. Freeze-drying came to the fore during the second world war because it helped to preserve the quality of blood plasma and penicillin needed by the medical corps without the need for refrigeration. In the modern context, freeze-drying is exploited in numerous manufacturing processes, e.g. astronaut's rations, vaccines and biologicals produced by the pharmaceutical industry and chemical synthesis.

Ensuring cryogenic temperatures are maintained from the bench to the bedside is of paramount importance for many cellular therapies. Cold storage and cold chains are used to transport a wide range of perishable biomedical/agricultural products around the world. These cold chains may involve large, refrigerated containers that rely on an external power supply, or individual static units, dry-ice packages or dry-shippers that can hold LN_2 temperatures for 10 or more days. Dry-shippers have an absorbent lining which means that once they are charged with LN_2 they can hold their temperature without fluid sloshing about inside the container. This makes these devices safe for air travel.

Cold chains facilitate exchange of biological material across international boundaries, allowing products to be moved in bulk and health-screened ahead of shipment. However, developing countries often have difficulty in maintaining cold chains, particularly to remote areas where refrigeration plants and cryogenic gases are in short supply. Many healthcare facilities in the poorest parts of the world do not even have a reliable electrical supply. This presents an acute problem for the distribution of medical supplies, such as getting effective vaccines to SARS-CoV-2 and other infectious agents to billions of people around the world.

The medical profession often use cryotherapy or cryosurgery (cryoablation) in their patient care. These procedures involve exposing the body to extremely cold temperatures (cold shock) for a short period of time. Cryotherapy may involve localised cooling or whole-body cooling. Devices designed for localised cooling may use LN_2 sprayed directly onto the skin or a stream of CO_2 gas that has been rapidly expanded and cooled by the Joule–Thomson effect. These devices can be used to treat skin lesions, particularly pre-cancerous lesions on the skin and cervix, or to provide pain relief.

More invasive applications can be used to destroy tumours within the liver, kidneys, bones, lungs and breast etc. This type of invasive cryosurgery most often involves the use of imaging guidance, temperature monitoring and a needle-like probe (cryoprobe) to guide the placement of liquefied nitrogen or argon (−186 °C or 87 K) gas. Advances in cryosurgery technology have dramatically reduced the long-term side effects once associated with the treatment, but like all medical interventions, more studies are needed to optimise these procedures.

Whole body cryotherapy or therapeutic hypothermia may have a neuroprotective effect in babies with mild hypoxic ischaemic encephalopathy or patients who have received a traumatic brain injury. Alternatively, cryo-spas may be used as part of a wellbeing programme. In these devices, the patient stands in a cryosauna filled with extremely cold (around −60 °C to −132 °C) air for 1 min to 3 min. During this procedure the peripheral blood is drawn into the body's core. At the end of the treatment the peripheral circulation is resupplied. Studies have shown that this treatment offers many benefits to the body, such as boosting metabolism, reducing inflammation, increasing energy, managing pain, and even treating anxiety and depression.

8.1.3 Concluding remarks

During cryopreservation by vitrification or freezing, more than 50% of the intracellular water is eventually replaced by the CPA. In essence, cryopreservation is the replacement of water by other molecules that do not freeze. Bearing in mind how critical water is to maintain the shape of cells and the configuration of proteins and other molecules, it is hard to believe that these delicate structures can survive the rigours of this process. And yet cells can withstand several freeze/thaw cycles without any apparent detrimental effects. Cryobiology has come a long way since 1949 when the cryo-protective properties of glycerol were first discovered. We can look forward to many more exciting discoveries in the future.

8.2 Cryogenic techniques in biological electron microscopy

David T Clarke and Marisa L Martin-Fernandez

8.2.1 Electron microscopy sample preparation and the motivation for the use of cryo-fixation

The electron microscope (EM) sample stage is a hostile environment for biological samples. The presence of gas causes scattering of the electron beams, so with the exception of certain specialist machines employing differential pumping techniques, the sample stage is under high vacuum. In addition, the sample is subject to bombardment by a highly damaging beam of charged particles, with increasing electron doses required for high resolution imaging [7]. The biological electron microscopist is therefore faced with the challenge of how to preserve biological samples under these difficult conditions, while keeping sample structure as close to the native state as possible. The traditional approach has been a combination of dehydration and some sort of chemical fixation. This approach varies depending on the type of microscope and the nature of the sample. Transmission EM requires thin samples through which the electron beam can pass. These can be individual biological macromolecules, which are loaded onto a thin support film and stained, usually with a heavy metal compound such as uranyl acetate [8]. Thin sections (~100 nm) from, for example, cells and tissue can also be imaged. In this case, the samples are usually infiltrated with an epoxy resin, producing solid blocks that can then be cut into sections using an ultramicrotome [9] or milled into thin lamellas using a focussed ion beam (FIB) [10]. Scanning EM is a surface technique and so does not necessarily require sample sectioning, but dehydration and fixation of the sample is still required. This usually involves chemical cross-linking of proteins with glutaraldehyde or formaldehyde, sometimes followed by fixation of lipids using osmium tetroxide. The samples are then dehydrated using solvents at a series of concentrations [11]. Scanning electron microscopy (SEM) can be used to produce 3D volume images by serial sectioning and imaging of the sample, either mechanically or by FIB [12].

These sample preparation approaches have been in use since the invention of electron microscopy, and studies based on these methods have revealed many secrets of cell and molecular biology. However, the fact remains that even if the correct procedures are followed rigorously and carefully, dehydration and fixation still changes the sample and introduces artefacts that are visible at the resolution of the EM. The most well-known of these is probably the so-called 'mesosome' in bacteria, first described in the 1970s but later shown to be a fixation artefact [13]. A number of types of artefact is possible, and their description is beyond the scope of this section, but more information can be found in the literature [e.g. 8, 14]. The potential for chemical fixation techniques to introduce artefacts therefore required the development of an alternative method to keep samples stable in the EM stage, leading to the birth of cryo-EM.

8.2.2 The development of cryo-EM and the current state-of-the-art

Freezing is an obvious method for the preservation of biological material. It should therefore be an ideal way of achieving the same result as chemical fixation, i.e. stabilising the sample for high resolution imaging, and allowing it to withstand the harsh environment of the electron microscope sample stage. If the sample can be maintained at a low enough temperature, water will not evaporate and the sample's hydrated state will be maintained. The potential for cryo-fixation to reduce electron beam damage was first recognised some time ago [15]. Also, the use of freezing as part of the sample preparation process for EM was already common, with methods such as freeze-drying [16], freeze-fracture [17], and freeze-substitution [11], although the final product was always dehydrated. Before cryo-fixation could become routine for the imaging of frozen hydrated samples in the EM, a problem had to be overcome: damage to biological structures from the formation of ice crystals during the freezing process.

In fact, it had been known for some time that frozen water can adopt a non-crystalline 'vitreous' state, and that this state could be achieved in principle by cooling the water very rapidly [18]. However, achieving the goal of vitrifying biological samples proved to be elusive for decades. The breakthrough occurred in the early 1980s when it was discovered that thin films of aqueous solutions of biological molecules could be vitrified by rapid plunging into liquid ethane [19]. Ethane is a much more effective cryogen than liquid nitrogen because it avoids the Leidenfrost effect, which occurs when immersion of a warm object into a liquid with a much lower boiling point results in the formation of an insulating layer of gas around the object, slowing cooling. The use of liquid ethane allows the cooling rates necessary for vitrification (10^4–10^6 K s^{-1}) to be achieved. Once this barrier had been overcome, the pathway was open for the visualisation of unstained, frozen biological macromolecules in the transmission electron microscope (TEM).

Throughout the following decades an enormous amount of work went into developing and optimising cryo-TEM of biological macromolecules. There is insufficient space here for a full description of these developments, but a number of excellent review articles has been published [20–22]. Instead, we skip to a description of the workflow currently employed for the preparation and examination of vitrified biological molecules in solution, known as single-particle EM. In order to reduce background noise and improve contrast, the frozen samples are best viewed in the absence of any background supporting material. This is achieved through the use of carbon support films containing holes of various sizes, which are supported by a standard TEM specimen grid. These films can be prepared in the laboratory [23, 24], and are also available commercially. More recently, support films with arrays of regularly-sized holes have become available (Quantifoil, Hatfield, PA, USA), as shown in figure 8.1(a). In order to ensure even spreading of the sample solution, the grids are usually glow-discharged before use. A solution of the biological sample is then applied to the grid in the form of a droplet. The optimal sample concentration varies and must ultimately be determined by trial and error, but concentrations are typically of the order of 100 μg mL^{-1} or lower. The grid is then blotted, with the aim of leaving a thin film of the solution suspended in the holes of the grid, before being plunged rapidly into the cryogen (usually liquid ethane or propane). This can be done manually, using filter paper and a simple gravity-driven device that drops the sample into the cryogen. However, it is now more commonly done using specially designed equipment that achieves consistent blotting and freezing (e.g. the Vitrobot from ThermoFisher).

Following freezing, the sample grid is placed into a transfer device that allows it to be put into the EM column through an airlock. These devices keep the sample frozen using a dewar filled with liquid nitrogen, with heat being conducted away from the sample through a metal rod. When a sufficiently good vacuum has been achieved, the sample can be imaged in the TEM, using a phase contrast approach, and keeping electron beam exposure to a minimum to avoid radiation damage. A typical cryo-TEM image of a biological sample is shown in figure 8.1(b). In the early days of cryo-TEM, images were recorded on film, but now direct electron detectors [26] are used, a development that has revolutionised the technique. Cryo-TEM images can be used to obtain high resolution structures of the molecule being observed. Thousands of images of individual particles are aligned, classified, and software used to reconstruct the 3D structure at high resolution. Again, the development of cryo-TEM analysis is a story in itself, but the developments that have taken place over the years have moved it from a relatively low-resolution

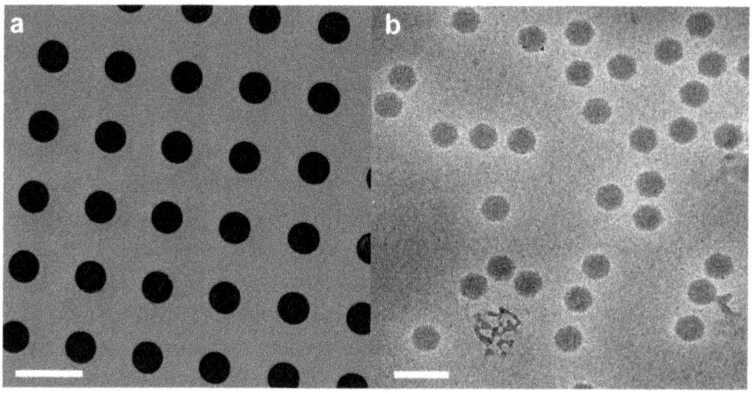

Figure 8.1. (a) Scanning electron microscope image of a Quantifoil support film (courtesy of Benji Bateman). Scale bar 4 μm. (b) Cryo-TEM image of micrograph of adenovirus type 2 suspended in vitrified ice. Scale bar 150 nm. Reprinted from [25], copyright (2008), with permission from Elsevier.

technique (tens of Angstroms) to one competitive with x-ray crystallography [27]. At the time of writing, a search of the Protein Data Bank [28] reveals the highest resolution structure currently determined by cryo-TEM to be that of Apoferritin, with a resolution of 1.25 Å [29]. Around 500 structures with a resolution better than 3 Å have been recorded. This is a remarkable achievement, recognised through the award of the 2017 Nobel Prize for Chemistry to three pioneers of cryo-TEM: Jacques Dubochet, Joachim Frank and Richard Henderson. Single-particle cryo-TEM is now the method of choice for structure determination of viruses and protein complexes, and the number of structures solved using this method is growing at a remarkable rate.

The single-particle work described above has received much attention in recent years, but its development was paralleled by the development of cryo-TEM techniques for the study of cells and tissue. In particular, the development of cryo-electron tomography (cryo-ET) has provided a great deal of highly valuable information on cellular and sub-cellular structure. In this technique, a thin (<500 nm) slice of frozen biological material is imaged in the TEM at a series of angles with respect to the electron beam [30]. Images from these 'tilt series' can then be combined computationally to produce a high-resolution 3D image of the sample. The challenge with cryo-ET is to vitrify 3D volumes of cells or tissue, and then to obtain sufficiently thin sections to allow imaging with electrons. Because of the increased mass and volume of these types of samples, the simple plunge-freezing approach cannot achieve freezing rates rapid enough to avoid ice crystal formation. As an alternative, high-pressure freezing techniques are employed. With this method, the sample is exposed to the cryogen at high pressures (~200 MPa), which prevent the formation of crystalline water ice [31]. Initially, sample sectioning was achieved by cryoultramicrotomy, in which the frozen samples are cut into thin slices using a cold diamond or glass knife [32]. Although this technique can provide good results, it is difficult and can lead to the formation of artefacts, such as sample compression. In more recent years, the use of focussed ion beams at cryogenic temperatures has become a valuable technique for the preparation of vitrified thin samples (known as lamellas) for cryo-ET. In this

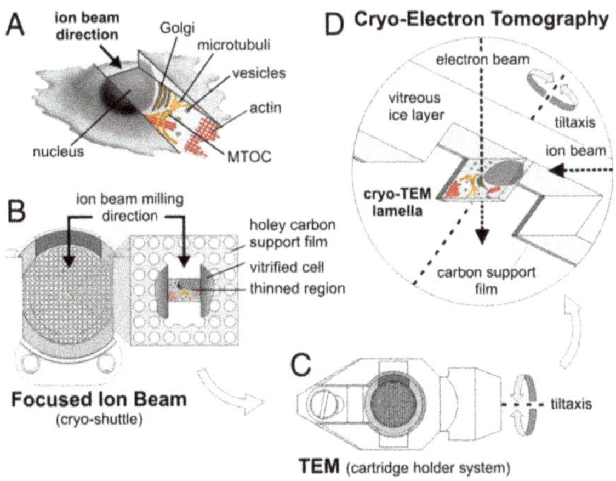

Figure 8.2. A schematic of the cryo-ET process. (a) Ion-beam thinning process. (b) Preparation of lamella on EM grid. (c) Grid placed in cryo-holder. (d) Tilt series collected from the lamella in the TEM. Reprinted with permission from [33].

method, a focussed beam of gallium ions is used to ablate the frozen sample, producing a thinned region which can then be imaged in the TEM [33, 34]. A schematic of the process is shown in figure 8.2.

The motivations for using cryogenic techniques in SEM are the same as those for its use in TEM: preserving sample integrity in the microscope stage while retaining as close to native structure as possible. The principles of sample preparation are also similar, in that the sample must be frozen rapidly to ensure structure preservation. For cryo-SEM, samples are frequently frozen by plunging into liquid nitrogen slush. They are then mounted on a cold stage, where further preparation steps take place before imaging. These typically include sublimation of surface ice and coating with a metal (e.g. platinum) to improve conductivity, minimising charging effects from the electron beam. The samples may also be freeze-fractured to reveal areas of interest within the sample. Cryo-SEM has found a wide range of applications in the characterisation of biological samples (e.g. [35, 36]).

8.2.3 The use of cryogenic techniques in correlative light and electron microscopy

Although EM can provide extremely high-resolution images, it remains a challenge for EM in cells to identify and locate specific regions of interest. Conversely, optical microscopy, and in particular fluorescence microscopy, while generally offering lower resolution, does enable the specific labelling of molecules of interest within the cell. These complementary characteristics have given rise to the technique of correlative light and electron microscopy (CLEM). In this method, fluorescence imaging is used to identify regions of interest (ROIs), which are then imaged at high resolution in the EM [37]. In simple terms, the CLEM workflow involves labelling the samples with a fluorescent marker, which can be achieved either by chemical labelling of molecules that target the protein of interest (specific ligands, antibodies, and antibody-like molecules), or by the expression of fluorescent proteins through genetic modification [38]. Fluorescence microscopy images are recorded, identifying the ROIs. The sample is then prepared for electron microscopy, and the previously identified ROIs are imaged at high resolution.

Cryogenic techniques are already playing a significant role in CLEM. In this case, the typical workflow involves cells being imaged at room temperature by fluorescence microscopy, followed by freezing on EM grids, fluorescence microscopy at cryogenic temperatures, and finally imaging in the EM [39]. This approach has been invaluable in targeting specific ROIs in the EM, but is still limited by one factor: the resolution mismatch between fluorescence and electron microscopy. This limits the technique to the identification of fairly general ROIs rather than specific sub-cellular structures, molecular assemblies, and molecules. The development of sub-diffraction limit or 'super-resolution' fluorescence imaging methods [40] has the potential to close the resolution gap between the two techniques. There are, however, barriers to super-resolution fluorescence microscopy at cryogenic temperatures; firstly, the requirement for high numerical aperture (NA) oil immersion objective lenses, which are incompatible with low temperatures. Secondly, super-resolution fluorescence techniques usually require relatively high levels of sample illumination, which can result in localised ice melting and devitrification. In spite of these difficulties, cryo-CLEM workflows using super-resolution fluorescence have been described, using in-air objectives and lowering illumination intensities [41, 42], the latter employing fluorescence microscopy on a SEM sample stage, eliminating the requirement for transfer of the sample between microscopes. An alternative approach that shows promise is the use of solid immersion lenses (SILs), which provide very high NAs without the requirement for immersion fluids, and have been demonstrated to work at cryogenic temperatures [43].

8.2.4 Challenges and future perspectives

Perhaps the ultimate goal in structural and cellular biology is to determine atomic resolution structure from biomolecules in their 'natural environment' in the cell. One way to achieve this would be to combine the structural determination of single-particle cryo-TEM with the imaging of cells and cellular structures obtained by cryo-ET. The barrier to this is the highly heterogeneous and crowded environment of the cell. In single-particle cryo-TEM, only one molecular species is present, and structure determination depends simply on acquiring images of a sufficient number of molecules to achieve adequate resolution. In the cell, there is a large number of different molecular types, and the resolution of a single TEM image is typically insufficient to identify specific molecules by structure alone. We therefore need a technique which will allow the identification and annotation of specific molecules, so that only those molecules are averaged from the tomograms and used for structural analysis. It is possible to conceive of a 'super-CLEM' approach in which, instead of identifying general regions of interest by fluorescence microscopy, specific molecules are identified. To date the resolution of even super-resolution fluorescence methods has been insufficient to enable the identification of individual molecules within the cell. However, a recently described fluorescence microscopy technique, 'MINFLUX' can achieve 1–3 nm resolution in 3D [44]. If MINFLUX could be used at cryogenic temperatures, it could in principle be used to annotate individual molecules for averaging and structural analysis from cryo-electron tomograms. This would be truly revolutionary, bringing together structural and cell biology and giving us structure within, rather than isolated from, the cellular environment, with unprecedented insight into the structure–function relationship.

A second area for development is in the instrumentation for cryo-CLEM. In most cases cryo-CLEM workflows require the fluorescence and electron microscopy to be carried out in separate instruments, using fiducial markers to locate the same areas in both microscopes. This limits the precision of correlation and introduces the possibility of distortion and contamination during the transfer process. Combining both fluorescence and electron microscopy in a single instrument would avoid these limitations. The sample chambers of SEMs are relatively spacious, and it is therefore reasonably easy to incorporate basic optics for fluorescence microscopy. An example is the 'PIE-scope', which integrates an optical microscopy set-up into a FIB-SEM [45]. Achieving super-resolution fluorescence imaging in such a set-up is more difficult, because of the need for the objective lens to be close to the sample. The SIL-based approach described by Wang *et al* [4] may bring benefits here. Combining fluorescence and TEM in the same instrument is extremely challenging, because of space limitations. Such microscopes have been described [46], but more work is required before they will be generally available.

In conclusion, it is no exaggeration to say that the use of cryogenic methods has already transformed the field of biological microscopy, in particular with the huge advances in structure determination from single-particle cryo-TEM. However, in some ways it can be said that the work has only just begun. Exciting times are ahead, with the potential for some truly revolutionary developments.

8.3 Cryosurgery

Louise Cuff

8.3.1 History of cryosurgery

Cryogenic surgery, or cryosurgery, is the controlled destruction of tissue by freezing. The benefits of cold have been recognized for thousands of years; the Egyptians used cold to treat injuries and inflammation as far back as 2500 BCE. Today this process is widely practiced in the medical, cosmetic, sports and wellbeing industries.

The past couple of centuries have seen the greatest advancements in ways of working with cold. Through the development of new applications, breakthrough technology and innovative design of medical devices, it is now possible to achieve extremely cold temperatures capable of treating some of the most complex medical conditions. Cryosurgery today is vastly different from where it began.

8.3.2 Pre surgical device applications

- The first person to use extreme cold for the destruction of tissue was English physician James Arnott in the mid-1800s. He used a mixture of salt and crushed ice for palliation of tumours and his salt solutions reached temperatures of −18 °C to −24 °C.

- Twenty-five years later, the technology capable of compressing and liquifying atmospheric gases was developed. Meaning it was possible to achieve extremely low temperatures.
- By the end of the 19th Century, the use of liquid air (−194 °C) or liquid oxygen (−182.9 °C) was also deemed useful to treat a large range of ailments such as warts and leg ulcers.
- In the early 1900s, William Pusey popularized the use of carbon dioxide snow. This acted as an alternative to liquid air as it was easy to produce using the Joule–Thomson principle. As the temperature for this process was limited to −79 °C it could only be used for surface treatments, such as the removal of warts and moles with low level scarring.
- After the Second World War, liquid nitrogen was in abundance and was considered safer to use than liquid oxygen. In 1950, Dr Ray Allington introduced cotton swabs dipped in liquid nitrogen, to treat a variety of skin diseases. However, heat transfer between swab and skin was insufficient for tumour treatment.

8.3.3 Early surgical device applications

- In 1913 Dr Irving S Cooper designed a liquid nitrogen probe that achieved temperatures of −196 °C. By freezing the thalamus, he treated previously inoperable brain tumours, as well as Parkinson's disease and other movement disorders.
- In 1961, Cooper and Lee contributed enormously to the evolution of cryosurgery, by designing the first automated cryosurgical apparatus cooled by liquid nitrogen.
- In 1965, Dr Douglas Paul Torre developed the first liquid nitrogen spray. Two years later his design was adapted into a hand-held device called the Kryospray—a basic device used for the application of liquid nitrogen, directly onto moles and skin cancers.
- In 1967, a new approach to cryogenic therapy was emerging. Torre used an experimental argon gas system, which could achieve temperatures as cold as −185 °C, via the Joule–Thomson effect.
- SP Amoils developed a gas-based closed cryoprobe. With this type of device, closing the probe tip allows gas and liquid to cool the tip. He wrote, 'It is the first cryoprobe that can be applied to thc lens at ambient temperature (room temperature) having almost instantaneous freezing and rewarming characteristics.' This kind of probe is more in line with modern day gas-based cryosurgical probes (figure 8.3).

8.3.4 Modern day cryosurgical devices

Modern day cryosurgical devices predominantly fall into two categories, a gas-based system or a liquefied gas system. The two types have different characteristics and are therefore generally used for different surgical procedures.

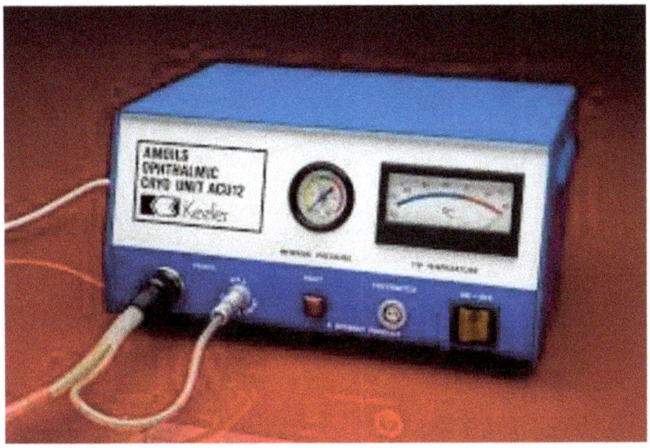

Figure 8.3. Early ophthalmic cryounit (image: KeelerAmoils).

Gas-based systems typically produce freeze temperatures of around −60 °C to −70 °C, and as a consequence will have smaller ice growth compared with liquefied gas systems. These devices are well suited to delicate procedures where the freeze zone must be carefully controlled to avoid harm to the patient.

Liquefied gas systems are capable of achieving much lower freeze temperatures with the patient-applied part reaching temperatures as low as −196 °C with a liquid nitrogen system. Consequently, ice growth is considerably larger than a gas-based system, making them very suitable for the treatment of cancerous tumours which require very low temperatures to ensure the destruction of cells.

8.3.5 Liquefied gas cryosurgical applications

Sophisticated liquid nitrogen systems, using closed cryoprobes, are used for the treatment of various cancers such as, breast, liver, prostate and kidney. The process is known as cryoablation. A needle is inserted into the tumor and the probe is then inserted percutaneously, with the needle acting as a guide. A freeze cycle is then initiated, and the destruction of unwanted tissue begins (figure 8.4).

Liquid nitrogen spray guns are commonly used to treat a range of dermatology conditions, including skin tags, moles, skin cancer and warts. The spray gun is a simple device that works by submersing a metal rod into the liquid nitrogen which causes the nitrogen to boil and generate pressure in the vessel. The liquid nitrogen can then be sprayed directly on the treatment site by activating a trigger.

8.3.6 Gas-based cryosurgical applications

Cryosurgery is widely used in ophthalmic surgery and is most frequently used for the treatment of retinal tears or full retinal detachment. Cryosurgery retinal attachment works by irritating the tissue causing it to grow and form a strong seal along the tear (figure 8.5).

Cryosurgery can also be used to treat Trichiasis, a condition of abnormal eyelash growth where an extra set of lashes grows behind the normal lashes. These extra

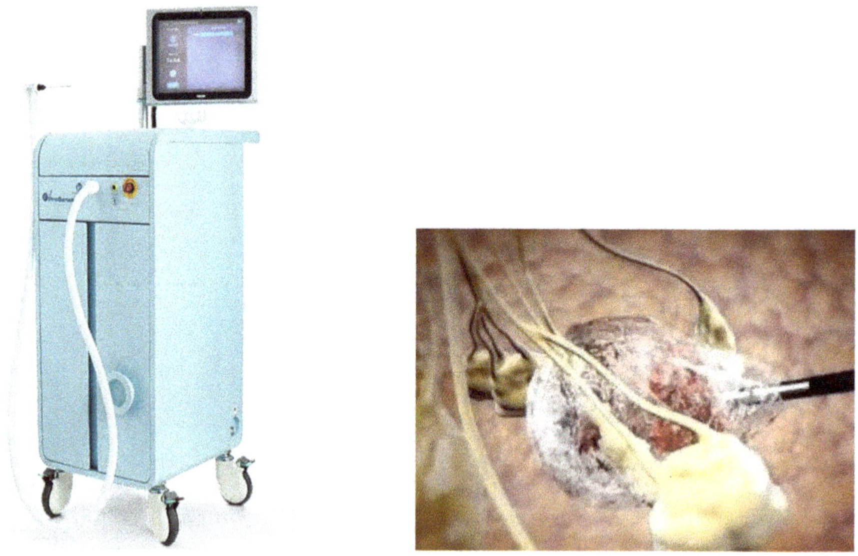

Figure 8.4. Liquid nitrogen system (image: IceCure ProSense).

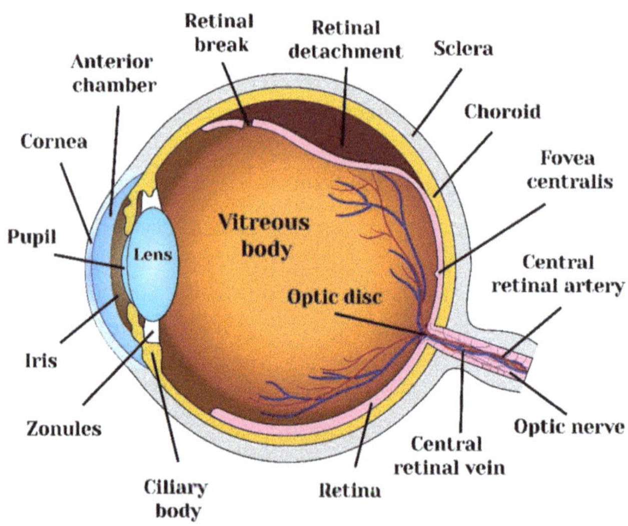

Figure 8.5. The eye (source: Mrs_Bazilio/Shutterstock).

lashes grow inward towards the eye causing irritation and possible damage to the eye. The unwanted eyelashes are removed by freezing the follicles with a cryoprobe.

Another ophthalmic surgery that uses cryo is the removal of cataracts, often referred to as cataract extraction. The cryoprobe is frozen to the surface of the lens so that it is firmly attached prior to extracting the lens with the cryoprobe.

Gynaecologists use cryosurgery as a means to kill abnormal cells, to remove genital warts and to treat cervical dysplasia, the growth of pre-cancerous cells on the cervix.

Cryosurgery is used to treat chronic pain. Devices consist of a combination of nerve stimulation and cryotherapeutic technologies, within a single medical device. Treatment with one of these devices enables drug-free pain relief, as temporary destruction of nerves block pain for approximately six months. Percutaneous, minimally invasive procedures are the preferred method for this type of chronic pain. To achieve electrical stimulation of the nerve structures, electrodes or probes are placed on the target nerve (figure 8.6).

Podiatrists use cryosurgery for the treatment of Plantar Fasciitis by applying extreme cold to inflamed tissues which very quickly reduce pain and inflammation. For the treatment of Morton's Neuroma, the cryoprobe is used to destroy nerve tissue.

Heart arrythmia or irregular heartbeats are successfully treated with cryosurgery. Surgery can be performed as an open procedure, or by using catheter cryoprobes which are long and thin and can be inserted directly into the artery in the groin. The cryo-catheter probe is then guided through the artery up to the heart.

Lung biopsies can be carried out using cryo-catheter probes. A long thin probe is guided to the biopsy site and the freeze cycle activated; this causes cryoadhesion to the tissue which can then be extracted. In a similar way, this process can also be used to remove foreign bodies from the lungs (figure 8.7).

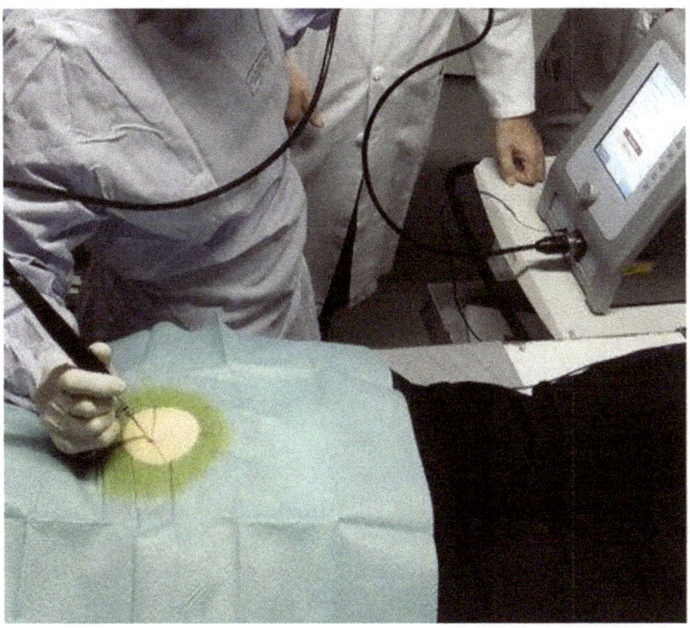

Figure 8.6. CryoSystem for nerve freezing (image: Inomed C3).

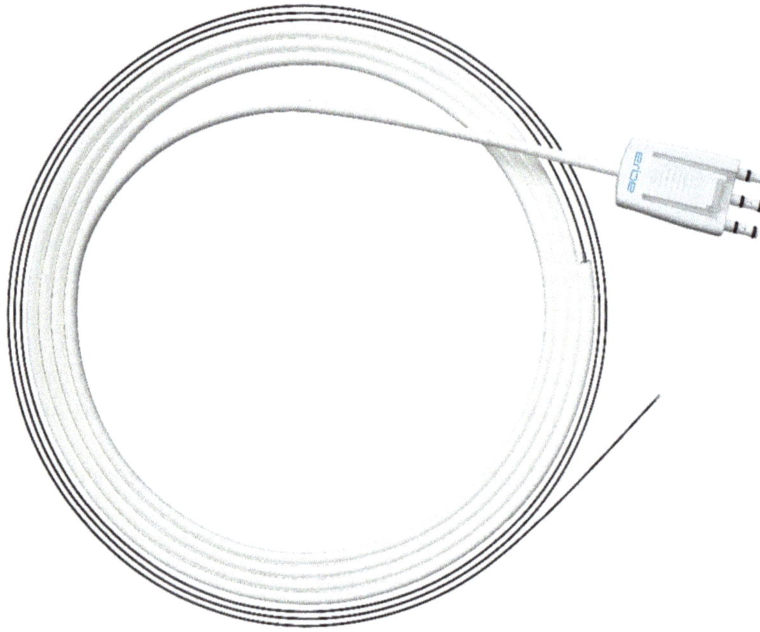

Figure 8.7. Flexible cryoprobe for lung biopsies (image: ERBE).

8.3.7 Liquefied gas cryosurgical devices

Liquefied gas systems consist of a probe and a console, they use liquefied gas such as nitrogen, helium or argon, liquid nitrogen being the most frequently used. The probe tip is cooled by circulating the liquefied gas through the probe tip, with a heat exchanger incorporated inside the tip to facilitate the boiling of the liquid gas at the freeze area. The console is the means of storing the liquefied gas, delivering it to the probe and controlling the flow and pressure (figure 8.8).

Technological advancements in imaging and keyhole surgery techniques have made it possible to perform many different types of surgeries using this technique. This has driven the need for cryosurgical probes to become smaller in diameter whilst increasing the freeze performance. This can present additional challenges in the design and development of cryoprobes and systems. In order to prevent unwanted freezing during keyhole surgeries, probe stems are vacuum insulated.

Liquefied gas systems tend to operate at lower pressures than gas-based systems and much lower temperatures are achieved with liquefied gases (−196 °C for nitrogen).

In order to achieve optimal performance of liquefied gas systems, it is essential to achieve nucleate boiling at the tip (patient-applied part) and to prevent liquid boiling in other parts of the system. Two types of boiling can occur, nucleate or film boiling.

Nucleate boiling—bubbles form on the surface but there is still fluid in contact with the surface. Liquid transfers cold to the surface, therefore producing much lower external tip temperatures; typically this can be within the region of −150 °C to −190 °C on the outside of the tip (figure 8.9).

Figure 8.8. Liquid cryogenic probe (image: IceCure ProSense probe).

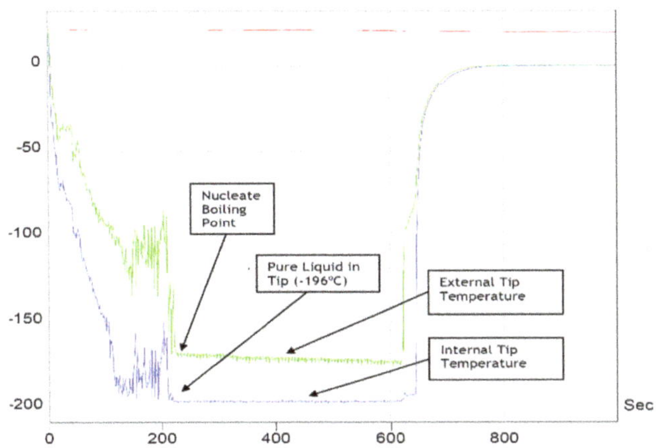

Figure 8.9. Boiling process.

Film boiling—the surface it comes into contact with is relatively hot so that all liquid in contact with the surface boils. This creates an insulating layer of vapour between the surface of the tip and the liquefied gas. This prevents cold transfer from the liquid flowing into the tip to the tip surface, resulting in poor external tip temperatures and a poor surgical outcome.

Nucleate boiling in cryosurgical probes is characterised by a sudden drop in both internal and external tip temperature as shown in this graph (figure 8.9).

8.3.8 Gas-based cryosurgical device

Gas cryo systems consist of a probe and a console and use cryogenic gases such as carbon dioxide or nitrous oxide (figure 8.10). The probe is the patient-applied part

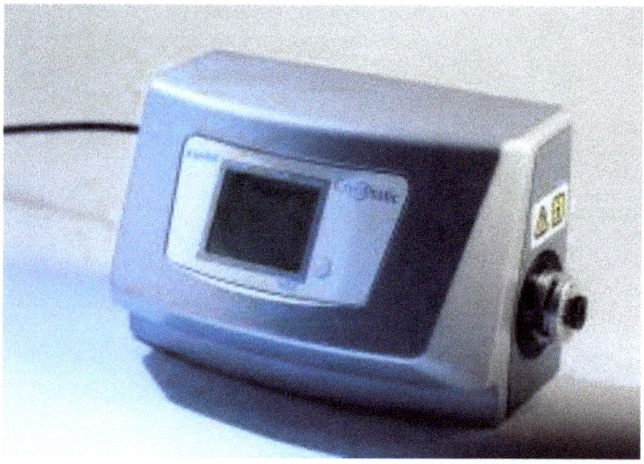

Figure 8.10. The latest technology (image: Keeler Cryomatic II).

and has an enclosed tip. The outside of the tip gets cold and is applied to the treatment area. The console is the means of connecting the probe to the gas bottle and sets the regulated freeze/purge pressure, activates defrost/freeze cycles and can offer other features such as a timer, alarm, pre-set freeze durations or temperature indicator.

Gas-based cryosurgical devices are based on the Joule–Thomson effect. A temperature drop is achieved in the tip of the cryosurgical probe by throttling the gas to achieve a sudden pressure drop. Temperatures in the region of −60 °C to −70 °C can be achieved at the tip by using carbon dioxide or nitrous oxide. Equalising the pressure results in an almost instant defrost.

Single-use probes are becoming increasingly more popular for gas-based systems as there is no reprocessing requirement and the risk of cross contamination is significantly reduced. Many hospitals and clinics do not have sterilising facilities on site, so reprocessing is outsourced, resulting in increased downtime and costs.

8.4 Cryotherapy

Agnieszka Piotrowska

8.4.1 Cryochambers

Although treating pain with cold (such as ice) is natural and has been well known for ages, there is no broad historical background for medical cryogenic chambers. They did not evolve from winter/ice swimming or low temperature rooms which might have been a more natural way. The idea of an extremely cold room used for medical treatment of pain was introduced more than 40 years ago. The first cryochamber (with an iron casting weighing more than 70 tons!) was fabricated in Japan in 1978 by Professor Toshiro Yamauchi and his team from Reiken Rheumatism Village Institute in Oita. The second unit was constructed in Germany in 1982 by a team led by Professor Reinhard Friecke, who developed the first standards for use of

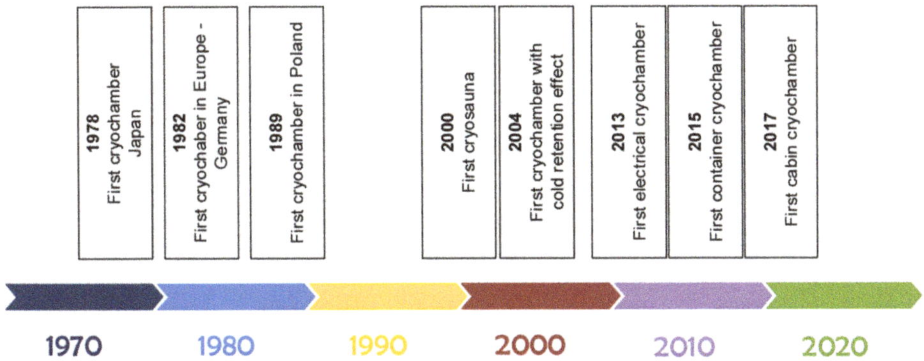

Figure 8.11. Timeline of cryochamber's evolution.

cryotherapy in medicine. The third cryochamber in the world was designed and constructed in Poland in 1989 by engineer Zbigniew Raczkowski. These three systems definitely opened a new path, creating whole-body cryotherapy (WBC)—therapy using temperatures below −100 °C to cause a natural defensive reaction in the human body. However, cryochambers were outside the interest of most medical doctors and engineers in the 80s and 90s. Only a few countries have been continuously developing, improving and testing cryogenic chambers from both a technical and medical point of view. Poland is the undisputed leader in cryotherapy and cryochamber technology right now. It's the only European country with more than 250 cryochambers in operation, located all over the country. Experience of several million cryotherapy sessions has been gained thanks to cryotherapy being reimbursed by national health insurance (the Polish National Health Fund). Since 1989, cryotherapy has become a recognized method in Poland supporting treatment of mobility disorders. Until 2010, cryogenic chambers were dedicated to medical applications only, so they could be found exclusively in hospitals or rehabilitation centres. Starting from 2015, the group of users extended. Professional athletes and their trainers gave attention to WBC as an alternative method of post-training recovery. Growing interest from sport, spa and wellness industries opened a new market for manufacturers. Cryotherapy devices have become available for commercial (non-medical) users.

Over the last 40 years, constructions and supply systems of cryochambers have been slowly evolving from industrial-type prototypes to desirable commercial products. The evolution is graphically presented in figure 8.11.

Currently, the commercial market offers different constructions and cryogenic gas or electrical supply system options which will be described in detail in the next section.

8.4.2 Definition and categorization of the cryotherapy devices

Going through the literature, it becomes clear that there is no standard definition of a cryochamber. Furthermore, the terms cryochamber and cryosauna are used interchangeably for cryotherapy devices. This misleads users and creates disorder in the literature. Which device can then be called a cryochamber?

The word 'chamber' suggests enclosed space, whereas the prefix 'cryo' indicates the fundamental role of temperature level. Cryotherapy treatments require temperatures below −100 °C. Despite (some) definition(s) of cryogenics starting at temperatures below −153 °C (120 K), temperatures from −90 °C can be considered cryogenic as well since they are achievable only by cryogenic technologies and fluids. They are below the range of standard refrigeration technologies and refrigerants.

Therefore, the term 'cryochamber' should be used only for cryotherapy devices that provide a minimum operating (treatment) temperature below −100 °C inside an enclosed space, allowing the entire human body to be exposed to cold. Sessions performed in cryochambers are called whole-body cryotherapy (WBC).

There are currently three types of cryochamber available on the commercial market: standard (also called classic or 'Wroclaw-type'), cabin, and cryochambers with cold retention effect.

Another type of cryotherapy device, often confused with the cryochamber, is the cryosauna. The cryosauna is an individual-size barrel-like enclosure, open at the top. The user's torso and legs are enclosed in the device and exposed to cold vapour of liquid nitrogen while the head remains above the enclosure at room temperature. The session should be called partial body cryotherapy (PBC) because the head, with thousands of cold receptors, is not stimulated by low temperature. Despite their fundamental differences and endless discussions of terminology, the cryosauna is often considered to be as effective and safe as a cryochamber, which is perhaps not so clear. The general categorization of cryotherapy devices is presented in figure 8.12.

The following sections will focus on standard LN_2, electric cryochambers and cryosauna. These three constructions have dominated the global market. Cryochambers with cooling retention effect account for less than 2% of the total number of cryochambers installed so far. The low number of installations results from the need for specific cellar-like construction (requiring space on two floors) and use of liquid synthetic air as the medium, which is inconvenient in use and can be unsafe.

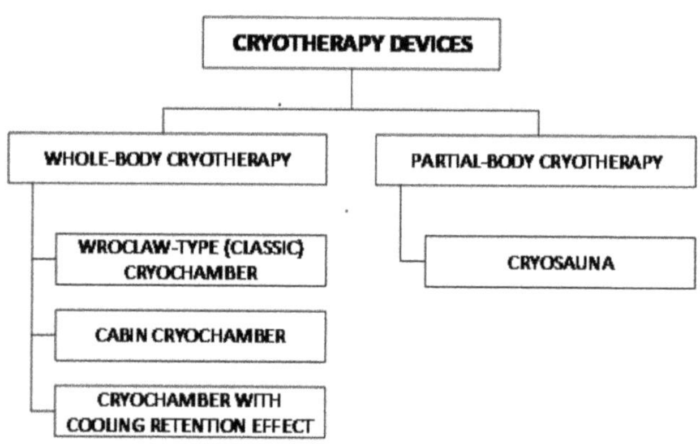

Figure 8.12. General categorization of cryotherapy devices.

8.4.3 The standard (Wroclaw-type) cryochamber

The standard cryochamber (also called 'Wroclaw-type' from the name of city where the first Polish cryochambers were constructed) consists of a minimum of two rooms: pre- and main chamber (figure 8.13). Cold is produced by evaporation of a liquid medium inside heat exchangers located in both pre- and main chambers. The temperature inside the pre-chamber is kept at −60 °C, while inside the main chamber from −110 °C to −140 °C. The pre-chamber plays a double role; for the user it is a place where he/she can adjust to the cold, and from a technical point of view it helps to keep low temperature inside the main chamber, reducing cold losses.

The cryotherapy session starts in the pre-chamber where the user spends 30 s. The user spends the remaining time inside the main chamber. At the end of the session, the user returns to the pre-chamber and then to the exit. The entire session does not exceed 3 min.

There are two types of supply system: liquid nitrogen and electrical (compressor driven). Synthetic air also used to be considered as a cooling medium. However, the construction of the supply system was basically the same as the nitrogen one, so it was not economically nor safely justified to use liquid air.

Most cryochambers are supplied by liquid nitrogen (because of low cost of LN_2). The general scheme of cryochamber supplied by liquid nitrogen is presented in figure 8.14.

Liquid nitrogen is transferred from the cryogenic vessel via vacuum or foam insulated transfer line to be evaporated inside the heat exchangers and safely released back to the atmosphere. This solution prevents direct contact between the user and liquid nitrogen. To maximize safety for both user and personnel, oxygen level is constantly monitored.

The alternative method to liquid nitrogen is the (cryo)refrigerator (also called electrical system). The construction of the cryochamber itself is similar to the liquid nitrogen unit. The general scheme of the electric cryochamber is presented in figure 8.15.

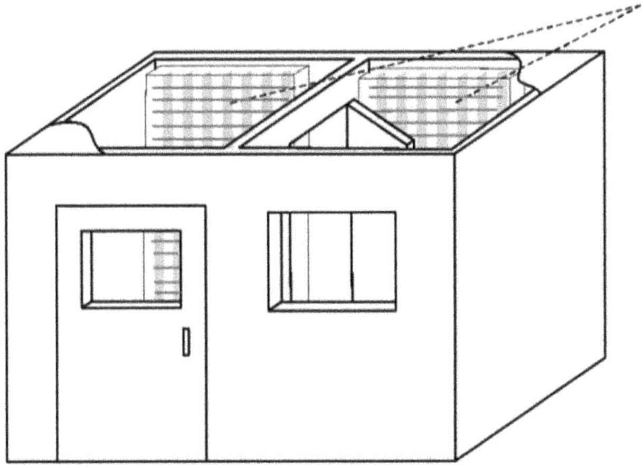

Figure 8.13. Standard cryochamber scheme.

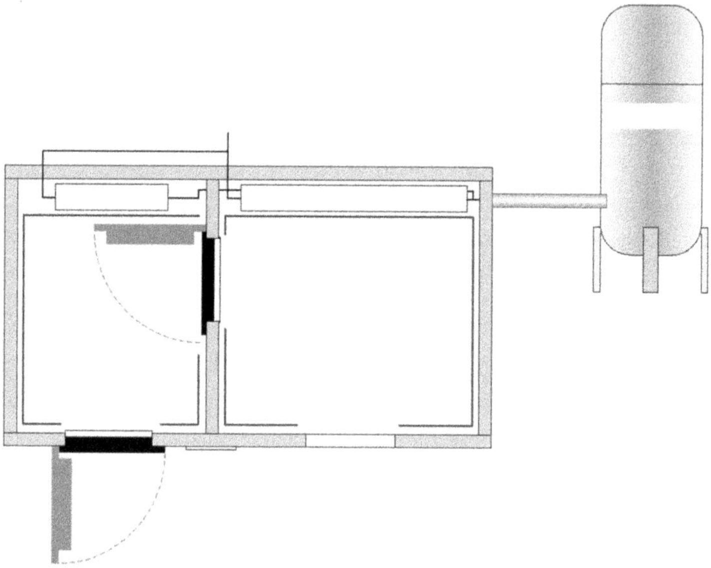

Figure 8.14. Scheme of cryochamber supplied by liquid nitrogen.

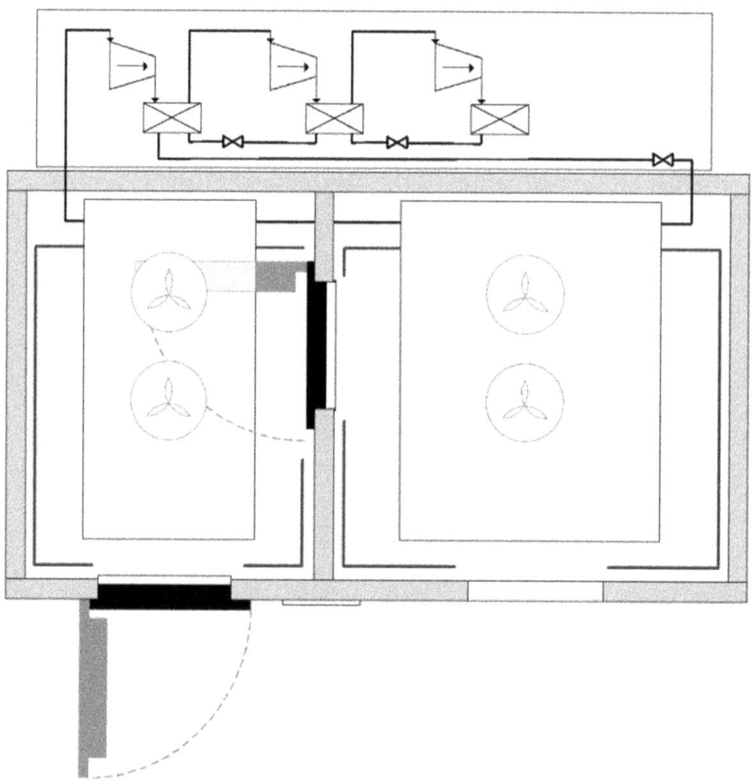

Figure 8.15. Scheme of the electric cryochamber (cryocooler, refrigerator system).

The cooling power required to decrease temperature is produced by a compressor cycle, usually three-stage cascade which assures slightly higher temperature (−110 °C) but doesn't require periodic delivery of cryogenic gas. The main advantages of this kind of system are the closed loop of the refrigerator (no need to replenish the refrigerant, no outflow of the used gas) and the need only for electrical power supply (generally available everywhere). Needing only electrical supply allows installation of systems in crowded areas (like city centres), hard-to-reach places (e.g. luxury spas on islands/in the mountains), places where delivery of liquid nitrogen would be impossible or LN_2 is not available. However, the investment cost can fundamentally limit the commercial market at the time of writing, being often more than double the price of a standard LN_2 cryochamber as a result of much more complex infrastructure. Professional compressors, heat exchangers and several other components increase the final price significantly. Nevertheless, electrical cryochambers (cascade, auto-cascade refrigerators, Joule–Thomson coolers) are currently under strong development.

The main differences between both systems concern session minimum temperature and temperature gradient. Cryochambers supplied with liquid nitrogen are cooled down with the process of natural convection. The system is not equipped with fans (fans may be used only in the cool down mode, without the user), so the speed of air is limited to the minimum to avoid any unpleasant feeling of cold wind inside. Heat exchanger(s) is/are placed at the back and side wall(s). Cold air flows down, filling the room from the bottom to the top. It naturally creates a temperature gradient (along the vertical axis). Temperature of −120 °C to −135 °C is measured at knee height (50–60 cm from the floor) while temperature at 1.5 m (where air is inhaled) is more like −40 °C to −60 °C.

The temperature inside the electric cryochamber (three-stage compressor cascade) doesn't go below −110 °C. However, heat exchangers are installed under the ceiling and the system is equipped with low-speed air fans. It allows reduction in the temperature gradient to 10–15 K of difference between floor and ceiling level. The temperature distribution is equal throughout the entire volume of the main chamber, which results in sensation of cold similar to the LN_2 system.

8.4.4 Cryosauna

The general scheme of a cryosauna system is presented in figure 8.16.

The system consists of an open top cylinder with platform height adjustment in the space where the session takes place, with technical components like heat exchanger, valves and fans are placed in the elsewhere in the construction.

Liquid nitrogen is transferred from a vacuum-insulated vessel to the heat exchanger where it is evaporated. Cold vapours flow directly to the cryosauna cylinder. The user's area of the cryosauna is fully filled with nitrogen vapour which from an operational point of view assures very low temperature of the session (according to the manufacturers down to −160 °C) while from a safety point of view, this can be a source of oxygen deficiency hazard (ODH). The main issue is the atmosphere which surrounds the user. Cold and heavy nitrogen displaces all oxygen from the cylinder. The user has direct contact with the cryogenic medium in

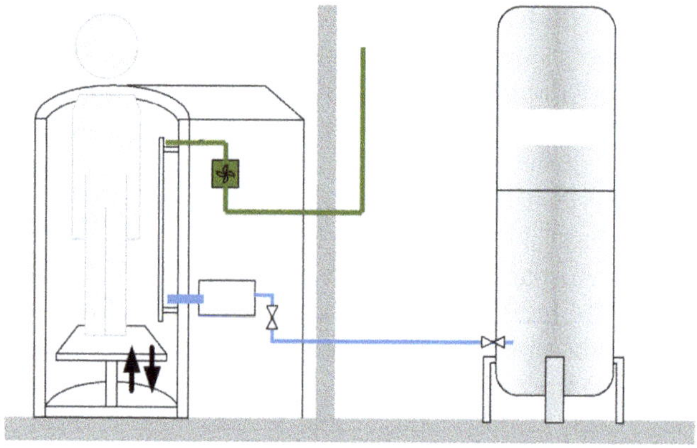

Figure 8.16. Scheme of cryosauna.

standard operating mode (which is not possible in cryochambers) and is submerged into an atmosphere with oxygen content 0%. Therefore, the risk for the user arises not only from damage to technical components, but also from human error.

Temperature gradient is negligible because the session space is relatively small, so it is easy to fully fill with cold vapour. However, most available units give the temperature of cold gas entering the cylinder as the session temperature. In the case of cryosauna, operational parameters like temperature distribution or regeneration time (time to cool down the system required between two consecutive sessions) are secondary issues. The main issue remains the hazard of oxygen deficiency in the standard operation mode.

8.4.5 Safety

Until 2010, cryochambers were installed mainly in hospitals and rehabilitation centres. Cryotherapy was prescribed by a medical doctor to support the basic treatment, the session was preceded by a medical examination and conducted under the supervision of physician and operating personnel (usually with medical preparation). Cryochambers have been certified according to the requirements of directive 93/42/EEC. The group of users was significantly enlarged approximately five years ago. Professional athletes appreciated cryotherapy as a method to speed up post-training recovery or increase the body's efficiency. Further, the anti-ageing effect of low temperatures was presented in several commercial articles and publications. Consequently, cryotherapy devices appeared in training centres, gyms and spa facilities. Without any construction or operational changes, they suddenly began to be treated as commercial, sports or spa equipment. The absence of clear rules enabled a random element to the operational staff, some trained by the manufacturer and some with no qualification. However, the risk has remained the same, requiring safety issues to be addressed.

The first lethal accident in a cryotherapy device was reported in the USA in October 2015. The 24-year-old manager of a Spa centre was found dead after she became trapped in the low temperature chamber overnight. She broke the principal of safety by taking a session without any supervision. Strong discussions concerning safety of cryotherapy devices began, resulting in the FDA statement (U.S. Food and

Drug Administration) released on 5 July 2016. The two-page document titled 'Whole body cryotherapy (WBC), a cool trend that lacks evidence, poses risks' [52], covered general information on WBC, the available devices and potential risks for users. The contents of this document proved that the technology of cryotherapy devices required alignment, with definitions and recommendations or standards for manufacturers—though no new rules appear to have been introduced since.

Analysing standard cryochambers supplied with liquid nitrogen, electric cryochambers and cryosaunas, the following conclusions can be formulated:

Design, manufacturing and operation of the compressor systems (cascade, Joule–Thomson, etc) are much more regulated by law than nitrogen systems (both standard and cryosauna). The installation, commissioning and maintenance service of the compressor refrigerator must be performed according to the applicable regulations and standards, only by qualified and certified personnel, whereas (in Europe) nitrogen systems must be designed and manufactured according to sound engineering practice (SEP) which results from Pressure Equipment Directive 2014/68/EU categorization (nitrogen—fluid of group 2, low pressure approximately 3 bar and relatively small volume of the piping and system components). It gives a lot of freedom to manufacturers. Sometimes too much freedom. The problem is that almost the entire system (heat exchangers, piping) is enclosed within a small space (approx. 50 m^3) and directly connected to the cryogenic storage tank, capacity starting from 120 litres to 1000 litres or more. In case of any leak (volume of nitrogen increases approximately 600 times during evaporation), the closed space of the cryochamber can be filled with cold vapour of N_2 immediately.

The cryosauna is an example of an open cryogenic vessel with regulation of liquid nitrogen flow operated in confined space. Even in standard operation mode it can create a local atmosphere with zero oxygen content.

Looking at the growing interest in cryochambers and cryosaunas located in gyms or spas, we should expect increasing numbers of systems in operation each year which can be operated by staff with no proper qualifications or by the user themselves.

In the author's opinion at the time of writing, clear rules and standards for production, test and commissioning should be introduced immediately. Attention is required from engineers, gas companies and notified bodies.

8.5 Cryogenics in proton and heavier-ion radiotherapy

Richard Amos and Russell Thomas

8.5.1 Rationale for proton radiotherapy

Radiotherapy, an important component of comprehensive cancer care, is well demonstrated to improve overall survival and improve quality of life for patients [47]. The goal of radiotherapy is to induce tumour-cell death by causing irreparable damage to DNA. The efficacy of radiotherapy relies on the accurate targeting and delivery of a therapeutic dose of ionizing radiation to the tumour volume whilst minimizing dose to the surrounding healthy tissue, thus reducing the risk of radiation-induced toxicity.

Proton beam therapy (PBT) is an advanced form of radiotherapy that is able to target the tumour volume whilst sparing more of the surrounding healthy tissue than conventional x-ray-radiotherapy (XRT).

For therapeutic x-ray beams, beyond a build-up of dose at shallow depths, energy deposition decreases approximately exponentially as a function of depth. Any x-ray beam thus penetrates the entire body, depositing dose along its path, with the maximum dose delivered near the surface.

Consequently, the delivery of a therapeutic dose to deep-seated tumours with a single x-ray beam requires the irradiation of tissues both proximal and distal to the target, with the proximal tissue receiving a higher dose than the tumour. In clinical practice this is mitigated by targeting the tumour with beams from multiple angles. This allows for the prescribed dose to be delivered to the tumour volume, but at the cost of increasing the volume of healthy tissue receiving low-dose irradiation with an associated increased risk of inducing secondary malignancies some 20–30 years post-radiotherapy. Reducing this risk is of particular significance for paediatric and young adult patients cured of their disease who may live long enough for such effects to manifest. It is worth recognizing that modern techniques for XRT delivery strive to reduce this 'low-dose bath' as much as possible.

The physical characteristics by which proton beams deposit their energy offers the potential for an improved therapeutic ratio. As a proton beam (or any charged-particle beam) enters a medium, such as a patient, the protons continuously lose energy along their path as a result of Coulomb interactions. As protons lose energy, they lose momentum, slowing down and becoming more densely ionising before eventually stopping. This gives rise to the characteristic low entrance dose and relatively flat dose plateau, before rising sharply to a maximum, known as the Bragg peak, and then falling to zero at the end of the beam range (figure 8.17). The depth of

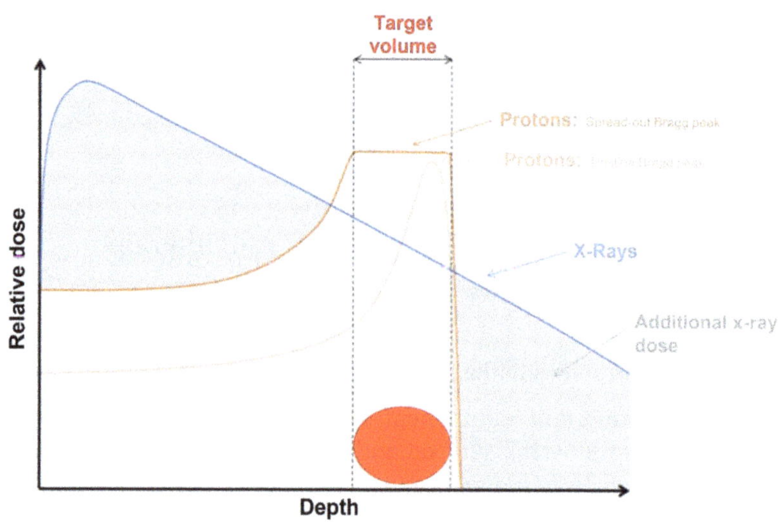

Figure 8.17. Dose deposition characteristics for x-ray and proton beams. (Credit: adapted from [53].

the Bragg peak is dependent upon the composition of the tissue through which the beam is travelling and the energy of the protons. The incident energy of the therapeutic proton beam is chosen such that the depth of the Bragg peak is coincident with the tumour. For the clinical delivery of proton radiotherapy, the so-called pristine Bragg peak of a monogenetic beam is typically too narrow to adequately encompass the target (figure 8.18). To achieve a high-dose region that encompasses the target along the depth direction, the range is modulated such that multiple weighted pristine Bragg peaks at various depths are delivered in superposition resulting in the so-called spread-out Bragg peak (SOBP) (figure 8.17). The

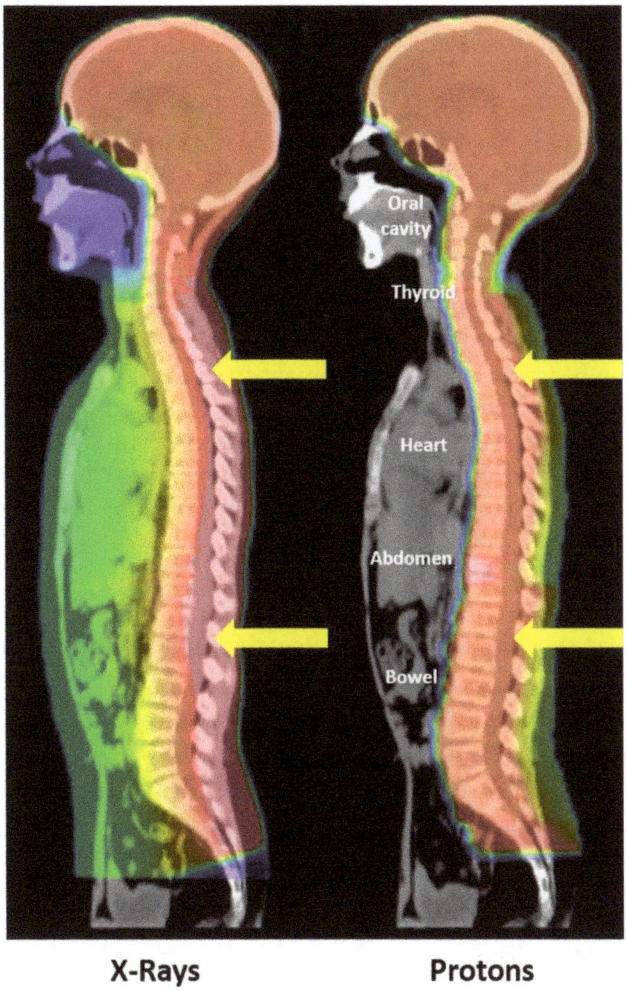

Figure 8.18. Comparison between x-ray and proton irradiation of the craniospinal axis to treat medulloblastoma in a paediatric patient. Radiation field directions are indicated by the yellow arrows. Protons avoid critical structures anterior to the spine, thus reducing the risk of radiation associated toxicity. Reprinted from [54].

reduction in low-dose exposure to healthy tissue with PBT compared to that of XRT is clearly demonstrated in figure 8.18.

8.5.2 Early history of proton radiotherapy

The potential benefit of using proton beams for radiotherapy was first recognized in 1946 [48], and the first treatment of human patients took place in the mid-1950s at the University of California Lawrence Berkeley Laboratory (LBL). This is perhaps both fitting and unsurprising since LBL is where the cyclotron particle accelerator was first developed in the 1930s. There followed a period of development of PBT utilizing particle accelerators in a number of nuclear physics laboratories around the world, where a small number of patients were treated with what could be regarded as an experimental or niche technique. This continued until the late-1980s–early-1990s which saw the advent of hospital-based PBT facilities.

The Clatterbridge Centre for Oncology [Liverpool] in the United Kingdom opened the world's first hospital-based PBT facility in 1989. This facility uses a low-energy, approximately 60 MeV, cyclotron proton source exclusively for the treatment of ocular disease. Just one year later in 1990, the world's first hospital-based PBT facility for the treatment of deep-seated tumours opened at Loma Linda University Medical Center (LLUMC) in California. The LLUMC facility was developed in collaboration with the Fermi National Accelerator Laboratory (FermiLab) in the United States and consisted of a 250 MeV synchrotron serving multiple treatment rooms, three of which contained large rotating isocentric gantries to enable proton beam delivery to the patient from any angle.

The rotating gantry configuration is similar to that used for XRT. However, in order to accommodate the bending magnets required to steer the proton beam around the gantries and towards the beam delivery system focused on the patient, the diameter of each gantry in the LLUMC facility is 13 m. The general design and layout of the LLUMC facility set the mould for many subsequent hospital-based PBT facilities.

8.5.3 Cryo-cooled superconducting technology in modern proton radiotherapy

The number of hospital-based PBT facilities worldwide has grown significantly over the last two decades [49]. The majority of facilities contain a single accelerator (cyclotron, synchrocyclotron, or synchrotron) generating proton energies up to 230–250 MeV, each serving between three and five treatment rooms. The rationale for a single accelerator serving multiple treatment rooms is based on cost; the accelerator is the most expensive hardware of the entire PBT system. However, the cost of each treatment room is not insignificant. The space required to accommodate each large isocentric gantry is essentially a three-storey vault constructed with walls designed for radiation shielding. The recently developed PBT service for the National Health Service (NHS) in the UK was backed by government funding of £250 million. This was to develop two separate facilities, each consisting of a single accelerator and three treatment gantries. Costs associated with the development of PBT facilities such as these are prohibitive to many, so despite the steady increase in patients

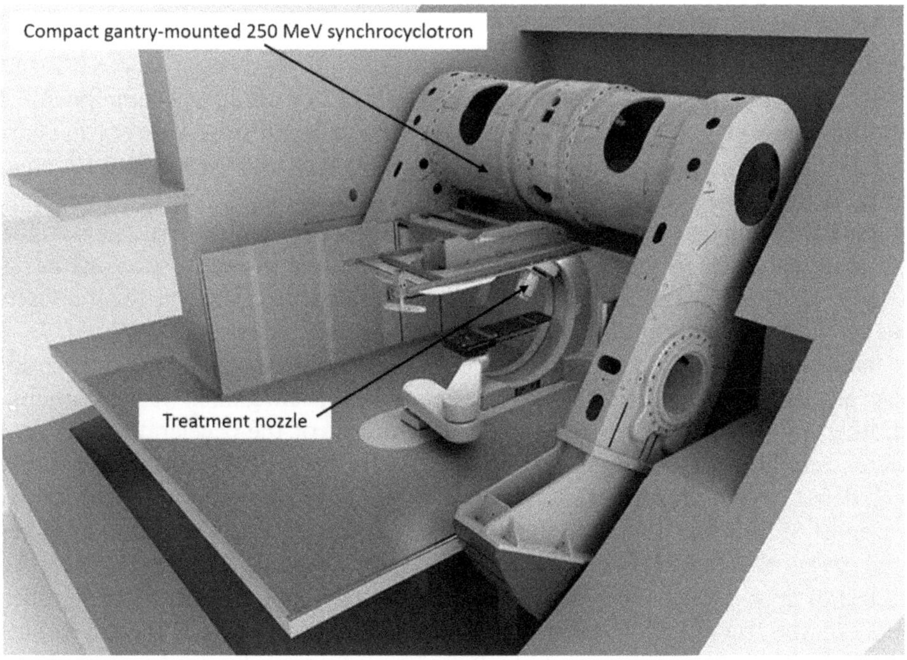

Figure 8.19. Compact gantry-mounted 250 MeV superconducting synchrocyclotron made by Mevion. (Credit: image courtesy of Mevion Medical Systems, Inc. All rights reserved.)

treated with protons (220 000 worldwide [50]), PBT still only accounts for a tiny fraction of radiotherapy provision available to cancer patients. In order to democratize this advanced treatment modality, costs associated with the development of these facilities need to reduce further. To achieve this, recent efforts have focused on reducing the size of both the accelerator and gantry systems using superconducting magnets. By using smaller cryo-cooled superconducting bending magnets on the gantry system, manufacturers are able to develop compact gantry systems that require significantly smaller vaults. Likewise, many contemporary PBT facilities recently developed, or currently under development, use a commercially available compact superconducting isochronous cyclotron or synchrotron.

Figure 8.19 shows a gantry-mounted 250 MeV superconducting synchrocyclotron, developed by Mevion, currently in clinical operation in a number of facilities. This system was the first commercially available compact single-room PBT system. By using superconducting technology to reduce the size of PBT systems, thereby reducing the cost to develop a facility, PBT will become available to a greater number of cancer patients who may benefit from the favourable dosimetric characteristics exhibited by proton beams. Not only is this a potential benefit to a greater number of individual cancer patients, but also the long-term financial burden on a healthcare system in treating post-radiotherapy related toxicities will be reduced.

8.5.4 Heavier-ion radiotherapy

As previously described, PBT offers dosimetric advantages compared to XRT due to the Bragg peak. However, research into the efficacy of heavier-ion beams for radiotherapy is ongoing. Like protons, heavier-ion beam dose delivery is characterized by a Bragg peak, however the Bragg peak is typically sharper due to reduced scattering of the heavier particles from their direction of travel. Furthermore, there is a potential biological advantage to heavier ions as they are more densely ionising than protons at the Bragg peak with a corresponding greater chance of causing DNA damage in the cancer cells. This may be particularly useful for a subset of patients with particularly difficult to treat radio-resistant tumours.

Studies into the efficacy of heavier-ion therapy include helium- and oxygen-ions, however, the majority of clinical experience has been gained from carbon-ion radiotherapy (CIRT). The increase in size and cost of the accelerator and gantry systems required for CIRT compared to PBT is very significant. There remains much debate about the cost-effectiveness of CIRT [51], and until more clinical data in support of CIRT is obtained, this will likely continue. The paradox is that some initial investment in CIRT must be made in order to obtain sufficient clinical data. The use of cryo-cooled superconducting magnet technology is having a significant impact on the size of CIRT synchrotrons and gantries, resulting in the development of more facilities around the world to join the existing small number of facilities in conducting the necessary clinical trials. If CIRT is proven to be highly efficacious for a significant subset of patients, then demand for affordable and compact CIRT clinical systems will increase.

References

[1] Polge C, Smith A U and Parkes A S 1949 *Nature* **164** 666

[2] Glenister P H and Lyon M F 1986 *Fertility. Transf.* **3** 20–7

[3] Bill H 2020 *The Biologist* **64** 16–21

[4] Wang X, Chen H, Yin H, Kim S S, Lin Tan S and Gosden R G 2002 *Nature* **415** 385

[5] Campbell B K, Hernandez-Medrano J, Onions V, Pincott-Allen C, Aljaser F, Fisher J, McNeilly A S, Webb R and Picton H M 2014 *Hum. Reprod.* **29** 1749–63

[6] Fahy G M, Wowk B, Pagotan R, Chang A, Phan J, Thomson B and Phan L 2009 *Organogenesis* **5** 167–75

[7] Egerton R F 2013 *Ultramicroscopy* **127** 100–8

[8] Ayache J, Beaunier L, Boumendil J, Ehret G and Laub D 2010 *Sample Preparation Handbook for Transmission Electron Microscopy: Techniques* (Berlin: Springer)

[9] Gnaegi H, Studer D, Bos E, Peters P and Pierson J 2008 Ultramicrotomy in biology and materials science: an overview *EMC 2008 14th European Microscopy Congress (1–5 September 2008, Aachen, Germany)* ed M Luysberg, K Tillmann and T Weirich (Berlin: Springer)

[10] Giannuzzi L A and Stevie F A 1999 *Micron* **30** 197–204

[11] Feder N and Sidman R L 1958 *J Biophys Biochem Cytol* **4** 593–602

[12] Denk W and Horstmann H 2004 *PLoS Biol.* **2** e329

[13] Nanninga N, Brakenhoff G J, Meijer M and Woldringh C L 1984 *Antonie van Leeuwenhoek* **50** 433–60

[14] Malhotra S K 1963 *Nature* **198** 611–12

[15] Knapek E and Dubochet J 1980 *J. Mol. Biol.* **141** 147–61
[16] Rice R V, Kaesberg P and Stahmann M A 1955 *Arch. Biochem. Biophys.* **59** 332–40
[17] Severs N J 2007 *Nat. Protoc.* **2** 547–76
[18] Luyet B J and Gehenio M P 1940 *Life and Death at Low Temperatures* (Normandy, MO: Biodynamica)
[19] Dubochet J and McDowall A W 1981 *J. Microsc.* **124** RP3–R4
[20] Murata K and Wolf M 2018 *Biochim. Biophys. Acta* **1862** 324–34
[21] Fernandez-Leiro R and Scheres S H W 2016 *Nature* **537** 339–46
[22] Bai X-c, McMullan G and Scheres S H W 2014 *Trends Biochem. Sci.* **40** 40–57
[23] Grassucci R A, Taylor D J and Frank J 2009 *Nat. Protoc.* **2** 3239–46
[24] Stokes D L, Rice W J, Hu M, Kim C and Ubarretxena I 2010 *Methods Mol. Biol.* **654** 187–205
[25] Taylor K A and Glaeser R M 2008 *J. Struct. Biol.* **163** 214–23
[26] McMullan G, Faruqi A R and Henderson R 2016 *Methods Enzymol.* **579** 1–17
[27] Callaway E 2020 *Nature* **578** 201
[28] Berman H M, Westbrook J, Feng Z, Gilliland G, Bhat T N, Weissig H, Shindyalov I N and Bourne P E 2000 *Nucleic Acids Res.* **28** 235–42
[29] Yip K M, Fischer N, Paknia E, Chari A and Stark H 2020 *Nature* **587** 157–61
[30] Plitzko J and Baumeister W P 2019 Cryo-electron tomography *Springer Handbook of MicroscopySpringer Handbooks* ed P W Hawkes and J C H Spence (Cham: Springer)
[31] van Donselaar E, Posthuma G, Zeuschner D, Humbel B M and Slot J W 2007 *Traffic* **8** 471–85
[32] Vanhecke D, Studer L and Studer D 2007 Cryoultramicrotomy *Electron Microscopy. Methods in Molecular Biology*™ vol 369 ed J Kuo (Totowa, NJ: Humana Press)
[33] Alexander R, Bäuerlein F J B, Villa E, Eibauer M, Laugks T, Baumeister W and Plitzko J M 2012 *Proc. Natl Acad. Sci.* **109** 4449–54
[34] Rigort A and Plitzko J M 2015 *Arch. Biochem. Biophys.* **581** 122–30
[35] Walther P and Müller M 1999 *J. Microsc.* **196** 279–87
[36] Fujikawa S and Endoh K 2014 *Methods Mol. Biol.* **1166** 99–116
[37] de Boer P, Hoogenboom J and Giepmans B 2015 *Nat. Methods* **12** 503–13
[38] Suzuki T, Matsuzaki T, Hagiwara H, Aoki T and Takata K 2007 *Acta Histochem. Cytochem* **40** 131–37
[39] Fu X, Ning J and Zhong Z *et al* 2019 *Sci. Rep.* **9** 19207
[40] Vangindertael J, Camacho R, Sempels W, Mizuno H, Dedecker P and Janssen K P F 2018 *Methods. Appl. Fluoresc.* **6** 022003
[41] Tuijtel M W, Koster A J and Jakobs S *et al* 2019 *Sci. Rep.* **9** 1369
[42] Peddie C J, Domart M-C, Snetkov X, O'Toole P, Larijani B, Way M, Cox S and Collinson L M 2017 *J. Struct. Biol.* **199** 120–31
[43] Wang L, Bateman B and Zanetti-Domingues L C *et al* 2019 *Commun. Biol.* **2** 74
[44] Gwosch K C, Pape J K and Balzarotti F *et al* 2020 *Nat. Methods* **17** 217–24
[45] Gorelick S, Buckley G, Gervinskas G, Johnson T K, Handley A, Caggiano M P, Whisstock J C, Pocock R and de Marco A 2019 *eLife* **8** e45919
[46] Agronskaia A V, Valentijn J A, van Driel L F, Schneijdenberg C T W M, Humbel B M, van Bergen en Henegouwen P M P, Verkleij A J, Koster A J and Gerritsen H C 2008 *J. Struct. Biol.* **164** 183–89

[47] International Cancer Control Partnership 2018 Radiotherapy: seizing the opportunity in cancer care https://www.iccp-portal.org/news/%E2%80%9Cradiotherapy-seizing-opportunity-cancer-care%E2%80%9D

[48] Wilson R R 1946 *Radiology* **47** 487–91

[49] PTCOG – Facilities in Operation https://ptcog.ch/index.php/facilities-in-operation

[50] PTCOG – Patient Statistics https://ptcog.ch/index.php/patient-statistics

[51] Beltran C, Amos R A and Rong Y 2020 *J. Appl. Clin. Med. Phys* **21** 6–9

[52] https://www.fda.gov/consumers/consumer-updates/whole-body-cryotherapy-wbc-cool-trend-lacks-evidence-poses-risks (accessed October 2022)

[53] Vanderwaeren L, Dok R, Verstrepen K and Nuyts S 2021 Clinical progress in proton radiotherapy: biological unknowns *Cancers* **13** 604

[54] Vanderwaeren L *et al* 2021 Clinical progress in proton radiotherapy: biological unknowns *Cancers* **13** 604

Chapter 9

Industrial applications

Jon Trembley, Joseph Boehm, Jack Cahn, Eleni Fitsiou and Jarek Nowak

9.1 Introduction

John Vandore

Given the classic, ubiquitous, enabling nature of cryogenics, this chapter could be endless. Air products, where one of this chapter's authors Jon Trembley is employed, lists around 60 industrial applications, ranging from cooling concrete to making snow! And perhaps this should be no surprise—by lowering temperature (and in the process lowering energy state and molecular vibration), resulting generally in physical contraction and changes of state (liquefaction and solidification), as well as the phenomena of biological cryopreservation and of superconductivity, the opportunities to exploit cryogenics really are without end. Out of this extraordinary choice we have taken VOC (volatile organic compound) recovery as one example from this constellation, described by the highly experienced Jon Trembly, Technology Manager for Cryogenic Applications, Air Products.

Additional applications covered in this chapter include a thorough review of superconducting magnetic separation by Josef Boehm—amongst other things, this technique can contribute to the purity and colour of porcelain! Jack Cahn gives an expert explanation of deep cryogenic treatment. And Eleni Fitsiou and Jaroslaw Novak from the University of Lancaster discuss cryogenics in the food industry—yet another vital and substantial industry in itself, dependent on cryogenics.

9.2 Cryogenic condensation for solvent recovery

Jon Trembley

Cryogenic condensation for solvent recovery. VOC recovery, or solvent recovery as it is commonly known, is the application of a technology to recover volatile chemicals from process gas emissions to prevent the release of these volatile chemicals into the

doi:10.1088/978-0-7503-2757-2ch9

Figure 9.1. Installation in the USA.

environment. It is called solvent or VOC recovery because organic solvent fumes, most of which are classified as harmful VOCs, tend to be the most common volatile chemicals being recovered. Although the main driving factor is typically environmental pollution prevention enforced by the government, there are times when solvent recovery is often completed for the sake of recovering valuable volatile chemicals that would otherwise be vented off and lost.

Cryogenic solvent recovery is a non-destructive way to control release of volatile chemicals from industrial sources. The chemicals to be recovered are not destroyed in the process as can be the case with alternative technologies like incineration, a common alternative to VOC recovery, and other similar destructive technologies for VOC abatement. Common solvent recovery technologies include adsorption, non-reactive absorption, membrane separation, and condensation.

For industrial gas suppliers, cryogenic VOC recovery has a more specific meaning: it is an application that uses the cold energy potential of liquid nitrogen (LIN) for condensing and recovering volatile chemicals from gas emissions. LIN is used to cool down volatile chemicals to cryogenic temperatures to condense them into liquid for recovery, cutting down emission of these volatile chemicals to the atmosphere by up to 99% or more (figures 9.1 and 9.2).

History: The first cryogenic solvent recovery systems were installed in the late 1980s, the primary market being the converting industry—facilities operating coating lines and drying ovens to apply paints and adhesives dissolved in organic solvents to films and other flat surfaces (e.g. in magnetic tape manufacturing); these systems were integrated with the drying process. Following early success, a new market developed with companies using ethylene oxide (EO) for medical device sterilization. These cryogenic solvent recovery systems were modified to allow EO to be recovered and recycled back to the sterilization process with the evaporated nitrogen gas utilised for purging and inerting the sterilization chamber.

Figure 9.2. Installation in Europe.

Since those early days of the technology's adoption, the potential for cryogenic solvent recovery systems has developed extensively into the fine chemical and pharmaceutical manufacturing industry, with the main drivers being: environmental —to recover regulated solvent emissions classified as hazardous air pollutants (HAPs), or ground-level ozone forming VOCs from chemical processes; economic —to recycle unreacted high value volatile raw materials, products, or solvents from chemical processes; and safety—to provide back-up refrigeration to protect vessels storing highly volatile raw materials or products from over pressurization

During the late 1990s, a new niche market for refrigerator recycling also emerged in Europe where cryogenic solvent recovery systems are used to recover regulated chloro-fluoro-carbons (CFCs) from shredding and recycling old domestic refrigerators; CFCs were used as refrigerants and also as propellants to create the foam insulation in old refrigerators. LIN is used to condense and recover the CFCs but more importantly used for inerting the whole process to safely operate under the lower explosive limit (LEL).

New markets continue to develop, with the latest being for organic liquids distribution terminals. In this industry sector, loading and unloading operations of solvents and other volatile chemicals for transportation create gas emissions that must be recovered for environmental reasons. Cryogenic solvent recovery systems are applicable here too.

Condensation theory: cryogenic solvent recovery systems rely on the simple concept of condensation, the physical change in matter from gas to liquid. By using LIN, with a boiling point of −196 °C almost any gaseous chemical can be condensed.

Condensation happens all around us. The most common chemical that condenses around us is water, moisture in the air condenses on the surface of your glass when you pour a cold beverage. Moisture in the air also condenses to form clouds, fog and rain when atmospheric temperatures get colder, so why does condensation occur? It occurs because of a change in the internal energy of the gas; as the internal energy goes down, the molecular attractive forces between the molecules overcome the vibrational energy that keeps the molecules in the gas state. Keeping pressure constant, gases generally condense into liquids as their temperature is lowered. If their temperature is further decreased, liquids start to freeze. There are times when gases can skip the liquid phase and change phase directly into solid, e.g., when the temperature change is fast, a process called deposition.

Different molecules condense and freeze at different temperatures because of different physical properties affecting the molecular active forces between them. Molecules that have higher attractive forces, such as water (because of hydrogen bonding), require higher internal energy to push them apart. In other words, molecules like water will melt and boil at higher temperatures than molecules with weaker attractive forces like acetone. For example, at a constant pressure of 1 atm, a sample of pure ice melts at 0 °C and boils at 100 °C. By contrast, a sample of pure frozen acetone melts at −95 °C and boils at +56 °C. Molecules with relatively strong intermolecular forces tend to have lower vapor pressures and are said to be less 'volatile' than molecules with relatively weak intermolecular forces.

When conditions remain the same, and a substance is no longer experiencing phase changes, it is said to have reached equilibrium. For pure substances, it is relatively easy to model the effects of temperature and pressure on its equilibrium. However, mixtures of substances are significantly more difficult to model because of the intermolecular influence or interaction these substances have on each other, which depends on the composition of the mixture as well as the temperature and pressure.

For general cryogenic solvent recovery scenarios, the VOC gas stream to be treated is usually a mixture of organic compounds and 'incondensable' gases like nitrogen, or air. As the temperature of this gas mixture is lowered, its equilibrium state will change, resulting in some of the less volatile (lower vapour pressure) compounds condensing out first while the more volatile or inert compounds remain in the gas phase. Generally, as the temperature continues to decrease, more and more of the VOCs will condense out of the gas phase. If temperature is decreased even further, even the 'incondensable' gases can start condensing—this requires temperatures low enough, say below −150 °C.

The concept of changing the equilibrium is easy to understand when considering a common mixture: air and water in the atmosphere. In this mixture, nitrogen can be considered 'incondensable,' with very high vapour pressure; engineers usually refer to such incondensable gases as 'carrier' gases. Water can be considered the least volatile compound in this mixture. As can be seen in figure 9.3, the maximum amount of moisture (gaseous water) in this gas mixture of water and air decreases (as it condenses into liquid water) as temperature is decreased. At around 85 °C, a maximum of 1 kg of water vapor can exist in 1 kg of air (this is called the saturation capacity); any excess water added to the system will exist only as liquid. By contrast, at around 55 °C, the saturation capacity of moisture in air is 0.1 kg per 1 kg of air. This means that if a gas sample of 1 kg of air and 1 kg of moisture at 85 °C is cooled down to 55 °C, about 0.9 kg of water will condense out of the gas mixture, a 90% reduction or 'recovery' of water. However, note that saturation capacity is also a

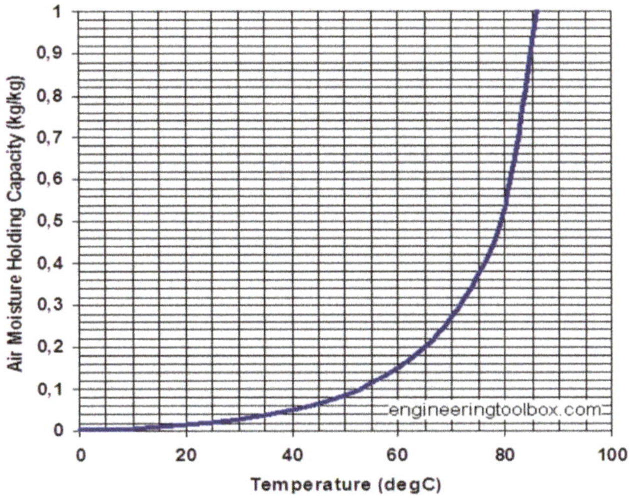

Figure 9.3. Saturation capacity of moisture in air at 1 atm.

function of pressure! Figure 9.3 shows the saturation capacity curve for water in air at constant atmospheric pressure (1 atm). This is the theory that governs VOC/ solvent recovery with LIN.

Just as rain is water condensing from a moisture/air mixture in the atmosphere, so can VOCs condense out of a gas mixture of carrier gases and VOCs. The question is: how much of the VOCs in the gas mixture do you want to condense? Once this question is answered, then one can use literature data [1–3], modelling programs (ASPEN and equivalent software [4]), or run an experiment to determine the temperature at which the desired amount of VOCs will be condensed out of the gas mixture. Once the start and end temperature are determined, then one can calculate how much cooling requirement (how much enthalpy must be removed from the gas mixture) is needed. The cooling requirement can then be met by a refrigeration process: ambient cooling, water cooling, mechanical refrigeration, or LIN cooling. LIN consumption can then be calculated by making some assumptions on the type of heat exchanger system to be used.

As an example, figure 9.4 shows the saturation capacity of ethyl acetate (a VOC) and water in a gas mixture with nitrogen as the carrier gas. In this case, the saturation capacity is shown as a recovery percentage of the initial amount of ethyl acetate and water. One can easily see the general pattern, that as temperature is lowered, the saturation capacity of the less volatile compounds decreases in the carrier gas and thus more of these less volatile compounds is recovered as liquid. Starting with a gas stream saturated with ethyl acetate and water at a temperature of 20 °C and a pressure of 1 atm, about 55% of the ethyl acetate and 70% of the water are condensed into liquid when the temperature is dropped to 0 °C. Environmental emissions limits in the US are usually expressed as recovery percentages, generally above 90%, but in Europe the limits are much lower. To recover at least 98% of the ethyl acetate, the gas mixture must be cooled down to about −40 °C or below. In this

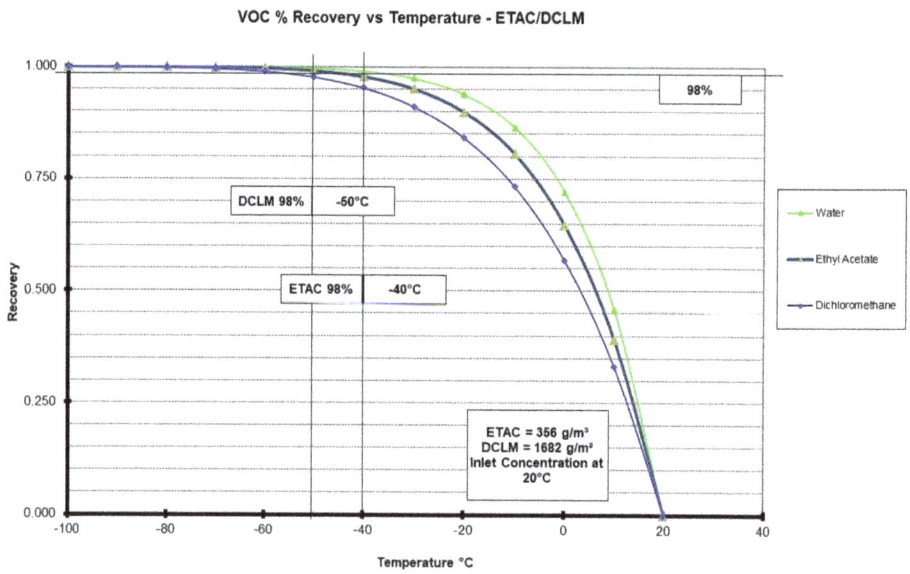

Figure 9.4. Saturation capacity of ethyl acetate (or dichloromethane) and water in nitrogen at 1 atm.

example you can also compare the same conditions for dichloromethane and can see that a temperature of <−50 °C is required for 98% recovery, although the TA Luft target is considerably lower.

Conclusion: Yes, there are other VOC emission control technologies as potential alternatives to a cryogenic solvent recovery system, however none offer a non-destructive way to control the release of volatile chemicals from industrial sources. There are other factors that can influence a decision on adopting the technology, but the benefits and flexibility of using a cryogenic based system can certainly influence the decision process. Couple that with the ability to recover and re-use the evaporated nitrogen then the economics can become very favourable [5].

9.3 Superconducting magnetic separation

Joseph Boehm

9.3.1 Magnetic separation, purpose and principles

In nature it is mostly the case that certain desirable materials are mixed with other, less desirable materials. A huge range of methods is employed to separate out useful components from an often diffuse mixture, literally separating the wheat from the chaff, or the diamonds from the gravel surrounding them. Gravity, density, colours, shape and much more all have an influence on the effectiveness of separation. The property of carrying electric charges is used to pull charged particles away from a mixture. The susceptibility to magnetic fields is just one of many ways to separate one compound from another.

Using a simple bar magnet can attract ferromagnetic materials (more later). Stronger magnet materials have been developed to enhance this method, in particular permanent magnets consisting of rare earth elements. These permanent magnets do seem to have an upper limit in their field strength though and this limit is in the region of 1.2 Tesla (the unit of magnetic flux density).

Sending an electric current through a conductor will result in a magnetic field around the conductor. Using this conductor wire in an arrangement with many turns, for instance in the shape of a solenoid, will amplify this field effect. The field strength from this wire is limited though, as the electric resistance of the wire will cause the wire to heat up and, ultimately, burn out if not cooled effectively. Fields of fractions of a Tesla can be obtained with this method without cooling. Using a field amplifier like a ferritic yoke around the conductor will concentrate the magnetic flux, and flux densities of around 2 Tesla can be obtained with cooled conductor and iron yoke. The power consumption of this type of magnet is considerable.

9.3.2 Basic concepts in magnetic separation

There are four main determinants responsible for influencing particles when exposed to magnetic fields:

- a magnetic field, H;
- the gradient of the magnetic field, $\nabla(H)$;
- the extent to which a particle can be magnetised by the magnetic field, $M = \chi H$, with χ being the magnetic susceptibility of the particle, or the medium it is in. The particle, or relative permeability, a related concept, to magnetic fields is written as $\mu = \mu_0(1 + \chi)$, μ_0 being the permeability of vacuum;
- the volume of the particle, V, which can also be written as mass m over density ρ.

It is assumed that the field over the volume of the particle is constant, which means that the particle is of small extent. One further assumes materials, consisting of spherical or granular shape which are around, or below 1 mm in extent. For these small particles the magnetic force F_m which acts on them is written as:

$$F_m = V\,\mu_0\,(\Delta M)(\nabla H) = \frac{m}{\rho}\mu_0\,(\Delta M)\,(\nabla H)$$

There are three main categories of magnetisable substances: ferromagnetic, paramagnetic and diamagnetic materials.

Ferromagnetic materials are easily magnetised, as their susceptibility is very high. Their values for χ can be 1000 or more which means that the susceptibility towards magnetic fields is 1000 times higher than the susceptibility of vacuum

Paramagnetic materials have a positive susceptibility (due to alignment of atomic spins) and diamagnetic materials have a negative susceptibility (due to the response of the electron orbit to a magnetic field). As an example, the (dimensionless!) paramagnetic volume susceptibility of aluminium is 2.2×10^{-5} (paramagnetic) and the volume susceptibility of silver is -2.3×10^{-5} (diamagnetic). Bismuth is an

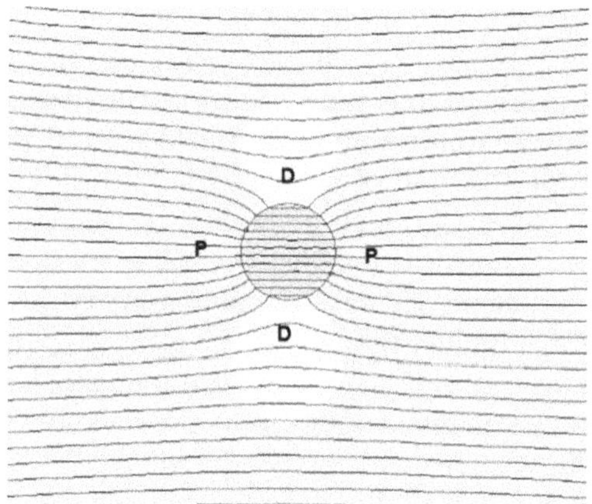

Figure 9.5. Field lines around ferromagnetic capture element: paramagnetic particles collect at P, diamagnetic ones at D. Field lines are concentrated by the ferromagnetic capture element, usually a thin wire. The attracting force is approximately proportional to the inverse of the wire diameter. The wire diameters used for magnetic separation are anywhere from 20 μm to several mm. The wire diameter must be chosen with the material to separate in mind.

exceptionally strong diamagnetic substance with a χ of -1.66×10^{-4}. Vacuum has a magnetic susceptibility of zero, meaning its permeability is μ_0.

Paramagnetic and diamagnetic susceptibilities are very small and little different from the magnetic behaviour of vacuum. This is the reason why one needs strong magnets and gradients to influence these particles, hence the use of superconducting magnets.

Paramagnetic particles are pulled into areas with increasing field, diamagnetic materials are pulled to areas with decreasing magnetic fields. Ferromagnetic materials should be removed from the material flow before any high field is applied as they simply would clog up the separation volume (figure 9.5).

Competing forces on particles are inertia (velocity), drag within the surrounding medium, temperature, the material of the ferromagnetic filter elements (if present) and possibly even inter-particle forces.

The main candidates for magnetic separation are materials that show paramagnetic behaviour. Magnetisation of these particles is linear with applied field, namely $M = \mu 0 \chi H$. Many metal oxides, or other metal compounds fall into this category and are the main target of magnetic separation.

When the creation of very high magnetic field gradients is realised by the presence of a ferromagnetic trapping centre, the separation method is called high gradient magnetic separation (HGMS).

Open gradient magnetic separation (OGMS) can pull materials of differing magnetic susceptibility apart in cases where, for instance during some manufacturing processes, high value metal grindings or swarf are mixed up and the economics justify employing special techniques to separate them again. The magnitude of the field gradient is very much smaller than in HGMS, but the separation process takes place in the open, continuously, without the need to remove trapped particles.

9.3.3 Permanent magnets, drums

Permanent magnets can be arranged in the shape of a drum, perhaps 1 m wide and 25 cm in diameter. Granular material like mineral sands are then fed tangentially to the top of the rotating drum. The magnetic fields emanating from these drums are then used to retain the more paramagnetic particles for longer than the less paramagnetic, or non-magnetic granules. One or more splitters, placed at the appropriate locations below the drum will then divide the particle stream in its variously magnetisable components. Magnetic separation has taken place.

9.3.4 Electric magnets, carousels, iron-enclosed pot magnets

Permanent magnets have only a limited capacity to create magnetic fields, are expensive and cannot change their magnetic fields. Electromagnets are able to increase the magnetic flux through circuits and these electromagnets are very widely used in all kinds of configurations. They are wound into solenoids and then embedded into ferromagnetic drums where the drums enhance the magnetic flux by concentrating it within the highly magnetisable iron yoke, resulting in designs that may be called 'pot magnets'. They can be configured as so-called C-type magnets where magnet gap is filled with a rotating filter carousel, containing ferromagnetic filter elements. Or they can be used as simple lifting magnets where one side of a pot magnet is left open and the field from that open side goes through a ferromagnetic component which then can be lifted, and released when the electric current is turned off. Magnets in scrap yards for cars are of that type.

All of these electric magnets are very flexible in their application but have one major disadvantage: they are large consumers of electric power. For instance, the pot magnet shown in figure 9.2, when used for the separation of kaolin, has a coil diameter of 2 or even 3 m. The larger of these magnets can consume up to 500 kW. It is also running through many cycles of energisation and this creates large stresses within the coil. The coils need to be cooled by large quantities of cooling water.

As long as these electromagnets are relatively small, and the materials they separate are economically viable, there is little advantage changing to a different technology. Electric power can be produced virtually everywhere and the maintenance of the magnets is relatively simple. From a certain size onwards, however, the cost of electricity will become a factor in the operation of a plant and very large quantities of power may be difficult to be provided in an economical way (figure 9.6).

9.3.5 Advent of superconductive magnets

Superconducting properties in solid mercury were discovered by Kamerlingh Onnes at Leiden University in 1911. Clearly, mercury would not be a suitable medium and metal alloys like niobium–titanium (NbTi) and niobium–tin (Nb_3Sn) have been developed as the main constituents of wires that can be wound into shapes suitable for producing magnets. These alloys are usually embedded within a matrix of copper, producing wire that looks like copper wire at first sight.

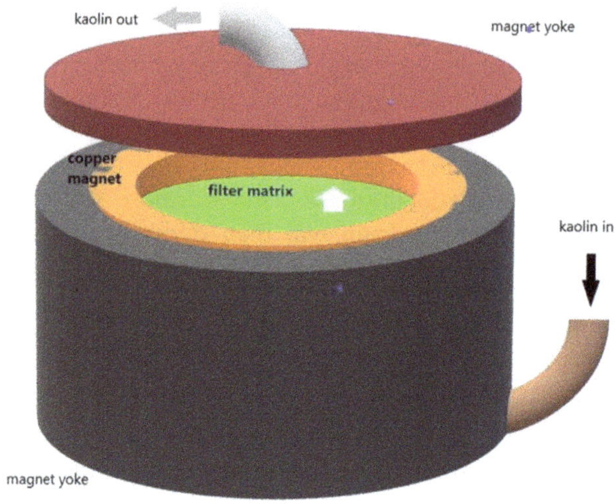

Figure 9.6. Pot magnet as used in the kaolin industry. The coils are up to 3 m in diameter and consume up to 500.kW.

9.3.6 Major market penetration with MRI magnets

Superconducting wire suitable for industrial use has been available since the 1960s, with large-scale applications in quantity emerging from application in magnetic resonance imaging (MRI). This happened from the late 1970s onwards. By the end of the 1980s thousands of large MRI magnets were built worldwide. Parallel to development of MRI devices came magnets for NMR too.

In discussing MRI and NMR magnets it is often overlooked that their existence and economic application is only possible by the presence of a sophisticated cryogenic infrastructure in which the magnet coils are embedded.

9.3.7 Application to magnetic separation

The case for magnetic separation in industry derives from commercial considerations. Material mixtures have to be found which are either relatively rare or valuable, or are present in large quantities where it makes commercial sense to use expensive technologies for their separation. A prime candidate for magnetic separation is kaolin, china clay, an aluminium silicate. Kaolin appears in very large deposits all over the world and has of course been mined for centuries. It is used extensively for the manufacture of clay pottery, but the main volume of kaolin nowadays is used for coating paper. Kaolin appears in platelet form which covers the fibrous structure of paper pulp very effectively and creates a smooth surface. Kaolin deposits are contaminated by a variety of minerals, hematite and titanium oxides are just two examples. As the contaminants are often paramagnetic the obvious candidate for kaolin purification is strong magnetic separation. By this method the brightness values of coated paper can be raised from, say, 85% to 92%.

The human eye can detect brightness differences of less than 1%. The price of paper goes up in line with its 'whiteness'. Magnetic separators that can produce white kaolin in quantities of up to 100 t/h make sense purely by the volumes they separate. One ton of kaolin is anywhere between $100 and $1000 in cost, the whiter the more expensive. Justification for the use of superconducting magnetic separation is therefore bulk.

9.3.8 First appearances of magnetic separators

9.3.8.1 Superconducting OGMS magnets

OGMS are without trapping areas, i.e. they have no filter elements, only a magnetic field and its gradient. The magnet coils can be arranged in either a linear fashion, i.e. a racetrack design with two current tracks in opposition, or in an annular fashion with two magnet coils energised in an antiparallel way. The racetrack approach for a superconducting separator is, at first sight, an elegant approach. It suffers, however, from a few design-inherent problems: the racetrack shape of the coil would like to adopt a round shape due to the huge repulsive magnetic forces. These forces will deform the coil, creating stresses and cracks within the epoxy matrix of the coil body. Any crack, or other movement of the wire structure will cause a quench of the superconducting coil thus limiting the magnetic field and field gradient to unnecessary low numbers.

Another disadvantage of this elongated design is the difficulty in cooling the coil. Submerging the coil in liquid helium will make the cryostat very bulky. Connecting a network of pipes, filled with liquid helium to the coil is feasible but not trivial as the flow of the liquid must be assured. The most elegant way to cool such a structure is to use a series of cold heads that are connected to the coil at intervals. The advantage of a racetrack separator is that it could theoretically be used with the separation surfaces on both sides. In practice this configuration may be difficult due to the cooling infrastructure and to the necessary coil arrangement that may work best when tilted.

The alternative to a racetrack coil is to use a split-coil arrangement with the coils shaped as solenoids and being energised in opposition. The magnetic field comes out radially from between the two coils, creating a large gradient field where magnetic separation (deflection) can take place. This design is scalable by simply increasing the diameter of the solenoids. The cooling arrangement is located on the inside of the coils, allowing any cooling method conceivable. The disadvantage of the design is that only the outer circumference of the coil body can be used for magnetic separation. If the magnetic surface should be tilted, then this angle needs to be designed into the structure of the coil and cryostat system (figures 9.7–9.9).

9.3.8.2 Superconducting pot magnets

Analogous to conventional electromagnets wound as solenoids, embedded within an iron yoke and field-ramped up and down continuously, a method was devised to replace power-hungry copper magnets with low power superconducting magnets. Direct *in situ* replacement of copper magnets was performed, mainly in Georgia USA;

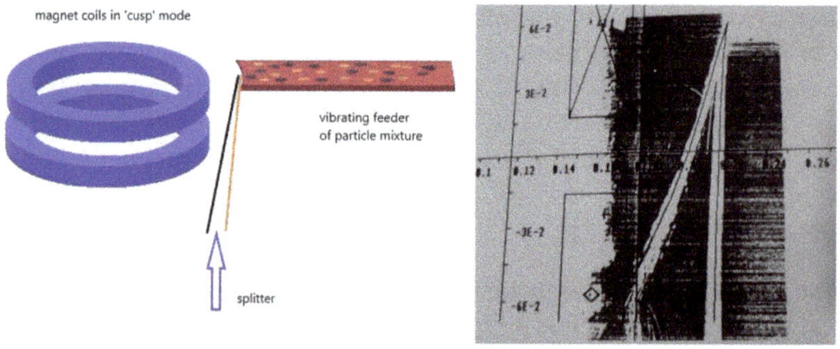

Figure 9.7. Schematic of OGMS separation using an annular magnet coil arrangement and photograph of real superconducting separation on annular separator (shown dimensions are in m). The materials to separate are magnesium oxide from silicate (sand). Particle sizes 200–400 μm, the shown lines in the photographs are superimpositions of magnet coils and calculated deflections (photograph by author).

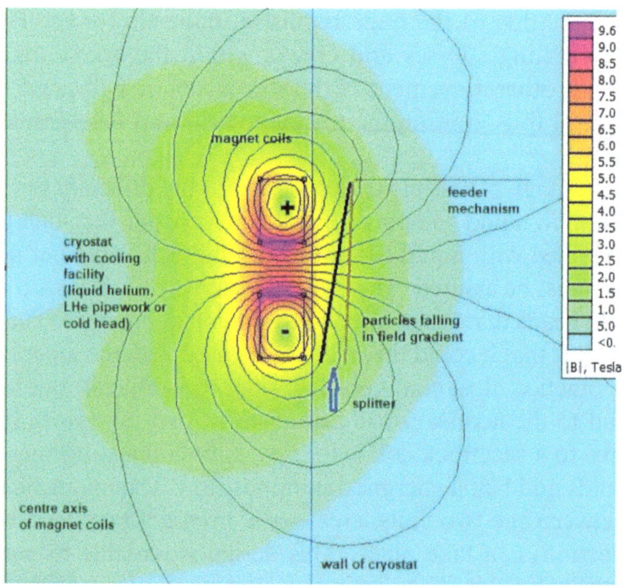

Figure 9.8. Coil arrangement and field profile for an OGMS magnet.

also, newly designed separators were built. The numbers for these superconducting pot magnets are not large, perhaps there are 10 in total, some have been dismantled.

The problem with large superconducting magnets that need to be power-cycled frequently is two-fold:

- They require a sophisticated and relative power-hungry cryogenic infrastructure. Joule–Thomson (JT) liquefiers are the most powerful devices that can create large quantities of liquid cryogen and this machinery is used for the constantly ramping magnets.

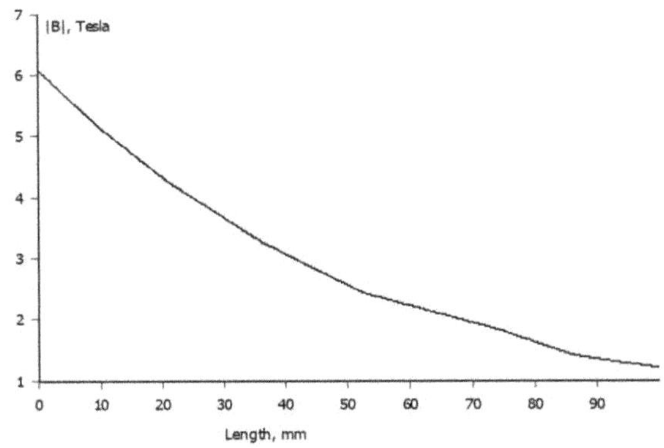

Figure 9.9. The field gradient of an OGMS magnet system.

- The magnets will experience a constant change of internal forces due to the magnetic fields. These forces will have a negative effect on the durability of the coil bodies.

The advantage of the pot magnets is their large volume of the filter (separation) bed and their simple mechanical structure, which is static.

9.3.8.3 Superconducting reciprocating separators

To overcome the complications with the cryogenic infrastructure required for constantly ramping pot magnets, reciprocating filter separators have been developed. Their design is focused on a much simplified cryogenic environment, at the expense of a more complicated mechanical requirement, and at the expense of a smaller filtration (separation) bed. The smaller filtration bed can be compensated by operating the magnets at a much higher magnetic field than is possible with large pot magnets. The magnets stay at field once they are energised and the field is not 2 T anymore, as with pot magnets, but 5 T or more. Instead of moving the field, one moves the filter. The filters are part of a train of magnetically balanced chambers. In figure 9.5, the filter sections are shown in blue and the magnetic balancing sections in yellow. Both filter and balancing sections must be packed with the same ferromagnetic material so that the magnetic field gradient at the end of the magnet coil will create identical forces on either side of the solenoid. In that way the moving train of the filter tube will only create eddy current interaction forces with the magnet coil, but no unbalanced magnetic forces. A well-balanced filter tube of 5 m length can almost be pushed through the magnet bore by hand. The filter tubes are of length 2.5–10 m and of diameter 0.25 m–1.0 m. The packing density with ferromagnetic capture material is only in the range of 5% to 7%.

Reciprocating separators (figure 9.10) can be kept cryogenically simple as they do not rely on the constant changing of the field. The working part of the separator has been shifted to the mechanical section. Cryogenically simple magnets can be installed in more or less any part of the world and the mechanical maintenance is

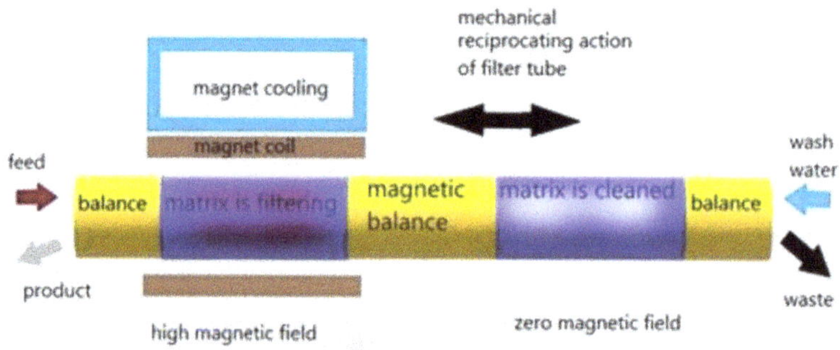

Figure 9.10. Reciprocating superconducting magnet separator.

open to any skilled mechanical engineer. The cryogenic service of such a separator can be limited to once per year when cryogenic engineers need to be available.

9.3.9 Cooling technologies for superconducting magnets, and new magnet materials

9.3.9.1 Various methods to cool separator machinery

There are a number of options to provide cooling for the magnet coil:

1. The original method of cooling superconducting magnets, with liquid nitrogen to cool the radiation shield and liquid helium to cool the magnet coil. This method is very wasteful as the cryogens need to be replenished often and the radiation shield is only cooled to 77 K (LN_2).
2. Cooling the magnet coil with the JT expansion method, and using a cold head (or expansion turbine) to cool radiation shield and pre-cool helium gas before expansion to liquefy the helium gas.
3. Cooling the coil with liquid helium but using a cold head to cool the radiation shield to 15 K and 50 K. This method was used by most MRI magnets until the advent of more powerful cold heads at the end of the 1990s. Large reciprocating magnet separators mainly use this method.
4. Cooling the radiation shield with a cold head and using a second stage of the cold head to re-liquefy evaporating helium gas. This method has—and is being used on superconducting separators, as it has two major advantages: (a) it is, in effect, a zero consumer of helium, and (b) in case of cold head failure there is plenty of liquid helium in the cryostat to keep the coil cold and potentially even allow continued operation of the separator. This method is now also the main choice for MRI systems. The volume of liquid helium within the cryostat can be sized to suit the application.
5. Cooling the coil and radiation shields directly with a cold head. This method dispenses with all liquid cryogens and is the simplest and most elegant way to cool a magnet. On the other hand it has a number of disadvantages:
   (a) The magnet takes a long time to cool down from room temperature. In case of power failure the magnet will immediately quench and heat up.

(b) More on magnet cool-down time (approximately 1 h kg^{-1} of cold mass from 300 K to 4.2 K when using cold heads of 1 W of cooling power). For large cold masses of hundreds or even thousands of kg this can be punitive.
(c) The temperature profile of the cold mass is not as uniform as in a liquid-cooled magnet which may limit the achievable field strength of the magnet.

For small magnetic separators this may be the best method of cooling.

The author only knows of one example where method 1 was used for a magnetic separator. It seems to have been on a large device in Russia. The cryogen consumption apparently was very high and it would have been very un-economical to operate in industry.

Method 2 (JT cooling) was used in larger numbers, both with cold head for pre-cooling the helium gas to below its inversion temperature and expansion turbine. JT cooling can produce large cooling power (in the region of 5 to 10 W @ 4.2 K) but it is maintenance-intensive. The compressed helium gas must be extremely clean from contaminating gases as the helium is forced through a pin-sized throttle behind which adiabatic expansion can take place. It is the small throttle that blocks frequently, making a JT cooling device less reliable than simple liquid or direct cold head cooling. The pot magnets, described earlier, were cooled by large JT-liquefiers in order to provide the cooling power required by the continually ramping super-conducting magnets.

Method 3, the combination of shield cooler and liquid helium was the crucial combination that made reciprocating separators possible. The technology was used in huge numbers for the cooling of MRI magnets. The top-up requirement with liquid helium was targeted to once per year. Service time was about 1 day during which the helium was topped up and the cold head was serviced (on the cold magnet!). In order to keep helium consumption low, the magnet was put into persistent mode, meaning that the magnet coil was electrically shorted at 4.2 K and the current leads were then withdrawn from the cryostat, leaving the cryogenic system with the magnet inside totally isolated from the outside world.

This combination of shield cooling and large helium reservoir also provides a crucial safety factor for industrial use: should the cold head stop or fail, the magnet does not stop immediately. There are days, possibly even a week available as a response time for taking remedial action. Remote locations for these separators do rely on this large buffer time and without this reserve many separators would not have been able to be installed.

Method 4 is a development of Method 3 made possible by the advent of cold heads that are capable of creating temperatures below the boiling point of helium. Cooling powers of 1 W and more can be achieved by these devices at 4.2 K. This cooling re-condenses the slowly evaporating helium from the cryostat of the separator and the condensate drips back into the reservoir. This results in two major improvements over the design of Method 3: (a) no helium will be lost as long as the cold head remains operational, and (b) it keeps the stability for operation with a reservoir of liquid helium, which provides a buffer against down-time.

Method 5, the direct cooling of the superconducting magnet body dispenses with a liquid helium reservoir. It provides no safe-time buffer against component failure. A superconducting magnet with a cold head failure will quench in a matter of seconds or minutes, the magnet coil will warm up and time-consuming re-starting of the process is required. There is also the disadvantage that, in most cases, the magnet cold mass needs to be warmed up during an annual service, or after a cold head failure. The time to warm up, repair the cold head and cool down the system again can be very long, depending on the mass of the magnet body.

This method 5 has also been made possible by the advent and application of high temperature superconducting (HTS) current leads. These current leads allow current to flow into the magnet coil, from temperatures of about 80 K onwards, without any electrical resistance and only minimal heat ingress. Magnets with direct cooling, and other magnets too, can therefore be operated with the power supply to the coil permanently connected, potentially allowing quick ramping of the magnetic field. Persistent mode operation of the coil is no longer necessary, although it might still be desirable in some applications.

9.3.9.2 The interdependence of cryogenic and magnetic technologies for magnetic separation

It is a safe assumption that the realisation and wide application of superconducting separators is intimately connected with development of cryogenic technologies. Initially, relatively inefficient structures required large cryogenic machinery (JT turbines) to allow their operation. Cold heads, be they Gifford McMahon devices with internally moving heat exchangers, or pulse tube coolers relying on the tuning of internal pressure pulses, allowed the helium consumption of separators at first to be much reduced (shield coolers) and then even eliminated (with re-condensing cold heads). Development of cold heads continues and cooling power is continually improved, allowing more powerful magnet applications.

The supply shortages of helium, which became very evident in the first decade of this century and multiplied the cost of liquid helium by factors of 3 to 5 made the use of low-temperature cold heads a necessity.

9.3.9.3 New magnet materials

The use of HTS for magnetic separators has been advocated since the early 1990s. However, the difficulties of working with these ceramic-type superconductors, and their cost of manufacture has largely precluded their application to industrial processes on a larger scale. Small HGMS separators, based on HTS technology were made in the late 1990s in order to prove the concept. Large-scale applications were prevented by the cost of the conductor material. Also, the advantage of HTS conductor, capable of operating at temperatures close to 77 K (the boiling point of liquid nitrogen) has been reduced by the widespread availability of cost-effective cold heads. HTS conductors can be operated at magnetic fields twice the value of more conventional niobium–titanium or niobium–tin coils. The materials on which these high fields would show an overriding advantage do not seem to be readily available.

Another material for the conductor of the magnet coil is more promising: it is magnesium-di-boride, MgB_2. The material is relatively cheap, being much closer to the cost of conventional niobium–titanium wire than to HTS ceramic conductor. The operating temperature of this material is below 40 K. Coils at 20 K could be operated with very powerful single-stage cold heads, showing 10s of Watts of cooling power (rather than the 1 W for the 4.2 K-capable cold heads). The conductor will not be able to produce the high fields of niobium–titanium wire. Applications for MgB_2 material would therefore be very large magnetic separators, like the 3 m diameter pot magnets of the 1980s but with negligible power consumption. Large separation volume certainly is an attractive proposition for HGMS separators, but also for OGMS devices.

9.3.10 Economic advantages/disadvantages

9.3.10.1 The case for superconducting separators

Superconducting separators have the advantage over normal electromagnets that their power consumption is minute. Devices that consumed 500 kW of electric power have been replaced by equivalent separators consuming 10 kW or 20 kW. The earlier large size of the separators has been reduced by use of a much higher magnetic field. The difficulties with location of power-hungry separators has been removed with power-efficient superconductors, and they are now found even in the middle of the Amazonian rain forest.

9.3.10.2 The case against superconducting separators

The high cost of production of a superconducting magnet must be put into relation to the product it is used on. There are not many materials that are valuable enough, and are used in large quantities, that would justify a significant investment in such a separator.

9.3.10.3 Outlook for various technologies

Superconducting pot or reciprocating magnetic separators (HGMS-type) are now used on a large scale and their technologies are well understood. The cryogenic environment in which they operate is relatively simple and efficient now with the widespread use of cryogenic cold heads. There are only a few producers worldwide that make these superconducting separators.

Superconducting OGMS are not used to that extent. This is probably due to the fact that there are other, cheaper methods allowing separation of mixed phases. There may be processes available which require separation, based on deflection. The purification of mixed metal waste is one example.

The emerging use of MgB_2 as a magnet coil material appears to be a good candidate for construction of large-scale separator devices where the need for very high fields is not absolutely necessary, but where high fields can be compensated by their large processing volume.

9.4 Deep cryogenic treatment

Jack Cahn

Deep cryogenic treatment (DCT) is a post-heat-treat, cold thermal process that increases mechanical strength and corrosion resistance of stainless steel, cast iron, aluminium, copper and tool steel by 20%–60%. DCT works by placing items in a vacuum insulated tank and slowly cooling from ambient to −196 °C, cold-soaking the items in a dry-nitrogen atmosphere for 18–30 h, then slowly returning to room temperature. Tempering steps then follow to remove hydrogen embrittlement (figures 9.11 and 9.12).

Figure 9.11. 52100 steel spur gears after DCT [6].

Figure 9.12. Production tanks at DCI in Lunenburg NS [7].

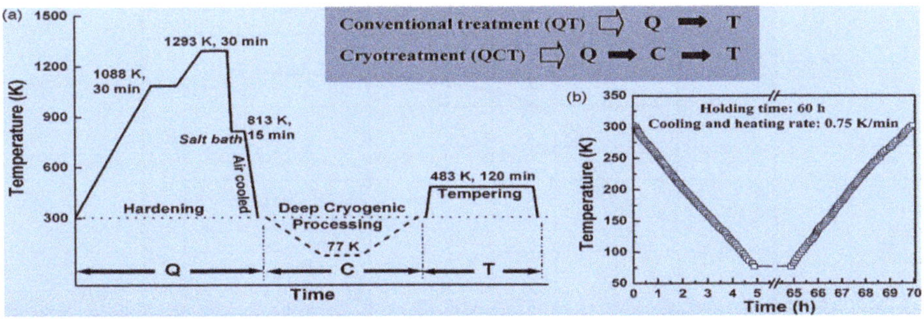

Figure 9.13. Typical sequence for deep cryogenic processing. The time-temperature curve shown is unique for each different material, desired improvement characteristic or performance application [8].

In steel, softer retained austenite is transformed into more durable martensite via a thermo-kinetic exchange. Fine carbides precipitate and nucleate at point-defects and grain-boundaries—cementing the metal matrix and yielding higher strain rates, greater resistance to abrasive wear and increased resistance to surface corrosion.

DCT takes 3–4 days to complete, costs roughly 12% of the original item, does not change dimensional part-size and allows multiple parts, weighing several thousand pounds each, to be simultaneously treated. The process is non-toxic, uses no chemicals and generates no environmental waste. It is scaleable to large industrial use and is supported by over 25 years of quantitative scientific research. DC-treated parts can be authenticated, tested and certified for industry use using existing ASTM standards and destructive/non-destructive test methods. Lack of industry awareness and, until recently, no large-size DCT tanks have been the key challenges for adoption by industrial end-users (figure 9.13).

Unlike hard facing, cladding, thermal barrier coatings or chrome plating, DCT is a through-material process, so it doesn't erode or peel over time. Evidence shows that DCT increases electro-mechanical affinity at the substrate–bond-coat interface. The process is complementary to all surface treatments—it is used in addition to heat treating and surface coatings, not to replace them (figures 9.14 and 9.15).

DCT improvements to stress–strain, tensile and yield strength are common, confirming that the process increases wear life and mechanical strength across most austenitic steels that transform to martensite. However, not all materials are improved by DCT and Rockwell C scale hardness, RC, frequently doesn't increase. Each material requires a unique treatment protocol to ensure optimized improvement. Items can be DC-treated as raw material, casting, forging or in finished state as supplied by an original equipment manufacturerer (OEM).

Commercial use

DCT has been successfully used in automotive racing applications, CNC machine tools, stamping and forging dies, brake rotors, drill bits, bearings and gears. The technology has tremendous value to increase the wear life of 3D printed items, marine components, jet engines, wind turbines and power generators. DCT also has

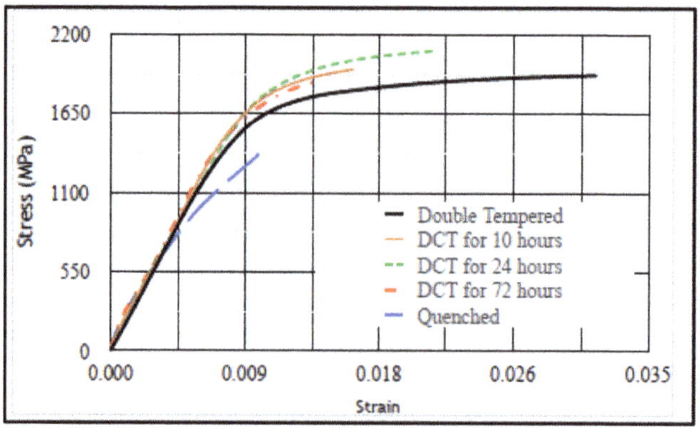

Figure 9.14. Stress–strain engineering curves [9].

Heat treatment processes	Mechanical Properties			
	Yield Strength (MPa)	Tensile Strength (MPa)	Elongation (%)	Elastic Modulus (GPa)
Quenched	1081.90	1371.85	0.95	213.21
Double Tempered	1692.20	1925.41	31.52	186.85
DCT for 10 hours	1720.70	1966.61	15.85	193.91
DCT for 24 hours	1818.30	2094.31	20.95	188.59
DCT for 72 hours	1781.39	1878.21	13.36	199.81

Figure 9.15. UTS of D2 steel in soft, HT and three DCT conditions [9].

the unusual effect of increasing the work-hardening capability of fully austenitic manganese and TRIP steels used in hard-rock mining and road construction.

Prior obstacles to technology adoption

Despite attempts using electron microscopy and nano-characterization, the science behind deep cryogenic phenomena is still relatively unclear. Although heat treatment has been around for over 10 000 years (when cavemen began fire-hardening spear points), DCT launched in the 20th century when nitrogen gas was separated, chilled and liquified. Like electric cars, the newness of this process may explain the recent adoption.

Industry qualification

Several qualification agencies have charted paths for industrial acceptance of this technology. Because DCT doesn't change source of supply, material type or dimensional tolerance, qualification time can be compressed in traditionally conservative industries. A key benefit is that DCT can be added to existing manufacturing processes without changing, modifying or eliminating any prior steps.

History and equipment
DCT has evolved greatly since WWII, when Clarence Zener experimentally poured liquid nitrogen on aircraft forging dies in primitive attempts to increase wear life. Zener knew he was onto something, but the war ended, and his project was mothballed. Forty years later the process re-emerged with technology advancements including digital-controlled LN_2 supply, use of dry-nitrogen vapour, PID optimization and *in situ* annealing capability. The deep cryogenic industry is still in an early development stage. Fewer than 150 firms worldwide provide DCT services—mostly to individual end-users. There is only one scientific research lab in the world (Canada) and both scale-up and full commercialization has yet to occur.

Key technology applications for deep cryogenic treatment are:

- Discrete components or systems with limited life due to abrasive wear or corrosion;
- Forged, cast, or 3-D printed items whose failure results in costly down-time;
- Key machined parts whose performance, when worn, lowers platform efficiency;
- Items whose upper performance ceilings are defined by short life.

9.5 Applications of cryogenics in the food industry

Eleni Fitsiou and Jaroslaw Nowak

Cryogenics finds many applications in diverse fields, being also an established technology in the food industry that produces high-quality nutritious food with increased storage life. The food and drink industries hold an essential place in the UK's manufacturing sector and the UK's economy; therefore, the impact of new technologies using cryogenics can be of significance.

The main aim of the freezing process is to control how and where ice crystals are formed, and it is generally accepted that fast freezing rates produce more and smaller ice crystals. Thus, any physical damage to the structure of the food is reduced. Usually, the goal is to achieve less mechanical stress during the formation of ice crystals, and small crystals with uniform distribution throughout the food. With cryogenic freezing, water molecules are frozen before they can press through the cell membrane and before they move on to the space between the muscle cells, and therefore, minimise the damage to the texture of foodstuff. However, there are arguments in the literature with regards to the benefits in the sensory quality of the products (e.g. for meat products) [10].

Currently, three freezing systems are industrially available: mechanical, cryogenic and impingement freezers [11]. There are also cryo-mechanical freezers, which combine the advantages of both mechanical and cryogenic freezers [12]. Mechanical freezers are very common and rely on heat exchange between air and food. They use a circulating refrigerant to convey heat from a low temperature region to another region at high temperature. Impingement freezers direct high-velocity jets of air on the food product and are better suited for flat-shaped products [10]. Cryogenic freezers can achieve the lowest temperature compared to the other two freezers by direct application of food-grade gases, such as liquid nitrogen or carbon dioxide onto the food. The main advantage of cryogenic freezing is the fast freezing rate achieved, which reduces moisture and yield loss, and minimises changes in the texture, colour and flavour of the product (enhances the sensory characteristics of food). Fast freezing technologies *lock-in* the aroma and quality of fresh food, preserving them efficiently for a longer period.

The advantages of cryogenic systems in the food industry can be significant, especially for food processing industries that deal with many or small size products, high value/quality products or seasonal products. The boost of productivity, the flexibility when handling different products, and the substantial operational benefits (e.g. higher throughput, quicker changeover rates) are also essential factors that need to be considered. Cryogenic systems are very flexible and can be used in-line or as a batch system. In some cases, they can be used to boost the performance of traditional freezers, e.g. in cryo-mechanical freezers the cryogen initially crust-freezes the product before moving on to the conventional mechanical freezing [13].

However, not all food is suitable for cryogenic freezing since it can cause physical and chemical changes (e.g. freeze-cracking). Common foods suitable for cryogenic freezing are those that contain considerable amounts of water, e.g. meat, poultry, seafood, vegetables, dough products and individually quick-frozen products. It is also noteworthy to mention some new applications, such as the cold grinding of spices and coffee beans (used to minimise the loss of important volatile compounds) [14, 15]. It is suggested that grinding the coffee beans at cryogenic temperature can affect particle size distribution. Additionally, the distribution becomes narrower as the beans are cooled, resulting in a more uniform grind. The characteristic aroma of the spices and coffee is better preserved when they are ground cold.

Food technology is under constant development to meet the high demand of today's society for fresh, safe and high-quality food. There are many novel and innovative food freezing techniques examined in the research, some of which are improvements of existing methods such as impingement and hydrofluidisation [13]. Yet only a few (including cryogenics) are commercially realised techniques [13]. Some innovative methods aim to either improve the quality of the product by rapid freezing and/or to control the way that ice is formed to improve product quality. It is challenging to access the benefits of new technologies since there are many different frozen food products with varying size and composition. The literature mostly focuses on measuring freezing rates and heat transfer coefficients, but the need to establish the relationship between the sensory properties of food and each technology requires further research.

With regards to food safety, which is of paramount importance, it is crucial to note that low temperatures do not kill all pathogens, but they deactivate some of them. Campylobacteriosis is a widespread infection caused by the *Campylobacter* bacterium. It has been suggested that liquid nitrogen can be used to reduce the presence of *campylobacter* bacterium on the surface of chicken [16]. Hence, rapid cooling could be a viable approach to reducing the above-mentioned foodborne pathogen.

The need for applying safety regulations (from 'field to fork' [17]) and to get acceptance of the proposed technique from consumers is imperative. New freezing methods are emerging, and research needs to continue to identify the pros and cons of each technique. The decision of which method is better suited for each product is not trivial and a lot of factors should be considered. Cryogenics is an established technology in the food industry, but further work is required to highlight and discover its advantages, considering its environmental benefits too.

References

[1] Silver L 1947 Gas cooling with aqueous condensation *Trans. Inst. Chem. Eng.* **25** 30–42

[2] Colburn A P and Hougen O A 1934 Design of cooler condensers for mixtures of vapours with non-condensing gases *Ind. Eng. Chem.* **26** 1178

[3] Sardesai R G, Shock. R A W and Butterworth. D 2007 Heat and mass transfer in multicomponent condensation and boiling *Heat Transfer Eng.* **3** 104–14

[4] Aspen HTFS+ Handbook, CM15, 16, and 18

[5] Environmental Technology Best Practice Programme. Good Practice Guide 12, 1996

[6] https://deepcryogenics.com/ *copyright* 2018

[7] Cahn J 2020 https://linkedin.com/in/jack-cahn

[8] Cahn J 2013 Deep cryogenic processing of weapons and armor *US Army Emerging Technology Proposal, Office of Sec Defense* 1

[9] Street T, Griffin D, Tawfik S, Russell Z and Plucknett K P 2019 Deep cryogenic treatment of high performance metals and related composites *NSERC Engage Programme, Dalhousie University* 24

[10] Evans J A 2009 *Frozen Food Science and Technology* (New York: Wiley)

[11] Kennedy C 2014 Cold feat: choosing the right freezing technology for your production line https://www.airproducts.co.uk/search#?query=kennedy&MediaTypes=&ContentTypes=&Industries=&Gasses=&SupplyModes=&Applications=&NewsTopics=&Countries=&SortOrder=Newest&Page=1&AvailableFilters=industry%2Ccontent_type&ShowChildFacets=False (accessed October 2022)

[12] Muthukumarappan K, Marella C and Sunkesula V 2019 Food freezing technology *Handbook of Farm, Dairy and Food Machinery Engineering* (Amsterdam: Elsevier) pp 389–415

[13] James C, Purnell G and James S J 2015 A review of novel and innovative food freezing technologies *Food Bioprocess Technol.* **8** 1616–34

[14] Uman E *et al* 2016 The effect of bean origin and temperature on grinding roasted coffee *Sci. Rep.* **6** 24483

[15] Sharma L *et al* 2016 Effect of cryogenic grinding on volatile and fatty oil constituents of cumin (*Cuminum cyminum* L.) genotypes *J. Food Sci. Technol.* **53** 2827–34

[16] Burfoot D *et al* 2016 Effect of rapid surface cooling on *Campylobacter* numbers on poultry carcasses *Food Control* **70** 293–301

[17] European Food Safety Authority 2012 Science protecting consumers from field to fork

www.ingramcontent.com/pod-product-compliance
Ingram Content Group UK Ltd.
Pitfield, Milton Keynes, MK11 3LW, UK
UKHW051900100926
8556IPUK00005B/708

9 780750 327558